Pest animals in buildings

Norman Hickin

GEORGE GODWIN
LONDON

George Godwin
an imprint of:
Longman Group Limited
Longman House, Burnt Mill, Harlow
Essex CM20 2JE, England
Associated companies throughout the world

*Published in the United States of America
by Longman Inc., New York*

© Norman Hickin 1985

All rights reserved; no part of this publication may be reproduced, stored in a retrieval system, or transmitted in any form or by any means, electronic, mechanical, photocopying, recording, or otherwise. without the prior written permission of the Publishers.

First published 1985

British Library Cataloguing in Publication Data
Hickin, Norman
   Pest animals in buildings.
   1. Household pests
   I. Title
   363.1'3   TX325

ISBN 0-7114-5644-5

Library of Congress Cataloging in Publication Data
Hickin, Norman Ernest.
   Pest animals in buildings.

   Bibliography: p.
   Includes index.
   1. Vector control. 2. Household pests–Control.
3. Pest control. 4. Animals as carriers of disease.
5. Household pests. 6. Pests. I. Title.
RA639.3.H5   1984   591.6'5   83-16246
ISBN 0-7114-5644-5

Set in 10/12pt Linotron 202 Bembo Roman
Printed in Great Britain at The Pitman Press, Bath

Contents

Foreword vii
Introduction ix
Acknowledgements xii

1. Carnivores – CARNIVORA 1
2. Rodents – RODENTIA 6
3. Bats – CHIROPTERA 23
4. Insectivores – INSECTIVORA 27
5. Marsupials – MARSUPIALIA 29
6. Birds – AVES 33
7. Snakes and Geckos – REPTILIA 44

ARACHNIDA 49

8. Mites – ACARI excluding METASTIGMATA 49
9. Ticks – ACARI METASTIGMATA 71
10. Spiders – ARANEAE 79
11. Sunspiders etc. – SOLIFUGAE *et al.* 85
12. Scorpions – SCORPIONES 88
13. Woodlice – ISOPODA 91

INSECTA 93

14. Beetles – COLEOPTERA 93
15. Woodwasps, ants, bees and wasps – HYMENOPTERA 146

16.	Fleas – SIPHONAPTERA	175
17.	Flies – DIPTERA	183
18.	Butterflies and moths – LEPIDOPTERA	225
19.	Thrips – THYSANOPTERA	237
20.	Bugs – HEMIPTERA	238
21.	Sucking lice – SIPHUNCULATA (=Anoplura)	251
22.	Feather, biting, chewing or bird lice – MALLOPHAGA	260
23.	Booklice or Psocids – PSOCOPTERA	264
24.	Termites – ISOPTERA	268
25.	Cockroaches – BLATTODEA	285
26.	Earwigs – DERMAPTERA	304
27.	Crickets – ORTHOPTERA	309
28.	Springtails – COLLEMBOLA	312
29.	Bristletails, silverfish and firebrats – THYSANURA	314
30.	Centipedes – CHILOPODA	320
31.	Millipedes – DIPLOPODA	323
32.	Tapeworms, flukes and roundworms – CESTODA, TREMATODA and NEMATODA	325
33.	Control methods and substances	341

References 367
Index 371

Foreword

It is now nearly forty years since the termination of the Second World War. At that time new chemical substances and novel techniques became available to combat harmful and injurious animals found in buildings. These pests had long been known to contaminate and destroy food, to tunnel into, weaken and finally to disintegrate wooden structures and artefacts of all kinds, indeed to render valueless most organic substances used by man in his economy within the environment of a building. In addition there are the truly parasitic animals on man, those that live both outside the skin, sucking his blood, and those that live wholly or in part within the human body.

For a period of twenty years or so great success attended the efforts to eradicate many noxious animals but since that time a number of factors have operated to modify the degree of the earlier success obtained. Resistance to what were thought to be "wonder" insecticides and rodenticides became apparent and the search for more sophisticated substances was intensified. At the same time the pollutant effect of some chemical substances which were in widespread use became recognized world wide. Safety to man as well as to wildlife rightly became paramount.

Other events were taking place simultaneously, notably the revolution in transport not only of people by air with their attendant parasites but the greatly increased volume of timber and livestock also carrying harmful organisms. In spite of international regulations and checks, many such harmful pests became established in lands far from their point of origin. Eternal vigilance has become the watchword for all associated with pest control.

Finally, in this context perhaps most important of all, attention must be directed towards the disturbed state of many areas of the world. During the last twenty years or so there have been many natural disasters of many kinds: drought, floods, earthquakes, and then civil strife, and wars of many kinds of origin but all are characterized by human misery arising mostly from the consequent unhygienic conditions. The importance of all this is that far from pest control becoming more simple in operation by the use of chemicals in large scale operations – the opposite has been the case. The target pest must be identified with precision; dosages of control substances measured with accuracy and used only in accordance with national and local regulations. No harm must come to wildlife. A special feature of this book is the association of many human diseases with animals in buildings.

The author Norman Hickin, a colleague of mine for many years, really needs no introduction from me. He already has a number of textbooks on pests and their control to his credit and indeed they have played an important part in training programmes in many parts of the world. He has specialized in wood-destroying beetles and termites but his present book covers all the groups of animals which occur in buildings as pests – on a world-wide basis. This would be considered a daunting task for most but Norman Hickin has stuck to the project in spite of difficulties and after many years work he has achieved his object.

I commend this book to you if you are engaged in pest control, Environmental Health, Buildings Inspection, Public Health, Public Administration, either as practical technician, technologist, administrator, manager or teacher.

W.H. Westphal
Chairman. Rentokil Group Ltd.

Introduction

This review concerns the animals which are considered by Man to be noxious when found in his buildings. Man uses building structures for a variety of purposes – for shelter in which to sleep – for protection against weather for storage purposes – and for special purposes, such as museums, art galleries and large commercial blocks.

The vast increase in population in recent years, together with immense changes in Man's way of life have meant that his buildings not only occupy a greater and greater area of land but tend to be of increasing complexity. In the more highly developed towns and cities a substantial proportion of buildings contain mechanisms for temperature and humidity control – systems which often tend to encourage the development of exotic animals.

Buildings are very diverse structures. In this respect it is a very loose term. On the one hand they are many-storeyed, mainly constructed of concrete, steel and glass, and generally occur in groups in the centre of large concentrations of population. The individual elements of the surrounding natural environment have little influence on the almost clinical sterility of the 'skyscraper'. On the other hand, and by far the most numerous, are the buildings generally of wood with a little stonework for a chimney. There is one floor only – on the ground. Such structures are situated in forest, steppe and savannah, and there is usually some invasion of animals into them from the adjacent countryside. At the extreme, we should mention that many indigenous peoples, living in tropical climates, construct their dwellings of four upright posts of tree trunks, and roof them over with vegetation, to make a thatch. Mud may be incorporated to form walls. Animal inhabitants of the surrounding forests, more particularly nocturnal ones, often take up residence side by side with the human occupants.

In between these extremes lies an almost indefinable diversity, and early in the course of the work on this present project, the difficulty of deciding where to draw the line was encountered. The decision was made not to include *all* animals that might invade the simplest types of shelter. Lions, tigers and leopards would have had to be described, as hundreds of records exist, in the recent past, of humans being dragged from their huts, killed and eaten!

In the environment of a building animals may be considered undesirable to Man for one or more reasons. They may be external parasites, biting and sucking his blood. Insects of widely differing orders, such as mosquitoes, body lice and bed bugs immediately come to mind, but mites and even mammals such as the Vampire bat are of importance in certain areas of the world. Many of the directly parasitic animals are of even greater importance because pathogenic organisms are often introduced during the attack. A number of these cause very serious cases of debility and disease. Some of the gravest epidemics the world has known, causing death on a vast scale, have been due to this cause and directly as a result of Man's association with animals in his dwellings. Bubonic plague caused by bacilli transmitted by the Rat flea *Xenopsylla cheopis* and malaria caused by plasmodia transmitted by mosquitoes are well-known examples. Less well known is the rabies transmitted by the bites of bats, especially the Vampire bat of Central and South America.

Man has two fairly distinct attitudes to animals, as to whether they occur in a building or outside it. It is true that he permits many domestic animals such as dogs, cats and other pets within his home but his relationship to these is of a special nature. He feeds them; he allows them to share his own home at night; he talks to them; he treats them much as he would treat members of his own family. Often he does this without any consideration for hygiene and certain sections of this book will show health risks which are run mainly in connection with the ectoparasites of household dogs and cats.

Some aspects of human diseases associated with buildings are given in more detail than is usually the case in books of this kind. As to the importance of this, it is left for the reader to make his own judgement. It will be difficult to underestimate this aspect of the undesirable nature of many animals in all types of dwellings.

One of the most important groups of undesirable animal species from an economic point of view are pests of stored products associated with the ability of Man to store his foodstuffs to tide him over the periods when they are not immediately available as a crop. This may be on a seasonal basis, or certain foodstuffs may be processed by drying so that they can be stored over one or more seasons when the crop fails. But unless knowledge is available concerning the habits of the undesirable fauna which would normally pass all or part of their life-cycle in the stored material, and appropriate measures takes, a substantial amount of damage can occur. Species of rodents, beetles, moths and mites are all of importance in this regard, and in some parts of the world their presence can make the difference between famine or plenty. Foodstuffs are obviously the most important cause for concern, but other stored products derived from vegetable and animal sources such as drugs, tobacco and hides can be seriously affected by pests.

INTRODUCTION

Strictly speaking, wood is a stored material of vegetable origin, the storage problems associated with it are distinct enough for it to merit a category of its own as wood-destroying insects are of great diversity. Wood is converted from the woody tissue of something like 18,000 different tree species, and many have infesting insects confined to one tree. On the other hand, some wood-boring insects attack a considerable number of species. Again, some are found boring into the sapwood of newly-converted timber, while others require an association with wood-decaying fungi. The object of this work is to survey and describe all the animals found in buildings which are noxious to man for one reason or another. This review endeavours to cover as much of the world as possible, although more information appears to have been published in Europe, North America and Africa than in Asia and South America. The reader, therefore, must be aware of some unavoidable geographical bias, although of course many of the animals discussed are of worldwide significance.

The chapters are in a descending order of classification, mammals being described first and invertebrates last. This taxonomic arrangement was thought to have advantages over a separation of species according to the type of damage caused. Each major group of animals is prefaced by a brief description and diagnostic features. Taxonomic information regarding nomenclature has been reduced to a minimum. However, many groups of animals are in a state of turmoil with regard to classification. It is, therefore, quite possible that some of the scientific names given in this book will be different from those known to the reader. This is unavoidable, but wherever possible the most up-to-date classifications have been used. The scientific names of the animals have been given first, followed by the well-known English or local name if available. No popular names in other European languages are given. The authority or author of the name is not given in the general text. They are all in widespread use. The author dislikes the use of abbreviations but, nevertheless, in a work of this magnitude some are inevitable. It is hoped that the reader will be unaware of, or will forgive those that occur!

Although this work ranges as widely as possible across the world it is likely that important problems concerning harmful species in some parts have been omitted, and the author hopes that readers will communicate with him so that future editions of the book may be amended accordingly. It should be mentioned that this work cannot hope to give complete keys for identification of all insects. Local keys are normally available at departments of health, departments of agriculture, museums and libraries in most regions of the world.

Acknowledgements

It is now a number of years since Mr W. H. Westphal suggested to me that I should write an account of animals, in or under buildings, that are harmful or injurious to human beings. It was to be in one volume and to include not only insects and other arthropods, but all animals. Mr Westphal thought that the book should not be entirely directed towards the United Kingdom but should be useful as a reference work throughout the temperate world. The task has now been completed and I must thank all those who have helped and encouraged me to carry it out.

First, then, I wish to thank Mr W. H. Westphal for his initiative and stimulation and for kindly agreeing to write the foreword.

As to the form and design which the book should take, Dr Peter Cornwell and Mr Robin Edwards gave much help and I am very grateful to them for their aid at the outset, when a number of time-consuming false starts were made.

Next I wish to thank my colleagues on the staff of Rentokil Ltd, for their help and advice over a number of years and the Company for allowing me the use of many facilities.

One of the most stimulating experiences in my professional life has been the encouragement and help given to me by many scientific friends and colleagues who have read the relevant manuscript sections. Their comments and suggestions have been invaluable and I am most grateful to them for all the trouble they have taken.

In this regard, I would like to thank Dr Michael Baker, Mr Alan Brindle, Dr Theresa Clay, Professor John Cloudsley-Thompson, Mr R. A. Davis, Dr John Freeman, Professor J. D. Gillett, Dr Victor Harris, Mr P. N. Lawrence, Professor Kenneth Mellanby, Professor T. G. Onions, Mr Brian Pitkin, Dr David Ragge, Dr Miriam Rothschild, Dr Philip Spear, Mr Gerald F. Thompson, Mr Harry V. Thompson, Professor Ian Thornton, Dr Gwynne Vevers and Mr E. J. Wilson.

I have drawn on the technical releases and other documents emanating from the National Pest Control Association of the United States, and I would like to record the unfailing help of Dr Philip Spear and other officers of the Association whenever their advice was sought. The period spent at their headquarters when it was in Elizabethville, New Jersey, going through their papers relating to this subject, was a stimulating experience and my thanks are tendered accordingly.

I have been privileged to receive abundant help with the illustrations and I particularly wish to thank the following: Dr J. P. Spradbery of the Division of Entomology of CSIRO, Australia, who kindly provided me with illustrations of wasps and hornets; Professor J. D. Gillett who allowed me to use 12 of Dr Judith Smith's beautiful drawings of mosquitos from his book *Common African Mosquitos* – unhappily, they had to be much reduced in size. It is a pleasure to record help I received from the Wellcome Foundation Ltd, which gave me permission to use what I wanted from its collections. The Centre for Disease Control of the Department of Health, Education and Welfare of the United States gave the free use of illustrations from its manual on household and stored-food insects of public health importance. The Deutsche Gesellschaft für Schädlingsbekämpfung MBH has allowed me to use its fine illustrations of stored product insects. These are indicated by *Degesch* in the text.

My thanks are also due to the United States Department of the Interior, Fish and Wildlife Service which helped with photographs; and to Musterschmidt-Verlag of Gottingen which gave assistance for which I am grateful.

Thanks are given to the Trustees of the British Museum who generously gave permission for a number of drawings to be reproduced. Dr Griffiths of the Pest Infestation Control Laboratory, Slough, of the Ministry of Agriculture, Fisheries and Food, kindly allowed me to use his electron-scanning micrographs of mites. By courtesy of the Shell Chemical Company USA, I have been able to use André Durenceau's beautiful drawings. These are identified in the text by *Shell Chemical*. I am also indebted to Dr A. G. Fisken of Shell Chemicals UK Ltd for obtaining this permission.

The range and scope of this work have required much correspondence and secretarial assistance, and I would like to thank Pamela Willis for her dedication to the many duties entailed. Hilda Maxwell, who has helped produced many typescripts for me in the past, continued undaunted to translate my handwriting into pages of type, for this I thank her. The final typescripts were prepared by Sheila Price and Jean Robinson and I am very grateful to them.

As for myself, after a number of years' gestation of the work, I shall miss the kindly questions from my colleagues as to how the book was going!

Chapter 1

Carnivores

CARNIVORA

The mammalian order CARNIVORA consists generally of flesh-eaters as the name implies. They are characterised by their specially adapted teeth and feet. The canine teeth are very large, sharp and curved, being used as the principal weapons for killing prey. The incisor teeth are reduced, particularly in the lower jaw, their function being to hold and pierce the prey. The cheekteeth are usually reduced in size and number but are pointed with sharp cusps; two on each side of each jaw (usually the fourth premolar of the upper jaw and the first molar of the lower jaw) are adapted for shearing flesh. These are known as carnassials and are most highly developed in the FELIDAE (cats). No teeth are used for mastication as flesh is highly digestible and is swallowed in large lumps.

Many carnivores, however, have a varied diet and are herbivorous to a greater or lesser extent, feeding on berries, nuts and fruits. Jackals, foxes, dogs, badgers and bears are very general feeders, often scavenging for food around human habitations.

Carnivores have five toes on the front feet and either four or five on the hindfeet. In some species, however, the inner toes of the front feet may be high off the ground. All toes bear sharp claws and some are extremely sharp. CARNIVORA are notable for their great variation in size. Some weasels weigh only 40 g or so, while some bears are known to weigh up to 800 kg.

Very few members of the CARNIVORA take up residence in buildings, although some will pursue their prey into them – such as weasels and stoats after rats and mice. Such behaviour, however, cannot be considered harmful to Man. Even more important in this regard are the species of the family VIVERRIDAE, known as the mongooses. Some are well known as controlling rats and snakes in the vicinity of buildings. They are native to the whole of Africa, the Mediterranean region generally, Madagascar, and South and South-East Asia. One of the most widely distributed species, the small Indian mongoose, *Herpestes auropunctatus*, has been introduced into the West Indies and Hawaii with the intention of reducing the number of rats and snakes, but it has also taken toll of small native mammals and birds, as well as of domestic poultry. The Egyptian mongoose or Ichneumon, *Herpestes ichneumon*, is found throughout Africa. It is about 60 cm in body length with a tail 46 cm long, and is a well known snake killer. All species of mongoose are, to a great degree, immune to snake venom, and as their reflexes are of lightning speed they are almost always more than a match for snakes up to 2 m in length.

MEPHITIS MEPHITIS   STRIPED SKUNK

In North America the Striped skunk, related to the badger, is one of the best known of all animals. This is on account of its distinctive coloration combined with a highly efficient and foul-smelling scent produced by anal glands. It is about the size of a domestic cat, head and body being 45 cm with a slightly smaller tail. It weighs about 2 to 4 kg. The body is black with a narrow white stripe up the middle of the forehead and a white area on the nape which continues backwards on each side in a V-shape to the base of the tail. These white stripes vary in width and extent. The tip of the tail may be white also. This pattern can be considered an example of warning coloration. Before ejecting its scent, the animal usually stamps its feet, then turning, aims the gland contents with a fair degree of accuracy to about 4 m.

Feeding
The food of the Striped skunk consists of insects, fruits, berries and mice. About one-sixth of its total food is made up of the last, so that it is generally a beneficial animal and, in some states, is afforded a degree of protection. Nevertheless, a large number are killed on the roads by motor vehicles.

Breeding
Mating takes place in early spring, and the 4 to 10 blind and naked young are born 51 days later. They are suckled from 6 to 7 weeks, the eyes opening after 3 weeks. When the young are well developed the male may rejoin the family.

A skunk weighing 1.6 kg can get through a hole 7 cm in diameter. It is when it takes up residence under a house that the intolerable odour makes its removal essential. It is now second only to the Red fox, *Vulpes vulpes*, as a wildlife source of rabies.

Distribution
Found throughout temperate and semitropical North America, except the coastal region of western Canada.

SPILOGALE PUTORIUS   SPOTTED SKUNK

A second species, the Spotted skunk, is widely distributed in North America west of the Mississippi (except Montana and most of Wyoming). It occurs also along the Gulf coast and the southern

Fig. 1.1 *Mephitis mephitis*, Striped skunk. (Victor B. Scheffer. US Fish and Wildlife Service)

Appalachians. There are four discontinuous white stripes along the body and it is much smaller than *M. mephitis*. The anal scent is just as effective, however, as in the latter species. The animal feeds on rats, mice and insects. It is also affected by rabies.

FELIS CATUS   DOMESTIC CAT

Domestic cats, in the family FELIDAE, often become feral in large buildings such as warehouses, hospitals and airports and thus constitute a hygiene hazard. They feed on discarded food and garbage as well as rats and mice. Their value as controllers of rodents, however, is outweighed by their undesirability in such situations. Young may be found any month of the year depending on food supply. Feral cats are extremely timid, sometimes excessively so to the extent that although a number of them may be present in a large building, they may rarely be seen. They flee at approaching footsteps and dart into dark hiding places. A search should be made for their sleeping quarters. These are usually located in some textile material – such as an old mattress, cloth or old newspapers in an attic or in a roof void where there is no human interference. These 'lairs' are almost entirely breeding places for the Cat flea, *Ctenocephalides felis*. In industrial localities often members of the workforce put out food scraps for these cats and so help to perpetuate unhygienic situations.

Rabies

Rabies is a disease of the central nervous system of mammals, including Man. It is acutely infectious and is caused by a virus which is almost always transmitted from the infected animal when saliva enters a wound caused by a bite or by the licking of a scratch. In rare cases the virus has been shown to be present in a fine mist of urine and other secretions in a confined space – such as in certain caves where large numbers of bats congregate and where rabid individuals are present.

The disease, which is characterised by a profound dysfunction of the central nervous system, is one of the most feared of human diseases and is the cause of many deaths each year (although many more people are killed by influenza, for example).

The incubation period varies from several days to over 12 months, and in dogs there is firstly a period of fever, fainting, hyperaesthesia and a change in the tone of the bark. A change in the general disposition of the animal is also apparent. This lasts from a few hours to several days and is followed by an excitation phase when the animal is restless and agitated, often with general tremors. It growls and barks incessantly and will make vicious bites at any animal encountered. Often there are convulsions or general paralysis may set in – usually starting with paralysis of the jaw which is accompanied by excessive salivation. Sometimes dogs die suddenly without showing any signs of the illness.

In Man, the prodromal phase is characterised by fever, general malaise, nausea and sore throat. There is tingling or intermittent pain at the site of the infected wound, accompanied by extreme stimulation of the general sensory system which is shown by acute sensitivity to sound, light, temperature change and draughts. Dilation of the pupils and excessive salivation may occur. Later, the muscles of the mouth, larynx and pharynx contract when drinking and, at a still later stage, even the sight of liquid causes this reaction. Hence the common name hydrophobia – meaning the fear of water – for this disease. The pulse is extremely rapid and periods of irrational behaviour, often maniacal, alternate with those of responsiveness. Paralysis of the throat muscles may lead to hoarseness or loss of voice. While in some cases the excitation phase predominates until death ensues, in others paralysis occurs shortly before death. Sometimes the excitation phase is absent altogether, the disease being characterised by ascending paralysis without hydrophobia. This is generally the case in Trinidad when rabies is brought about by Vampire bat infections.

There are no therapeutic treatments available for persons who already show signs of the disease and the victim is almost invariably condemned to die in agony and distress. If possible exposure to the virus has taken place (a bite from a mammal in an area where rabies occurs), then a protective treatment involving antirabies vaccine should be administered (this is very unpleasant and may not be successful) after thorough cleansing of the wound.

Except for Australasia and Antarctica, the disease occurs in most countries of the world. It is widespread in Europe and has become increasingly important since the Second World War. In Asia and some parts of Africa, stray dogs are the principal means of rabies distribution but in the more recent, western European extension of the disease, *Vulpes vulpes*, the Red fox, is the principal carrier, although Man falls victim usually through the bite of a dog or cat which has already been bitten by a rabid fox.

Other mammals living within the curtilage of a building and known to transmit rabies virus are: in North America, *Mephitis mephitis*, tbe Striped skunk; *Spilogale putorius*, the Spotted skunk; in Central and South America, *Didelphis marsupialis*, the Virginian opossum; *Herpestes edwardsi*, the Indian grey mongoose. Vampires and other bats carry the rabies virus in several countries.

Chapter 2

Rodents

RODENTIA

Most species of this very abundant order are small, nocturnal and secretive animals. Their economic importance cannot be overstated, and *Rattus norvegicus, R. rattus* and *Mus musculus* can be considered commensals of Man. A number of other species, including *R. exulans* and *Bandicota* spp. are also important in the Pacific region and tropical Asia respectively. An important anatomical feature of rodents is the degree of specialisation of the teeth. They are much reduced in number and consist of a few cheekteeth which are folded, or looped, and possess grinding surfaces, and a single pair of incisors in each jaw which are adapted for gnawing. The wide gap between the incisors and the cheekteeth is known as the diastema and is especially long in rodents. The incisors grow throughout life and this, together with constant grinding, has the effect of continuously presenting a fresh, chisel-like edge. It is this feature which is thought to have contributed most to the great biological success of these animals.

Rodents are divided into three main groups according to the widely differing structure of the jaw muscles. Individual species, however, even in the same group, may differ considerably in external appearance.

The hystricomorphs are generally ground-dwelling or burrowers, usually having blunt snouts. The Old World and New World porcupines are examples, but, in addition, there is a large number of South American species including the guinea-pig, the chinchilla and the largest rodent, the capybara. Included in this group also is the coypu, or nutria, of South America which is now feral in Britain – established only in parts of East Anglia. It will occasionally enter farm buildings where root crops are stored.

The sciuromorphs generally possess rounded heads and are either squirrel-like, living in trees, or are burrowers. The family containing squirrels, marmots and chipmunks is widely distributed and some species will take up residence in roof voids and other little-frequented areas of a building, often becoming a nuisance. The New World pocket gophers, pocket mice, kangaroo mice, kangaroo rats, spiny pocket mice, as well as the beaver, are also sciuromorphs.

The myomorphs have a long and more or less cylindrical skull with a pointed snout. This is by far the largest group, containing more than 1,000 species distributed throughout the world. The greater proportion of this group never come into contact with Man or his agriculture but some of the species do harm to growing crops. However, the family MURIDAE contains the true mice and rats, some species of which have been introduced into the greater part of the world by Man. The edible dormouse in the family GLIRIDAE often takes up quarters in buildings.

RATTUS

This Old World genus consists of more than 500 named species and subspecies, although it is likely that there is much synonymy. Some species are commensals of Man, reliant on him for food, harbourage and distribution. Two species of rats are perhaps the most undesirable of human-associated animals and are of cosmopolitan distribution. *Rattus* spp., together with another myomorph (*Mus*) and the bats, are the only mammals known to reach remote oceanic islands and to colonise them successfully. In most cases, Man has accidentally carried them by sea and by air: goats, rabbits and other mammals, however, have been wittingly established by Man.

Rats are injurious to Man in buildings on account of their damage to materials such as wood, plastics, and even soft metals; their destruction of food and its contamination by hairs, urine and droppings; their ability to bite, causing a painful wound (especially to children); and their ever-present potential danger as carriers of pathogenic organisms. This latter may be transmitted by biting, by contamination of foodstuffs and by their ectoparasites such as fleas and mites. In Indonesia, *Rattus exulans* is an important plague carrier.

RATTUS RATTUS

This species is known by a number of common names, chief of which are Roof, Ship, Black, House, Alexandrine and Fruit rat. On account of the degree of colour variability, the term 'Black rat' is probably not a good name. Although 74 subspecies have been granted various degrees of recognition, their systematic position is still under discussion. There are also a number of forms given specific rank which may ultimately be included in *R. rattus*. This rat is thought originally to have been indigenous to the Indo-Malayan region, extending eastwards as far as southern China. It was during the Middle Ages, when large ships from western Asia and Europe first began to explore and trade with the East, that *R. rattus* colonised areas around the world.

Rattus rattus is predominantly coastal in distribution but is sometimes found inland, as in Johannesburg. In the USA, it is found mainly on the west coast and, in a wider area, along the south-east coast. Inland its distribution is sporadic. In the British Isles, it is largely confined to ports,

8 PEST ANIMALS IN BUILDINGS

Fig. 2.1 *Rattus rattus*, Ship, Roof or Black rat. Studies in posture. Its agility compared with *R. norvegicus* is well illustrated.

the population probably being maintained by immigrants. There is some indication that their numbers are increasing in such situations. In some tropical areas, especially inland, the species lives distant from buildings and is not necessarily commensal. It is the most common species on ships.

Rattus rattus is now the only rat species on Ascension Island in the South Atlantic where it lives both commensally and ferally. It also burrows in soil or in cultivated terraces. On St Helena which is further

Fig. 2.2 Distribution of *Rattus rattus* in North America and Mexico.

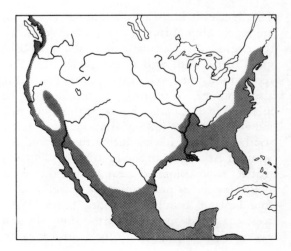

south, the two important species of rats are present over the whole island.

Origins
The first rat to reach western Europe was *R. rattus*. It was not known to the Romans or Greeks but by the twelfth century it was present in considerable numbers. It reached North and South America in the ships of the early explorers during the sixteenth century. In many areas it has been displaced by the later invasion of *R. norvegicus*.

Description
Three forms, probably only colour phases, previously classified as subspecies, are found in the greater part of the distribution range. The so-called 'Black rat', *Rattus rattus*, is black above and grey beneath and is the commensal form. *Rattus rattus frugivorus*, is tawny-yellow (agouti) with a white belly, and is the predominant wild-living form in India, New Zealand, Cyprus and New Caledonia. *Rattus rattus alexandrinus*, is tawny-yellow (agouti) but is greyish below. The latter colour is due to the slatey base of the white hairs. This is the predominant form in the Middle East.

Rattus rattus is more lightly built and rather more slender than *R. norvegicus*, and the coat is sleeker or less shaggy than the latter species. The great variation in coat colour is indicated above but, in addition, there is sometimes some white spotting on the head and chest. The tail is slightly longer than the head and body together; the size of the full-grown adult varies considerably, but an average length for head and body of the male has been given as 190 mm, with the tail 220 mm, while corresponding figures for the female are 170 mm and 200 mm. The weight is also variable but is usually about 200 g. The skull is lighter in weight than that of *R. norvegicus*, and there are pronounced supraorbital ridges which, curving outwards, give a pear-like shape to the cranium.

Other distinctive characters of *R. rattus* compared with *R. norvegicus* are the pointed, rather than blunt, muzzle; the slender, rather than stout, tail and the ear pinnae, which are relatively hairless, translucent and large, while those of *R. norvegicus* are hairy, thick, opaque and small. A valuable character in separation of the two species is that the tail of *R. rattus* is always unicolorous, whereas the tail of *R. norvegicus* is always lighter below (nearest the ground).

Habits
In wild populations it is generally aboreal, while commensal animals often inhabit upper storeys of buildings, travelling along cables and wires. The crescent-shaped, greasy smears from fur where rats habitually pass under cross-beams is characteristic. The rat is nocturnal in habit. Faeces are sometimes diagnostic, being about 10 mm long and 2–3 mm wide and curved. Faeces are usually easily distinguished from those of *R.*

Fig. 2.3 *Rattus norvegicus*, Brown rat. Studies in posture.

norvegicus, which are fatter and more solid, measuring approximately 17 mm by 6 mm. The range of movement of *R. rattus* is rarely more than 100 m. Although preferring fruit and other vegetable products, it will eat almost anything if pressed.

Breeding
In the United States, *R. rattus* is sexually mature at 3–4 months at a weight of only 90 g. About 4 litters are born annually. Gestation lasts for about 21 days and the number of young varies from 5 to 10 per litter. Breeding usually takes place throughout the year but peaks occur in summer and autumn. The annual mortality rate is from about 91–97 per cent.

RATTUS NORVEGICUS

Common names of *R. norvegicus* in Britain and other English-speaking countries are Brown, Norway, Common, Sewer and Wharf rat; sometimes Grey rat in the United States. It has a vast distribution, ranging from the Arctic to the Antarctic and is found in an extraordinarily wide range of commensal situations.

Rattus norvegicus is said to have originated in temperate Asia but did not invade Europe apparently until early in the eighteenth century. This was several years after *R. rattus* had arrived in Europe. The spread of *R. norvegicus* was probably very rapid and it is thought to have reached England between 1728 and 1730. It soon displaced *R. rattus* probably because it was more powerful and pugnacious but the successive decline of timber-framed and thatched houses has also been put forward to explain the virtual disappearance of the latter, at least from inland regions. *Rattus norvegicus* arrived at the North American seaboard around

1775 where, again, it quickly supplanted *R. rattus* generally in the northern latitudes.

Description
Considering the vast geographical range of the species, there is generally little colour variation. It is brownish in colour but paler underneath (see *R. rattus* for a comparative description). A small amount of albinism occurs, however, as does melanism which is much more widespread. The latter is present to the extent of 1 or 2 per cent, although up to 20 per cent has been observed in populations which are expanding rapidly. The full-grown male measures about 230 mm along the head and body while the tail is another 200 mm. The female averages about 240 mm along head and body and the tail is another 190 mm. The weight is variable but can (rarely) exceed 500 g. It is thus about two and a half times heavier than *R. rattus*.

Habits
Rattus norvegicus is essentially an obligate commensal of Man over the greater part of its range. It is associated with buildings, especially warehouses and farm-buildings, rubbish tips and sewers. In summer months it spreads from such situations along hedges and banks, notably river banks, and stone walls. It burrows extensively, sometimes using rabbit burrows but it always lives on or under the ground. Its range of movement is usually only about 30 m but can be considerably more.

Rattus norvegicus is stated to be one of the most damaging of pests with which Man has to contend. Although generally a cereal feeder, of necessity it can be omnivorous but, under special circumstances, it can subsist entirely on animal food, as when infesting meat stores. *Mus musculus* is killed and eaten and the everted skin is a sign of the presence of *R. norvegicus*. However, not *all* rats will kill mice, even when starved, and if they do they will not necessarily eat them. When feeding on food provided by Man, apart from that eaten much is damaged or contaminated.

Often populations are very dense and more than 500 rats have been recorded in a single corn rick. When populations reach saturation point, high mortality of young in the nest occurs and can be at the rate of 99 per cent per annum. A full-grown female can be aggressive when defending young, becoming a formidable opponent of dog or even of Man.

Breeding
Sex ratio is about 50 per cent. In some constant environments breeding may be almost continuous, with about 30 per cent of all females pregnant at any one time. Under other conditions there may be one or two peaks annually. At about 80 days old (at about 115 g), females become mature.

The number of embryos varies with the weight and age of the mother, being about 6 for the small animals and about 11 for the largest. The number of litters per year varies also according to conditions and is from three to five with the gestation period about 24 days. Young leave the bulky nest three weeks after birth.

RATTUS EXULANS   POLYNESIAN RAT, PACIFIC RAT

Found widely throughout the Pacific region from Malaya and Hawaii to Australia and New Zealand, it has colonised almost every island in the region, having been transported by Man. In New Zealand it is called the Kiore and was introduced by the Maori settlers.

Fig. 2.4   *Rattus exulans*, Polynesian rat. (R. E. Marsh)

Description
Very much like *R. rattus* in body structure and in its sleek and graceful appearance, *R. exulans* is, however, considerably smaller in size, generally weighing less than 90 g. The tail is less than 110 mm in length extending to, or nearly to the snout. The scales are fine and black. The muzzle is sharp and pointed and the ears can be pulled over the eyes which are large. The hind foot length is only 24 mm compared with 44 mm in *R. norvegicus*, and 35 mm in *R. rattus*. On the female the number of pectoral teats is two pairs, whereas in *R. norvegicus* it is three pairs and the number of inguinal teats on the male is two pairs, whereas in both *R. norvegicus* and *R. rattus* it is three pairs. The colour of the belly fur is

often difficult to determine accurately due to staining or soiling, but there is a similarity with *R. norvegicus* as it is white with grey underfur.

Feeding
Rattus exulans is similar to *R. rattus* in preferring natural plant food such as cereals, nuts, fruits and vegetables and can cause extensive damage in coconut groves.

Breeding
Whereas *R. rattus* and *R. norvegicus* generally have large litters (from 6 to 12 young), *R. exulans* usually has from three to six young. The main breeding season appears to be in late summer.

Rat-proofing
The physical abilities of *R. norvegicus* and *R. rattus* are fairly well known due to Man's continual efforts at control. The following represent the combined maximum abilities of both species which it would be wise to assume when building a rat-proof structure or when rat-proofing an existing one.

(a) Gain entrance through any opening larger than 1.25 cm square.
(b) Climb both horizontal and vertical wires and cables.
(c) Climb the inside of vertical pipes which are 4–10 cm in diameter.
(d) Climb the outside of vertical pipes and conduits up to 7.5 cm in diameter.
(e) Climb the outside of vertical pipes of any size if the pipe is within 7.5 cm of a wall or other continuous support for the rodent.
(f) Crawl horizontally on any type of pipe or conduit.
(g) Jump vertically as much as 1 m from a flat surface.
(h) Jump horizontally 1.2 m on a flat surface.
(i) Jump horizontally at least 2.4 m from an elevation of 4.5 m.
(j) Drop 15 m without being killed or seriously injured.
(k) Burrow vertically in earth to a depth of 1.25 m.
(l) Climb brick or other rough exterior walls which offer footholds to gain access to upper storeys of structures.
(m) Climb vines, shrubs and trees, or travel along telephone or power lines to gain access to upper storeys of buildings.
(n) Reach as much as 33 cm along smooth, vertical walls,
(o) Swim as far as 0.8 km in open water; dive through water plumbing traps and travel in sewer lines even against substantial water currents.
(p) Gnaw through a wide variety of materials, including lead sheeting, sun-dried adobe brick, cinder block and aluminium sheeting.

MUS MUSCULUS   HOUSE MOUSE

Generally known as the House mouse, *Mus musculus* now has a worldwide distribution. It is thought to have originated in the Asian steppes. Four subspecies are recognised.

Description
Grey to grey-brown above and grey to silver-grey below, with the belly almost invariably lighter in colour than the back. The muzzle is pointed, serving to identify all mice from voles, but not nearly so long and pointed as in the shrews. Species of voles, shrews and *Apodemus* enter houses on occasions. The odour of mice is musky and unpleasant, being similar to that of impure acetamide. The size of *M. musculus* is very variable but an average length of head and body of males has been given as 79 mm with the tail another 78 mm. The female measurements average 78 mm head and body with the tail another 78 mm. Males weigh about 15.5 g and females 16.5 g or more. (The sample from which the latter figure was derived contained a number which were pregnant.)

Fig. 2.5 *Mus musculus*, House mouse. With litter.

Habits
The House mouse is a common inhabitant of buildings, dwellings, warehouses (especially where food is stored or prepared), restaurants and kitchens, cold stores, farm-buildings used for corn storage or where spillage has occurred. In addition to buildings this mouse may be found

widely on arable land: these feral forms tend to stay outdoors all the time, rarely mixing with commensal mice. In some areas such as in corn-growing districts of North America and Australia it sometimes reaches plague proportions.

Mice will feed on a wide variety of materials but serious depredations have, in the main, been in connection with grain. An average consumption of about 3.5 g dry weight per day has been estimated, and damage of 16 per cent to grain stacks has been given. About 80 droppings are produced per day, particularly when feeding. *Mus musculus* infestations have been recorded in many unusual situations, such as refrigerated stores where they are able to survive in the wall insulation.

Domestic cats and occasionally *R. norvegicus* are responsible for an uncertain degree of control in dwellings and warehouses.

Breeding
Reproduction is continuous throughout the year but varies from an average of 10.2 litters in corn ricks to 5.5 in urban dwellings. A nest is made of any available material which is then shredded. Average litter size is 5.6. The gestation period is 19–20 days and weaning takes place at 18 days. Females are fecund at 7.5 g bodyweight and males at 10 g.

BANDICOTA   BANDICOOT RATS, MOLE RATS

Two species of bandicoots are of economic importance particularly in India and neighbouring areas of South-East Asia. *Bandicota bengalensis* (Lesser bandicoot) is more common than *B. indica* (Greater bandicoot), and has tended to oust *Rattus rattus* as the commonest commensal rodent in some cities of Burma and India. In 1907 in Bombay, *B. bengalensis* constituted only 1 per cent of the total rodent population, whereas in 1956 this figure had risen to 49.2 per cent. This is particularly important as bandicoots are more susceptible to plague than *R. rattus*.

They are thickset animals with a short tail and coarse, dark brownish-grey fur. *Bandicota bengalensis* is slightly smaller than *B. indica* with head and body length about 200 mm. Full-grown adults rarely exceed 300 g in weight. They are excellent burrowers, hence the name mole rat, and they can even burrow through poorly made concrete.

Bandicoots feed mostly on field crops, vegetables, fruit and stored grain and their range of movement rarely averages more than 30 m.

In buildings bandicoots breed throughout the year and one author claims the annual productivity of females to be the highest for any species of commensal rat. The reported mean litter size is 6.2.

NEOTOMA   WOOD RATS

The eight species of *Neotoma* cover practically the whole of North America. Sometimes they are referred to as Trade rats and pack rats. They are about the same size as *Rattus norvegicus* but instead of a scaly tail and coarse fur, the tail is hairy and the fur is fine. Additionally, the ears are larger and the bellies and feet usually white. They are nocturnal and are rarely seen. Compared with R. norvegicus they are of minor significance but sometimes they invade farmhouses, dwellings and other buildings in search of food or to construct nests. They are known to carry plague and are under suspicion of harbouring other diseases of public health importance.

ERITHIZON PORCUPINES

Two species of Erithizon, *E. dorsatum* (Canadian porcupine) and *E. epixanthum* (Yellow-haired porcupine) sometimes cause damage in farm buildings and in unattended country homes in North America. The former species occurs in the north-east, south to Pennsylvania, in the region of the Great Lakes and then north to Alaska. The latter species is found from the Great Plain westward to the Pacific and from Alaska in the north to southern Arizona along the Rocky Mountains. It appears much larger than its 10 kg weight limit would suggest on account of its covering of quills of which about 30,000 may be present on an adult. Apart from the face, the body underparts and under the tail, the body is covered with them. The stoutest and heaviest quills are on the rump and tail and their pointed tips are black or brown and covered with very small, diamond-shaped scales which act as barbs. The quills can be raised or lowered but cannot be ejected. When menaced, the porcupine will either make a rolling lunge with its body or slap its club-shaped tail, and

Fig. 2.6 *Erithizon dorsatum*, Porcupine. Damages perspiration-stained wood and leather in farm and other buildings in North America. (US Fish and Wildlife Service)

a number of quills may then become stuck in the flesh of its opponent and when the barbs hold fast the quills are pulled from the porcupine. The bead-like eyes are black and at night do not reflect light. The small ears are partially hidden in the fur. The incisor teeth are wide, heavy and have yellow outer enamel. They are used as chisels during winter when the animal feeds on the inner bark (cambium and phloem tissues) of a number of forest trees. In spring, summer and autumn it feeds on herbaceous plants and fruits.

Although never attacking, the porcupine often leaves quills in the tongue and on the face of inquisitive domestic livestock as well as in dogs. When entering buildings it is particularly attracted to woodwork stained with perspiration which it will chew and eat. Farm toolhandles, saddles and canoes are often treated in this manner. It has been known to break and eat plastic covering on a motor car steering wheel.

Human diseases associated with rodents

The capability of transmitting diseases to Man is shared by many rodents. Commensal species, however, are of the utmost importance – not only because of their close association with Man and his food but also because of their large numbers, wide distribution, ability to be carried in trains, motor transport, aircraft and ships and because of their potential role in the transmission of infection from other wild-living, rodent species. The infecting of Man with rodent-borne disease may be directly by bite or by Man coming into contact with the body of the rodent, dead or alive. It may occur indirectly through food contaminated by rodent faeces and urine or through blood-sucking insects or arachnids.

The order in which to discuss diseases transmitted by rodents is difficult to ascertain. Plague, for instance, kills more people than does food poisoning but, on the other hand, the latter is far more common. In the following account the bacteria are, therefore, in alphabetical order followed by the rickettsias and a roundworm.

Leptospiroses

A number of diseases are caused by the various species of *Leptospira*. Rats, mice, dogs, cattle, pigs and, no doubt, other mammals may become infected by Leptospirae and, thereafter, suffer a chronic, infective condition. Such animals excrete the organisms in their urine over long periods of time. Human infection occurs by contact with the leptospirae in the urine. Exposure to this hazard is most pronounced in those working in sewers, digging ditches, farming and slaughterhouse occupations. Swimming in contaminated pools has also given rise to human leptospirosis.

A number of different species of *Leptospira* are involved. Some of those known to cause human infection are *L. icterohaemorrhagiae* (Weil's disease), *L. canicola* (canicola fever); *L. pomona* (swineherd's disease); *L. autumnalis*, (Harvest sickness); *L. grippotyphosa* (swamp fever); and *L. hebdomadis*, (seven-day fever).

Portal of entry
It is not certain how the organisms enter the body, although it is thought that it is through the nose, mouth and cuts and abrasions in the skin. Thereafter the course of the disease varies with the species of *Leptospira* which has gained entry.

In young people mild forms of the disease usually develop but in those over the age of 30 the more severe Weil's disease is more common. The organism usually concerned in the case of the latter is *L. icterohaemorrhagiae*, but more rarely, *L. canicola* is isolated.

Weil's Disease   Haemorrhagic Jaundice

There are 8 to 12 days of incubation before the onset of one or more abrupt chills, followed by a fever. Prominent symptoms are headache, photophobia and muscular pains in the back and calves. In addition, there may be nausea, vomiting, diarrhoea, sore throat, cough or cold sores around the mouth. The fever lasts up to seven days when hepatitis and nephritis commonly occur. Jaundice sets in but subsides by the fourteenth day. About 5 per cent of cases end fatally.

Plague

This rodent-borne disease ravaged the human race for many centuries, causing great loss of life and, indeed, altered the course of human history. Some conjecture must surround its presence in the earliest times but some authors claim that the first epidemic was among the Philistines in 1320 BC. The first pandemic of Bubonic Plague started about AD 542 and is called the 'Justinian Plague'. Better known are the pandemic of the fourteenth and seventeenth centuries which affected practically all the known world. The pathogenic organism concerned is the bacterium *Pasteurella pestis* which occurs in many species of wild-living rodents in areas where Man may or may not be present. These animals may be merely acting as carriers or may become diseased themselves.

Two types of disease (distinguished in the main by the species of rodent from which Plague is transmitted), are also recognised – Urban or Murine Plague and Sylvatic, Campestral or Wild Rodent Plague. Urban Plague is transmitted from commensal rats via the bite of the Oriental rat flea, *Xenopsylla cheopis*, and Sylvatic Plague is transmitted from wild rodents as well as from rabbits and hares by the bite of their fleas or sometimes through handling diseased animals or carcasses.

The role of the flea

Pasteurella pestis multiplies at an extremely fast rate in the blood of an infected rodent, producing an overwhelming septicaemia. When a rat flea feeds on such an animal it takes large numbers of the bacteria with its meal of blood. In a relatively small proportion of fleas (never more than about 12 per cent), the bacterium establishes itself in the gut. Here it multiplies rapidly until the gut is a solid mass of bacteria. Such a flea is known as a 'blocked' flea and shows signs of great hunger by feeding ravenously on any animal within reach, including Man, when its own host dies. When the blocked flea feeds on Man it sucks blood to the limit of the elasticity of the gut then it regurgitates some blood which is now infected with bacteria. In addition, the flea defecates at regular intervals and, again, the faeces contain bacteria. The latter may be rubbed into the bite wound. Blocked fleas can remain infected and alive for at least six weeks, although the first fortnight is the period of effective transmission.

Only 2 of 17 species of flea parasitising rats are constantly known to be vectors of the Plague. These are the Asiatic or Oriental flea, *Xenopsylla cheopis*, and *Nosopsyllus fasciatus*, of which the former is overwhelmingly important. The Human flea, *Pulex irritans*, may transmit *Pasteurella pestis* on occasion from man to man.

When large numbers of the susceptible species of rodents become diseased simultaneously, this is known as an epizootic and a large proportion of them die. (An epizootic in animals corresponds with an epidemic in Man.)

In Man there are three forms of Plague – the Bubonic variety is the one most commonly occurring. This is characterised by the presence of large, painful swellings of the lymph glands called buboes which are often in the groin. This form of disease almost never spreads directly from man to man. Transmission of the bacterium from rat to rat and from rat to man is brought about by the bite of an infected rat flea.

Pneumonic Plague is brought about by the infection invading the respiratory tract and in this form of the disease the coughing victim releases extremely infectious sputum droplets into the air but this form of the disease cannot persist in the absence of the Bubonic form.

Septicaemic Plague is the form which occurs when the disease affects the bloodstream.

Bubonic Plague is the form which has caused disastrous epidemics in the past. In the initial weeks the case-mortality rate appears to approach 90 per cent but as the epidemic subsides it often falls to as low as 30 per cent. From 1900 to 1960 there were 534 cases of Human Plague in the United States and of those 345 (or 65 per cent) were fatal. There have been reports of outbreaks in California as recently as 1976. Pneumonic Plague is probably the deadliest of bacterial diseases of Man and, in the past before the use of antibiotics, a case-mortality rate of practically 100 per cent occurred.

Tularemia

This disease is caused by the bacterium *Pasteurella tularensis*, which is related to that causing Plague. A large number of rodent species, (including commensal rats and mice), are infected by this organism. One account states that human infection is brought about by the handling of infected rodent carcasses or from drinking contaminated water. In North America, however, contact with infected rabbits during autumn, winter and spring is said to be the cause, as well as the bites of ticks and deer flies in summer. The species of ticks involved are said to be mainly *Dermacentor andersoni, D. variabilis* and *Amblyomma americanum* (see page 75). *Chrysops discalis* is the deer fly most commonly implicated.

Salmonellosis   Food poisoning

Salmonellosis is a general term describing any infection of Man or animal involving bacteria of the genus *Salmonella*. It is commonly called 'food poisoning'. Hundreds of *Salmonella* serotypes are recognised but the most common rodent-borne species are *S. typhimurium* and *S. enteriditis*. Both have been used in the past as rodenticides but with little success. Rodents transmit the bacteria to Man through infected droppings which contaminate foods in supermarkets, bakeries and restaurants.

Symptoms are characterised by vomiting, diarrhoea and abdominal pains. It is highly infectious from man to man but fatalities are rare. Effects of the bacteria are often mild and the association with rats and mice frequently goes unnoticed.

Spirillary rat-bite fever

This disease is caused by the spirochaete *Spirillum minus* and is recognised by a relapsing-type fever, together with a hard ulcer at the site of the rat bite. The causative organism is short, 2–5 μm in length, thick and spiralling into from one to three angular curves and with polar flagellae present. It moves rapidly with a. darting motion.

The rat-bite wound heals fairly rapidly unless there is a secondary infection, but after an incubation period of from 5 to 28 days there is a flare-up of the wound which is accompanied by chills, fever, headache and general malaise. The wound becomes swollen and purple and may ulcerate. After a few days the symptoms subside but several days later they reappear. Temperature rises and falls with the periodic fever and may continue for several weeks if treatment is not given. Early in the disease a sparse, spotty rash may appear on the legs, arms and trunk. A fatality rate of 10 per cent has been recorded.

Streptobacillary rat-bite fever

This disease does not invariably follow the bite of a rat – in some cases it may be acquired through eating infected food. It is caused by the organism *Streptobacillus moniliformis*, which develops in chains in several forms, but is generally 2–15 μm in length. The organism commonly occurs in the nose and pharynx of both wild and laboratory rats.

The incubation period of the disease is shorter than that for spirillary rat-bite fever, being from one to five days. Only rarely does an abscess form at the wound and usually normal healing takes place but chills, fever, vomiting, headache and severe pains in the back and joints occur abruptly. Within the first 48 hours a rash appears and the disease is similar to that of dengue; the fever often abating within two or three days, but one or more joints become swollen, red and very painful. This acute arthritis is persistent and is a prominent symptom. Subcutaneous abscesses containing the organism often occur. About 7 per cent of cases are reported as being fatal.

Rickettsialpox

The organism, *Rickettsia akari*, occurs in a number of house mice in the USA, South Africa and the USSR, and is transmitted to Man through the bite of the House mouse mite, *Allodermanyssus sanguineus*. A rash similar to that caused by chickenpox is produced as well as fever. The Tropical rat mite, *Ornithonyssus bacoti*, is known to transmit the disease under laboratory conditions.

Murine typhus

Murine endemic typhus fever is caused by *Rickettsia mooseri* (= *R. typhi*). In nature it is maintained as a mild disease of rats, transmitted from rat to rat by the Rat flea, *Xenopsylla cheopsis*, or by the Rat louse, *Polypax spinulosus*. The presence of the rickettsiae has no great effect on the health or length of life of the rat or its flea.

The acquisition of the disease by Man usually comes about through the bite of an infected Rat flea. It is possible also that food recently contaminated by infected rat urine or faeces of infected Rat fleas, if ingested, would result in murine typhus. In nature the cycle of rat–flea–rat maintains the disease, i.e. it is not obligatory for Man to be present. When Man becomes infected the disease does not spread from man to man. There have, however, been unconfirmed reports that murine typhus rickettsiae have been transmitted to Man by the Body louse.

The incubation period is generally 12 days but may vary from 6 to 14 days. Although similar to epidemic louse-borne typhus, *R. prowazekii*, in many respects, murine typhus fever is milder and shorter in duration, the rash being not only less extensive but also less persistent. There are fewer complications and the fatality rate is lower – at less than 5 per cent for all groups, with the great majority occurring in older patients.

Scrub typhus

This is a disease occurring over a large part of South-East Asia. The pathogenic organism is *Rickettsia tsutsugamushi*, which infects a number of rodent species including the commensals, and is transmitted to Man by the bite of the larval stage of mites of the genus *Leptotrombicula* (see page 66).

Trichinosis

Trichinosis is an infection of many species of animals, including Man, rodents and pigs. The causative agent is the small roundworm, *Trichinella spiralis*.

Man becomes infected by eating raw or improperly cooked pork, which in turn has become infected by the pig eating an infected rat. Rats can become infected in a number of ways, but usually by eating another infected rat or pig-remains at slaughterhouses. In Man, the disease rapidly produces a fever, gastrointestinal symptoms, muscular pain and eosinophilia. It is occasionally fatal, if untreated.

Chapter 3

Bats

CHIROPTERA

Bats are the only true flying mammals. They are capable of long, sustained flight and in the intricacy of their aerobatics they emulate birds. On the other hand, they are generally awkward and slow when on the ground. When resting, they hang by the toes of their hindfeet, which are so adapted that this may be done for long periods without effort.

Classification
There are about 800 different species of bats, making them one of the largest mammalian orders, and about one-seventh of all mammals are bats. Generally they are so little-known that it is certain that many species are still to be described, especially in tropical areas where they are most numerous.

Bats are divided into two suborders, separated basically by their size, although there is an amount of overlapping. The MEGACHIROPTERA or 'big bats' vary in weight from 25 to 900 g, the wing span is between 250 and 1,500 mm. They also possess large eyes and there is nearly always a claw on the first finger as well as on the thumb. Usually they have 'dog-like' faces and are often known as 'flying foxes'. Although by no means all of them feed on fruits, they are also known as 'fruit bats'. Usually they roost in trees (often in large colonies), but sometimes they use outside walls of buildings.

The MICROCHIROPTERA or 'small bats' range from 3.5 to 180 g in weight and the wing span is between 150 and 900 mm. They are divided into 16 or 18 families. No fingers possess a claw and there is an extraordinary range of facial appearance due to the shape and size of the ears and the membranous extensions of the nose, called the nose leaf.

The family DESMODONTIDAE contains the vampires which subsist only on vertebrate blood.

Description
Bats are externally characterised by the enormously developed forelimbs being formed into wings; a membrane of skin extends between the handbones to the forearm, side of the body and the hindleg.

Additionally, most bats possess a membrane connecting the legs and including the tail, known as the interfemoral membrane. A cartilaginous support for the free edge of the interfemoral membrane is known as the calcar and is fixed to the inside of the foot and extends out along the membrane edge. The keel is a definite extension of the free edge of the membrane beyond the calcar but if the latter lies along this free edge it is said not to be keeled. A leaf-like structure in the ear is known as the tragus.

Habits
All bats are nocturnal, hiding away during daytime in hollow tree-trunks, caves and rock crevices. Many have adapted themselves to using buildings for this purpose, especially roof voids. They often enter through a relatively narrow opening to find an ideal situation – dry, out of direct sunlight and free from predators. Many species of bats congregate in colonies which may number several hundreds or even thousands of individuals. In such cases a bat colony may cause considerable annoyance due to odour, faecal droppings (which may support undesirable insects), urine stains on ceilings, general unhygienic conditions as well as noise. More important, is the factor that many species of bats are known carriers of rabies which their sharp teeth can easily transfer to another animal, including Man. For this reason it is most unwise to handle bats, especially those which are obviously sick, unless strong gloves are worn.

No bats make nest or other habitations.

Feeding
Most of the MEGACHIROPTERA eat fruit or other vegetable material – such as flowers, pollen and nectar, but the MICROCHIROPTERA show a much wider variation in types of food. Most are insectivorous and in many areas are beneficial in removing large numbers of night-flying insects. Some bats in the family NOCTILIONIDAE catch surface-swimming fish, as does a Mexican species in the VESPERTILIONIDAE. Others, in the families MEGADERMATIDAE (the false vampires) and PHYLLOSTOMATIDAE, take vertebrates such as lizards, mice and other bats. Some species in the latter family feed on fruit, pollen and nectar. The blood-sucking vampires can drink large quantities of blood fairly quickly and become engorged, excess water being quickly excreted. The saliva contains blood anticoagulants.

Hibernation
Insectivorous bats in temperate climates obviously experience a period of insect scarcity during the winter. Such bats hibernate, although in some species it is often interrupted during periods of warm weather (even in mid-winter), and they may change hibernating roosts.

Atmospheric conditions during hibernation are very important to bats.

Temperature must be low so that body reserves are not exhausted too rapidly, but, on the other hand, if it is too low the bats may die. Atmospheric humidity is important too, if not high enough the bats will desiccate, so damp places are always chosen.

Flight guidance
The manner in which bats, especially the MICROCHIROPTERA, find their way about, avoid obstacles and hunt for food, has attracted a great deal of attention. No bat is blind, although there is variation in the size and performance of their eyes which are only of use in twilight. About 200 species of the MICROCHIROPTERA have been investigated and all have been found to emit sound pulses during flight and when active on the ground. These pulses are produced from a modified voice box and are nearly always ultrasonic in character, i.e. they are too high in pitch to be audible to the human ear. The sound pulses vary in duration from about 0.25 ms to as much as 60 ms. The echo-location system is of great sensitivity. Some species are known to be capable of detecting and avoiding wires only 0.1 mm in thickness. Flying nocturnal insects – such as moths – are found, followed and intercepted in mid-air. It is of great interest, however, that some groups of moths possess hearing organs which can detect bats' ultrasonic pulses and they will endeavour to evade the predator. Some other moths are able to produce their own ultrasonic pulses as a warning of their distastefulness.

Breeding
Mating takes place in the roosts inhabited during the day and there is little or no courtship. There are no pair formations, bats being entirely promiscuous. In temperate regions, mating occurs in the autumn but ovulation does not take place until the following spring on account of the interruption brought about by hibernation, during which time the sperm are stored within the female. Mating may take place again in the spring but the greater number of conceptions are brought about by the delayed fertilisation.

Young bats are born in colonies consisting of all females. One well-developed young is born at a time, although in some species twins are not infrequent. The female has two teats on the chest and in some there is an additional pair in the groin which do not lactate but which the young one clings to during flight. When the young bat becomes too much of an encumbrance it is left to roost while the female seeks food.

DESMODUS ROTUNDUS   VAMPIRE BAT

The Vampire bat, *Desmodus rotundus*, nourishes itself only by drinking blood of mammals, including man. This it is able to do by making a cut with its razor-sharp teeth then lapping the blood as it flows. Even a lightly-sleeping person may be parasitised in this way without waking,

Fig. 3.1 *Desmodus rotundus*, Vampire bat. Three-quarter view of face, showing large eyes and razor sharp teeth. (Christine Hawkey, Zoological Society of London)

even though the bat is resting on its victim. The Vampire bat is about 80 mm in length, grey to brown in colour, with short fur. Its eyes are large and its ears are sharply pointed. It can run swiftly – even up vertical walls – and can hop, quite unlike the usual movement of a bat on the ground. It breeds at any time of the year but has only one litter annually. The range of distribution of the Vampire bats, of which there are three species, is from Mexico to central Argentina. Vampire bats are the only parasitic mammals.

In Britain, under the Wildlife and Countryside Act 1981, it is illegal for anyone without a licence intentionally to kill, injure or handle a wild bat of any species.

Chapter 4

Insectivores

INSECTIVORA

The order INSECTIVORA is the most primitive of the placental mammals and consists of generally small animals with the following characters.

Most possess a fleshy muzzle or proboscis projecting in front of the teeth. With only a few exceptions the eyes and ears are small, frequently hidden by skin or fur and the number of toes on all feet is five, all projecting forwards. (Only four toes are present on the forefeet of most mice with which shrews may be confused.) The brain of insectivores is low and flat, not much expanded beyond the level of the forehead. The cheekteeth have sharp conical cusps enabling the animals to seize and crush insects.

The insectivores are classified into eight main groups, the taxonomic importance of which is in doubt. Only three of the families of INSECTIVORA concern us here and two of these only remotely.

The ERINACEIDAE includes the European hedgehog which often inhabits suburban gardens and sometimes hibernates in the garden shed, or among leaves in an outhouse.

The TALPIDAE includes the moles which, although only entering buildings rarely and accidentally, are otherwise a source of great annoyance because of their damage to lawns and gardens.

The SORICIDAE, however, contains the shrews. These are all small or very small animals (the smallest mammal is a shrew) with long slender and very sensitive snouts beset with vibrissae. They have small eyes, velvet-like fur and short legs.

There are about 200 species of SORICIDAE, and they are generally distributed throughout the temperate and tropical world, except for Australia and the greater part of South America.

In Britain, *Sorex araneus*, the Common shrew and *S. minutus*, Pygmy shrew enter houses, where the householder usually mistakes them for the House mouse, *Mus musculus*. *Sorex minutus* is a good climber and is sometimes found as high as the upper floors of a dwelling. In Europe, in rural areas, they often make their way into buildings, especially during autumn, where they may become a nuisance, but do little damage.

In South-East Asia, the House or Musk shrew (*Suncus murinus*) is found only in or near buildings and habitations. It is up to 140 mm long, with the tail accounting for about another 80 mm. This shrew has a long mobile nose, glossy brown or black fur and produces a strong scent which smells of musk. It will eat anything, but prefers larger insects, such as cockroaches or crickets. House shrews have been known to eat young rats. Often this species is considered beneficial.

Shrews are short-lived, usually breeding only for one season, and are very active. Although consuming large numbers of insects and other arthropods some species will eat vegetable matter such as seeds.

Chapter 5

Marsupials

MARSUPIALIA

Marsupials are primitive mammals showing anatomical characters, principally of the urinogenital system, embryonic development, teeth and foot arrangements, very different from those of higher mammals. Only the egg-laying monotremes, the platypus and the echidnas are more primitive, exhibiting a number of features associated with reptiles.

Marsupials are perhaps best known for their manner of giving birth. The period of gestation is always very short, from as little as 12 days to no longer than 38 days in the largest kangaroos. The young, when born, are at an early stage of development and are very small. Those of the Brush-tailed possum weigh only 0.2 g, which is about 0.013 per cent of the body weight of the mother, and those of the Red kangaroo, the largest living marsupial, still only weigh 1.5 g, which is about 0.003 per cent of its mother's weight. The young emerge from the single external aperture, the cloaca, and crawl upwards on the mother's belly to a group of teats. These are situated generally, but not invariably, in a pouch or marsupium. The young animal takes a teat into its mouth and the teat then enlarges, thus anchoring it firmly.

The young are suckled in the pouch until they roughly correspond in development with the young of higher mammals at birth. The length of the pouch-life may be as little as 8 weeks in some species to nearly 12 months in the case of the Eastern grey kangaroo. The young then leave the pouch but return to suckle for a period, or for shelter. There is no release of eggs during the pouch-suckling period, which prevents two litters occupying the pouch at the same time.

The most important characteristic of marsupials, however, concerns the urinary ducts, or ureters, which separate the sex ducts as they develop. This has the effect of doubling the uterus and vagina in the adult female. In the male, the paired vasa deferentia transporting spermatozoa from the testes to the penis lie to the outside of the urinary ducts, while in the higher mammals they lie between the ureters and loop over them at the descent of the testes. Marsupials generally possess more incisor teeth than placental mammals, having as many as five on each side in the upper jaw and three in the lower jaw. The cheekteeth consist of three premolars compared with four in placentals and four molars instead of three. There are also important differences in the provision of deciduous (milk) teeth: in marsupials the only deciduous teeth are the last premolars, whereas in the placental mammals the milk dentition is total with the exception of the molars. However, the marsupial embryo does possess a number of rudimentary teeth, the pre-lacteals, which are lost at an early stage of development, or do not even progress further than the germinal layers.

The cytology of marsupials is interesting in that the chromosome number ranges from only 10 to 32, the average being 18, while the average in placental mammals is 48, and for the monotremes it is 58. The brain of marsupials is comparatively smaller and of much simpler organisation than that of the placentals.

Only two marsupial species have association with buildings. In each case only shelter is sought, but the resultant unhygienic conditions and noise cause them to be undesirable occupants.

TRICHOSURUS VULPECULA   BRUSH-TAILED POSSUM

In the suburbs of Sydney, Adelaide, Perth and in various parts of Canberra, in addition to other cities in Australia, this widely distributed member of the PHALANGERIDAE (or phalangers), is an important invader of buildings. It often occupies the roof-space and causes disturbance as well as fouling the ceiling with its urine. It is usually live-trapped and released elsewhere – in a national park if possible – as Australian law forbids their being killed. The popular name of possum distinguishes the phalangers of Australasian distribution from the opossums of America.

Brush-tailed possums occur from North Queensland through New South Wales and Victoria, to South Australia, also in Western Australia and Northern Territories, as well as being serious horticultural pests in Tasmania and New Zealand. They are often seen dead on the road, having been killed by motor traffic.

Trichosurus is somewhat stouter in build than the domestic cat, the head and body being about 45 cm in length and the tail a further 30 cm. The muzzle is rather blunt and the ears are long and oval, giving a fox-like appearance. The head and body are silvery-grey, but the belly is a dirty yellow: there is, however, much colour variation. In Tasmania, where it was first introduced in 1837, the colour phases are most commonly black and grey. The fur is thick, soft and woolly and has been extensively exploited. The prehensile, bushy tail which is grey at the base with a black tip, gives some support when tree-climbing. The terminal half of the tail is naked on the underside. The forefeet each have five toes furnished with a sharp claw. The hindfeet have five toes, but the second and third are jointed, and the first toe which is clawless is opposable so that branches may be grasped.

Habits
Being nocturnal, it is seldom seen but is often abundant in dry forest,

Fig. 5.1 *Trichosurus vulpecula*, Tasmanian brush-tailed possum. (Australian News and Information Bureau)

open woodland and wherever there are sufficient trees in suburban areas. Even in open plains the possum penetrates along tree-lined river systems. It makes its nest in hollow tree branches, frequently at considerable height from the ground. Entirely herbivorous, the animal subsists on a variety of grasses, herbs and tree leaves. The male marks out its territory by rubbing a secretion from its chest on to trees and other objects.

The period of gestation is 17 days. The new-born young are only 13 mm in length and weigh 0.2 g, compared with the adult weight of 2 kg, that is 10,000 times heavier. The strong forelimbs bear sharp, inwardly-directed claws which are the animal's sole means of climbing through the mother's fur to her pouch. The claws are not shed in the pouch as in many other marsupials. Pouch-life is four to five months, and weaning is at six months. Sexual maturity is reached at 12–15 months. Generally only one young is born at a time, and either one or two litters, according to district, are reared annually.

DIDELPHIS MARSUPIALIS   VIRGINIAN OPOSSUM

Another marsupial which causes some fouling of buildings in which it sometimes takes up residence, is the Virginian opossum, *Didelphis marsupialis*, a member of the primitive family DIDELPHIDAE.

Originating in South America, it has colonised Central and North America within recent times, and has reached Canada. On the eastern side of the United States it occurs from Florida to the New England states. To the west it is found in Iowa, then south to southern California and Mexico. This animal has a total length of about 1 m of which approximately 40 cm is taken up by the nearly, or completely naked, prehensile tail. The poor quality fur is greyish to nearly white. Ears are nearly hairless, the eyes are dark and the snout is pointed. All feet are nearly black and have five separated toes, and one clawless toe is opposable on each foot. The total weight may be up to 2.3 kg and a 2 kg animal can get through a hole 7 cm in diameter. True hibernation does not take place, but times of bad weather are spent in its filthy den in a hollow tree trunk, in an underground burrow or in a building.

Although readily climbing trees, especially as an escape reaction, *Didelphis* spends most of its time on the ground. If surprised by a predator such as Man, it 'plays possum', pretending to be dead by assuming a state of tension. It will also take up a threatening posture and show its teeth, and may also hang from a branch by its tail. The opossum feeds both on vegetable and animal matter, including fungi, certain herbaceous plants, fruits, insects and other invertebrates, frogs reptiles, birds (including poultry) and small mammals. The nest of leaves and grass is placed in a burrow, a rock crevice, a hollow tree or a building. It is generally in a very unhygienic state.

In the southern states of the United States of America the Virginian opossum produces two litters annually, beginning at the age of eight months, but in the northern states one litter only is produced annually, commencing at one year old. Only eight hours after copulation, up to 25 egg cells are fertilised, and only 12 to 13 days afterwards the young are born. They are about 12 mm in length, and weigh only 0.16 g each.

The young make their way to the brood sac, but only the first 11 to 13 become attached to teats. The remainder perish. Their weight increases tenfold during the first week, and at four weeks they show their heads. At five weeks they may leave the pouch briefly, and at eight weeks they are fully independent. The opossum lives for up to eight years.

Chapter 6

Birds

AVES

Birds are probably the most easily recognised of all vertebrates. The adaptations to aerial flight are remarkable. The covering of feathers, the modification of the forelimbs as wings and changes in the structure of the breast-bone in order to accommodate the muscles actuating the wings, are but some of these adaptations. On the other hand, features showing reptile or dinosaur ancestry are the scaly legs, the horny beak and the laying of eggs. There are thought to be about 8,600 different species of birds, and although a large number of these cause damage to agricultural crops, relatively few are harmful in or on buildings.

Two bird species only can be looked upon as true commensals. These are the House sparrow, *Passer domesticus*, and the Feral pigeon, a variety of *Columba livia*. Both these birds are, to a large extent, dependant on Man for food, shelter and nest sites. In many large cities these species are found in great numbers where, in addition to enjoying their association with Man, they suffer little at the hands of predators. A number of other birds are partial commensals in that they seek the shelter of urban areas, often in immense flocks, for roosting at night. The Starling, *Sturnus vulgaris*, provides an example of this.

The harm done by birds to buildings is considerable, not only by fouling with faecal matter but by corroding the stone-work. It has recently been shown that certain fungi and bacteria flourish in pigeon excrement, and that it is the acidic products from these organisms that corrode the stone. In roosting areas the droppings are unpleasant and often dangerous, for example, by making pavements slippery. Commensal birds are potential disease transmitters both to Man and his domestic animals, although convincing evidence of actual transmission of pathogenic organisms to Man is not in any way substantial. Sparrows and pigeons are often found inside food warehouses, where they not only contaminate foodstuffs with their droppings but also peck open containers, causing spillage and wastage. Bird droppings and birds' nests also act as important reservoirs for many species of pest insects and mites, which may then enter buildings particularly after the young birds have left their nests.

Some species of woodpeckers (PICINAE) peck holes in roof shingles and other wooden structures (including mahogany lamp standards in one town) and create seasonal noise problems.

COLUMBIDAE

Doves and pigeons are included in this family. Most are attractive birds and they have always taken a place in Man's drawings and writings. Two species in particular, however, can be considered as pests when they occur in or near our buildings.

COLUMBA LIVIA   FERAL PIGEON

Of all the birds associated with Man and his buildings, the pigeon is certainly one of the most important. The history of its domestication goes back at least 3,000 years. There have been a number of theories put forward to account for the origin of domestic breeds, but that suggested by Charles Darwin in 1859 appears to be universally agreed. This gives the Rock dove (or Blue-barred rock pigeon as it is called in the United States), *Columba livia*, as the original progenitor. An inhabitant of rocky or cliff-girt coastlands, this species is widely distributed in Europe, Asia and Africa, a number of subspecies occurring in the different areas.

It seems most likely that domestication first took place in Asia, and a number of domestic breeds were introduced into Europe from that continent. The ancestral coloration of dark blue wing bars most commonly shows itself in domestic breeds. The pigeon of city streets is derived from domestic varieties which have assumed a semi-wild or 'feral' way of life. It is probable that there are few truly wild flocks of *C. livia* as there is usually evidence of interbreeding with feral and domestic birds.

Description
The Feral pigeon is so well known that a description of its appearance is unnecessary. At least 20 colour patterns (phenotypes) are known, which are largely derived from domestic stock. The wild Rock dove has bluish-grey plumage with two parallel, blue bars on the wings: this type colour is known as 'blue-bar'. In an estimation of polymorphism in pigeons in Syracuse, New York, this colour pattern occurred to the extent of 29 per cent. The pattern known as 'blue-checker' was only slightly less abundant at 28 per cent, with 'blue T-pattern' 21 per cent and 'spread black' 13 per cent. The remaining colour patterns were much less common, ranging from 3.4 to 0.2 per cent.

Habits
A population of feral pigeons is divided into a number of flocks whose individuals rest, roost and feed together with little interchange between

flocks. The territory occupied by the flock is relatively stable as is also the number of birds constituting the flock. Few birds venture more than 1,000 m from their usual flocking area. Nesting also takes place in colonies.

Breeding

Feral pigeons breed throughout the year. There is no season of non-breeding activity intervening between generations. Many young birds leaving the nest in January achieve breeding condition during the summer of the same calendar year. Often birds ringed as chicks pair and are incubating eggs at the age of six months. Birds younger than this can be distinguished by the presence of brown-tipped juvenile features.

The nest site is usually a ledge on a building where there is some protection from above by a parapet or overhanging masonry. A ledge under a bridge is frequently chosen. The nest consists only of a few twigs and stems of dried grass and is slight and almost flat. One or two eggs are laid, never more. They are blunt oval in shape, about 40 × 30 mm in size, and with a smooth white shell with a little gloss. Both sexes take part in incubation although the male's share is often meagre.

Incubation is about 17 days and for the first few days the young are fed with 'pigeon's milk', a cheesy substance produced in the crop of the parents.

STREPTOPELIA DECAOCTO   COLLARED DOVE

This small dove, which prefers to live close to Man, has spread from northern India across to Britain, where the first breeding occurred in 1952. Since that time, it has spread throughout most of the British Isles. The Collared dove is occasionally a pest, when its cries (in addition to cooing) disturb the occupants of nearby houses. Sometimes the birds frequent mills and may contaminate food with droppings and feathers.

PLOCEIDAE

The weaver birds, of considerable importance as agricultural pests, form a major part of this family. As pests of buildings, however, we are mostly concerned with sparrows.

PASSER DOMESTICUS   HOUSE SPARROW

This very common bird is closely associated with Man, on and around his buildings and cultivated land. It has been said to be the most successful city dweller of all birds.

In the United States it has a number of popular names, among which are English sparrow, European House sparrow, Gamin, Tramp, Hoodlum and Domestic sparrow.

Distribution

The House sparrow is found throughout Europe, with the exception of the greater part of Italy and some of the Mediterranean Islands. In these areas, as well as in North Africa and temperate Asia, distinct races occur. It does not occur in Finland or Russia. The bird has been introduced to many other countries and now occurs in North and South America, South Africa, Australasia (except Western Australia), Hawaii, etc.

Passer domesticus was first introduced into North America in 1850 when eight pairs were liberated in Brooklyn, NY. Since that time a number of further importations have been made and small groups are distributed throughout the United States. Today the House sparrow is found throughout the whole of the North American continent, wherever there are human settlements, with the exception of southern Florida. It is found also in the Bahamas, Bermuda and Cuba in the south, and Nova Scotia in the north.

Description

The House sparrow, like the Feral pigeon is so well known that it scarcely requires description. It may, however, be of value to give the salient features to assist with the identification of the sexes.

The cock sparrow has the upper parts chestnut-brown streaked with black. The crown is dark grey, the cheeks greyish and the throat black. There is a short but distinct whitish wing-bar. The underparts are rather dirty greyish-white, and the rump is grey. The female, as well as the immature male, lacks the chestnut-brown upper parts, grey crown and whitish wing-bar.

Habits

The House sparrow is gregarious and found much more commonly on the ground or on buildings, than perching in trees. Perhaps one of the factors contributing to the success of the House sparrow is its adaptability with respect to diet. In agricultural areas, corn can constitute three-quarters of its food, together with weed seeds and a few insects. Conversely, in urban areas, it subsists on street refuse, some insects and seeds.

The rapid dissemination of this bird throughout the North American continent has been attributed to its hardness, high rate of reproduction, great diversity in feeding habits, aggressive disposition and lack of natural enemies. Where introduced, the House sparrow is said to interfere with the breeding of some native birds by denying access to suitable breeding situations.

Serious damage may be caused by the birds transporting nesting materials into spouting, guttering and similar places on buildings, thus blocking them and fouling cisterns.

Breeding
The typical nesting site is in a hole in a building, but spaces under eaves, in spouting or hopperheads are common sites. Sometimes, large straggling, domed nests are constructed in high hedges. Dried grass stalks or straw are used for the nest, which is lined with feathers; more than 300 have been found in one nest. The clutch size is usually between three and five but up to eight eggs have been recorded. The eggs are variable in colour, usually greyish-white spotted with dark and light ash-grey, and brownish. One egg in the clutch is often lighter in colour than the rest. The average size of eggs in Britain is 22.5 × 15.7 mm.

In Britain, the breeding season commences in May and often continues until July or August. Two or three broods are usually produced. The eggs are incubated by the hen, with help from the cock, and hatching takes 12–14 days. The cock and hen feed the young from the beak and by regurgitation of crop contents. A constant source of insects is required. The fledgling period is 15 days.

PASSER MONTANUS   TREE SPARROW

In Asia, more particularly in the tropical areas, the place of *Passer domesticus* is taken by *P. montanus*. From central Europe to Japan it often occurs in extraordinary numbers especially in rice-growing areas where it does serious harm to crops. In spite of 5 to 10 million birds being sold annually for food, its numbers never appear to decrease. The Tree sparrow is a slightly smaller and more slender bird than *P. domesticus*. It resembles the cock House sparrow, and both sexes are alike. It does not, however, possess the grey crown, but it bears a black patch on the cheek.

Breeding
The nest is made in any sort of hole, in trees, sheds, haystacks, thatched roofs, cliffs, nests of larger birds and holes in buildings. The four to six eggs are smaller than those of *P. domesticus* and are darker and browner as well as having finer stippling. There are several broods annually.

STURNIDAE

This family of Old World origin contains 111 species of which *Sturnus vulgaris*, the European starling is the most familiar. Members of the family are active, aggressive, medium-sized birds with straight or slightly down-curved beaks. They generally posses a waddling walk and their flight is strong and direct. They are usually dark-coloured but often the feathers are shot with metallic sheens and in a number of species there are patches of white, yellow or red bare skin, and wattles are sometimes present.

STURNUS VULGARIS   EUROPEAN STARLING

Starlings are very common throughout Europe and are found in most cities and towns, often in very large numbers. They have spread to Australia and North America; other species, such as the Glossy tree-starling, *Aplonis panayensis*, of Singapore, take the place of the European starling in some parts of the world. All take advantage of our towns for roosting – a phenomenon of recent occurrence, starting only 70–90 years ago.

The starling was introduced into the United States in 1890, when 80 birds were set free in Central Park, New York. This was followed by another batch of 40 the following year. Five years later, they occurred in Brooklyn, and in another two years they were said to be well established in New York City. During this time they had spread over the first 40 miles of Long Island, along the Hudson River as far as Ossining, over much of eastern New Jersey and extended into Pennsylvania and Delaware. Today, starlings are found in most parts of the USA and southern Canada.

Description
The starling is most easily recognised by its quick, jerky walk and its bustling runs when searching for food on lawns and in farmyards. The swift flight with rapidly moving wings and the glides with extended wings are also characteristic. The plumage is iridescent and the feathers are white-tipped giving a spotted appearance. The bird feeds in flocks, probing turf with a quick jerky motion, searching for soil-inhabiting insects (particularly tipulid larvae). In summer, starlings take a large amount of fruit.

Habits
Besides the generally insanitary conditions brought about by the nesting of the starling in buildings, grave concern is often occasioned by the habit of communal roosting. Immense numbers of the birds frequently take part in this and although the roosts are generally in reed or river beds or dense coverts, in some districts they are situated in the centre of cities. They gather into flocks in the open country, sometimes as much as 30 miles away from the communal roost. The starlings then fly in, more or less directly, joining other flocks and often performing spectacular manoeuvres in the air before settling on and around every conceivable foothold on the roost. Thereafter movement goes on until dark, accompanied by constant twittering. The faecal matter deposited on buildings used as a roost creates a serious hygiene problem and the noise becomes intolerable.

Breeding
A large untidy nest of dried grass and feathers is fitted into any suitable aperture in buildings; gutters, hopperheads and downpipes are made use

of and if there is a hole large enough to admit the bird leading to a roof void or other cavity it is virtually certain to be utilised. The four to six pale blue eggs measure about 30 × 21 mm and they may be found from mid-April onwards. Incubation lasts 12–13 days and the young fledge in 20–22 days. At this stage the nest is most insanitary. In Britain two clutches are occasionally laid, although one is more common, but in some countries two is general and even more are sometimes produced.

Fig. 6.1 Nest of *Sturnus vulgaris*, the European starling. In a hopperhead.

LARIDAE

The most familiar of birds associated with the sea are the gulls. Most species, however, seldom venture far from land and some individuals are often encountered far from the sea; in their worldwide distribution they are absent only from deserts and permanently frozen areas. There are 43 species, and they are generally large, 30 to 80 cm in length, with fairly long, powerful beaks with a hook-shaped tip, long wings and webbed feet. Five species of gulls nest, or have nested, on buildings in Britain and whereas around 1940 this habit of nesting in areas frequented by human beings was almost unknown, at the present time it is widespread and increasing. Three species are mainly involved, the Herring gull, the Lesser black-backed gull and the Kittiwake. The first named is the most common and widespread.

Descriptions
Larus argentatus, the Herring gull, is the commonest gull in many parts of its circumpolar distribution, including Britain. It is generally silvery-white with pale, dove-grey wings which show a few white patches; the terminal part of the wing is black with white tips. The beak is yellow with a red spot near the tip of the lower mandible. Legs and feet are flesh-coloured.

Larus fuscus, the Lesser black-backed gull, is about the same size as the Herring gull (50 cm), but has slate-grey upper-parts and wings, and yellow legs. The bill is also yellow with a red spot. This species is much smaller and paler than the Greater black-backed gull, *L. marinus*.

Rissa tridactyla, the Kittiwake, is smaller, about 40 cm in length and of slighter build than the previous species. The grey wings have a black terminal portion, but there are no white tips. Legs are black. Flight is more graceful and the bird has a quicker wing action than the larger gulls.

Habits
There have been large increases in the numbers of gulls, including the species already referred to. This has been brought about by reduced persecution by shooting and egg-robbing and the increase in urban scavenging, especially in winter, at ports, rubbish dumps and sewage works. Another important factor is the increasing fish waste from boats in coastal waters. These factors have, in turn, brought about pressure on normal nesting sites which is probably the main reason for nesting on buildings; there is considerable scope for extension of this habit.

Many nests appear to be tolerated by Man, but in some districts, notably Dover, South Shields and Sunderland, vigorous measures have been taken for some years to destroy nests and drive the birds away. These efforts have, in the main, failed in their purpose. In 1969–70, the number of nests of the Herring gull on buildings exceeded 1,200. The largest colony, of 225 nests, was on the roofs in Dover in Kent, while that at Newquay, Cornwall (189 nests) was the second largest. Several other colonies were around 100 nests.

The overwhelming majority of nests were built on the roofs of houses, shops, offices and so on, often between the chimney pots. The localities were mainly coastal, but not invariably so. In north-east England, in the coastal towns of Sunderland and South Shields, there were 166 and 236 nests, respectively, in 1974. These caused widespread disruption in the town centres. Apart from the fouling and blockage of gutters, and general deterioration of the fabric of the buildings, during the breeding season attacks were made on anyone trying to approach the nesting sites. Streets had to be closed due to the large number of fledgling birds running about. There is also a considerable noise problem in the early morning.

The Lesser black-backed gull has been slower to adopt the habit of nesting on buildings. The first known instance was in South Wales in 1945, and in 1970 there were five colonies involving 62 pairs, almost all in the same area. Three colonies were sited on factory roofs.

The Kittiwake started to nest on man-made structures in 1931 on a harbour wall at Edinburgh, Scotland, and in 1934 it began to nest on a warehouse further along the coast at Dunbar. This is now the largest colony, with 163 pairs in 1969. Other sites are mainly in north-east England, except for one at Lowestoft in Suffolk. In 1969–70, however, there were only seven colonies with a total of 410 pairs.

Colony size merely follows trends in the wild populations, and is due largely to their low inter-nest distance, allowing more pairs to nest on a small area of buildings. It is of interest that the average colony size is over twice that of *Larus argentatus* and five times that of *Larus fuscus*. Most colonies are sited on warehouses and similar buildings, the nests being placed on window-ledges rather than roofs.

Nesting on buildings appears to be more prevalent in Britain than elsewhere, although the habit is widespread. In Bulgaria, the Herring gull has roof-nested since 1894 and in 1970 there were several hundred nests on the roof-tops of Varna and Burgas. In north-west Germany the same species has roof-nested in Bremerhavn since 1956, and there were 80–100 pairs in 1971. In Wilhemshavn a colony was started in 1961 and there were approximately 50 pairs in 1971. In Boston, Massachusetts, a colony of 150 pairs was found to be nesting on flat roofs in the harbour area and they were later found at Logan airport, Boston.

The Kittiwake has nested on window-ledges of warehouses in Norway for some years. A colony at Röst in the Lofoten Islands originated in 1928, and in 1963 there were over 130 pairs.

In New Zealand, *Larus dominicanus*, the New Zealand black-backed gull nests on the roofs of sheds on wharves and also on dwellings in the central city area of Auckland.

PICIDAE

This family of 210 species is distributed throughout the world with the exception of Australia and Madagascar, and is generally confined to wooded areas. The subfamily PICINAE are known as woodpeckers and they are characterised by their adaption to a life on tree trunks and branches. This is shown by their upright stance when clinging to the vertical bark. They have short legs and two toes only face forwards with very sharp, curved claws. Much of the body weight is taken by the pointed tail feathers which have strong, stiff shafts. The beak is hard, straight and pointed being used as a chisel for removing bark and wood when searching for insect larvae and making a nest hole. The tongue is extremely long and often sticky with backward-pointed spines at the tip. The skull is very thick and the muscles of the neck powerful.

Drumming

A number of woodpecker species cause considerable annoyance to the occupiers of buildings in many rural, wooded areas on account of the noise which they make in the early part of the breeding season which is generally February to early May and is known as 'drumming'.

This is produced by a rapid succession of blows made by the beak at a carefully selected spot. The duration is only for about 1 s but 8–10 blows occur in this period and they can be heard at a distance of 400 m or so.

In nature, drumming takes place usually at the end of a broken or dead branch which acts as a sounding board and causes a loud, vibrating noise, but it is when they transfer their attentions to buildings that noise problems become more acute. In North America and Europe corrugated iron roofs of buildings are often selected for drumming as well as cedar shingles and a large number of other situations, both wooden and metallic.

Drumming probably plays a part in courtship and/or in defining territory. Although causing considerable nuisance in this way it must be borne in mind that woodpeckers are protected birds in some countries. Also, the Great spotted woodpecker, *Dryobates major,* has developed the habit of opening milk bottles. However, as we shall see below, this is more usual in the tit family.

PARIDAE

Several members of the tit family regularly remove the tops from milk bottles to take the cream. Records for this habit go back to 1921 in the UK, but the incidence increased considerably in the late 1950s.

Tits also damage buildings by pecking away the putty from around windows, and by entering rooms to tear wallpaper, magazines and other papers. This usually occurs in years when the population of tits is high, thus forcing some birds to search for food in new habitats.

Psittacosis

This infectious disease was at first thought to be confined to birds of the order PSITTACIFORMES (which includes the parrots and parakeets), and through them transmission to Man. It is now known that birds of several different groups may be infected and may transmit the causative agent to Man. This has led to the suggestion that ornithosis may be more correct as a name for the disease.

The causative agent is a virus; the individual bodies, however, are among the largest viral organisms. Although pigeons, canaries, finches, petrels and chickens are known to have transmitted the disease, the most severe symptoms occur when members of the parrot order have been involved. The virus is present in the nasal discharge and the droppings of infected birds (which may not appear to be anything other than healthy) and contaminates their feathers and cages. The stable virus withstands drying, and enters the nasal passages of humans, presumably as a dust. An affected person may transmit the disease to another by coughing.

Chapter 7

Snakes and geckos

REPTILIA

REPTILIA are characterised by their skin which is more or less covered with scales, possession of simple teeth, and their production of relatively large yolky eggs. They are cold-blooded and thus are most abundant in tropical regions. This class of animals was of the greatest importance in the past but even so in some situations they are often one of the dominant groups. The chief living groups today are exemplified by the crocodiles and alligators, the snakes and lizards, the tortoises and turtles, and the tuatara (*Sphenodon*) of a few islands off New Zealand.

Squamata

The lizards and snakes, SQUAMATA, comprise the only order of the REPTILIA which concerns us here. Snakes have evolved from primitive lizards by elongation of the body and loss of limbs. Most people are able to identify a snake, although there are a number of legless lizards to cause confusion. Positive identification, however, is difficult and relies chiefly on the internal structure of the eyes and skull, as well as the relative positions of the thymus bodies, gall bladder and kidneys. The snake-like lizards also possess some trace of the shoulder-girdle, while snakes have none. The most important difference between snakes and lizards is to be found in the lower jaw. Whereas in lizards two rami are immovably joined together, in snakes they are joined by an elastic ligament which allows movement when the snake is swallowing a large animal, and extraordinary distension is possible. The total number of vertebrae in a snake varies from about 180 to 400. The ribs articulate freely and there is no sternum which might obstruct the passage of large prey. The eyes are without eyelids. The tongue is long, thin and bifurcated at the end and is concerned with a combination of the senses of taste and smell. When the tongue is protruded and waved about, it picks up extremely small quantities of chemical substances associated with the surroundings. When the tongue is taken into the mouth the microscopic particles are transferred to a sensory area lying in front of the palate known as Jacobson's organ. In this way the snake can detect the presence of its prey and the general features of its environment.

Serpentes Snakes

Many species of snakes are found in buildings, some habitually. They enter dwellings in pursuit of mice and rats and find nesting places in roof voids, among thatch or in underfloor spaces. Snake species occurring in buildings are, therefore, usually nocturnal in habit. The kraits, *Bungarus* spp. are typical examples. They possess extremely toxic venom although they are timid and seldom bite, except in self-defence. However, a Javan krait, *Bungarus javanicus*, is known to have bitten two men sleeping in a hut. One died within half-an-hour while the other, the younger, lived for only 16 hours.

In the VIPERIDAE, *Vipera russelli*, possesses the disconcerting habit of entering buildings after its prey and this has caused a high incidence of bites among humans.

Feeding
All snakes feed only on animals, which are swallowed whole, often being suffocated by constriction or poisoned by venom. Some species of small earth-burrowing snakes subsist only on termites and some swallow only other snakes and often their own species. Snakes are almost universally regarded with horror and abhorrence and this is almost certainly due to the possession by many species of a remarkably efficient venom-producing mechanism which can, on occasion, be turned upon man, often with disastrous results. Certain groups of snakes have developed specialised maxillary teeth through which venom is injected into the punctured wound which they cause.

Those snakes that inject venom after biting are divided into two main groups: the OPISTHOGLYPHA or rear-fanged snakes, the most dangerous of which is the South African boomslang. The short fangs are situated at the rear of the upper jaw and are grooved in order to convey the venom to the wound.

The front-fanged snakes are known as PROTEROGLYPHA. These include the cobras, among the most deadly of all snakes. The fangs are situated in front of the upper jaw and venom is conveyed to the wound by means of a channel. The Australian taipan, *Oxyuranus scutellatus*, which grows to a length of about 3.3 m is one of the world's deadliest of this group, as is the Death adder, *Acanthophis antarcticus*, which reaches a length of only 1 m and the Tiger snake, *Notechis scutatus*, with a maximum length of 2.4 m.

Venom

Owing to the great medical importance of snake bites in the tropical and subtropical world, many aspects of snake venom have been widely studied. Snake venom, stored in specialised and enlarged salivary glands, consists of a complex mixture of non-cellular protein. The venom is injected through hollow or grooved teeth when the glands are pressed by the contraction of surrounding muscles. There are two main types of venom protein. One has neurotoxic properties and the other is enzymatic in action. The cobras, for example, inject a complex of neurotoxins (nerve poisons). The respiratory muscles of the bitten person are paralysed due to the acute toxic effect on the nerves activating them. In addition, cardiotoxins are present which affect the heart muscle directly.

Many snake venoms contain powerful protein-digesting enzymes. The result of these is a rapid and serious necrosis surrounding the wound. Some of the enzymes have the effect of coagulating blood, others prevent coagulation, while a third type digest blood corpuscles (haemolysis).

Breeding

There is often an elaborate courtship when the sexes have been attracted to each other by scent produced from anal glands. In some species the males may chase each other in a frenzied manner and vigorously wrestle. The male crawls along the female's body stimulating her with his 'chin' and tries to lift her tail so that one of his paired penes may be inserted. It is then introverted and a seminal fluid flows along a groove running the length of the penis. Snakes are either oviparous when eggs are laid and incubated externally, usually among heat-producing vegetable matter, or ovoviviparous in which the soft-skinned eggs are retained in the oviducts for incubation and hatch as the eggs are laid.

PIT VIPERS

Snakes of the family CROTALIDAE are known as pit vipers and constitute one of the most important groups of venomous snakes. Their distribution includes eastern Europe, the greater part of Asia, the Indo-Australian Archipelago and North and South America, but they are absent from Africa. These snakes are characterised by the possession of organs of heat perception located in front of and slightly below the level of the eye. The so-called 'pits' are divided into two halves by a translucent membrane which is richly innervated. The external opening of the anterior part of the pit is often as large as, or even larger than the nostrils, while the opening to the posterior part is inconspicuous. The pit membrane is extremely sensitive to temperature changes and it has been demonstrated that the snakes can follow warm-blooded prey in total darkness and will strike only when the prey is within range. The erectile fangs in the upper jaw are long and curved and are replaced periodically. The genus *Bothrops* contains a number of species in South and Central America such as the Fer-de-lance, *B. atrox*, found in Brazil, Peru, Central America, Mexico and the West Indies. It reaches a length of about 2 m and feeds on rats and opossums and is to be found where these mammals abound.

The rattlesnakes, *Crotalus* and *Sistrurus*, are also pit vipers and the 30 or so species and 60 subspecies are distributed throughout the American continent. The rattle consists of a number of dried skin segments which fit into each other. When vibrated, a rattling or hissing sound is produced which serves as a warning to other animals. One of the most dangerous species is *Crotalus adamanteus*, the Eastern diamondback rattlesnake from the south-eastern United States. It reaches 2.5 m in length, but neither this species nor other rattlesnakes are particularly aggressive.

The largest member of the CROTALIDAE is *Lachesis mutus* the Bushmaster of tropical America which reaches 3.6 m in length. *Trimeresures flavoviridis* the Yellow-green pit viper or Habu, is found in a number of Pacific islands south of Japan, where it commonly occurs in buildings. The snake is about 2.5 m in length, is aggressive and nocturnal, feeding on rats and domestic chickens. *Trimeresures elegans* from the same area and *T. mucrosquamatus* from China, Burma, India, Vietnam and Taiwan are also known to frequent buildings.

VIPERA RUSSELLI   RUSSELL'S VIPER

Russell's viper, in the family VIPERIDAE, is widely distributed in tropical Asia. It reaches a length of 1.5 m and has a bold attractive pattern down the middle of the back, of large reddish-brown oval spots, each encircled by black, narrowly edged with white and there are adjacent spots sometimes fused into an undulating band. There are two dark marks on the top of the head from the eye to the back of the jaw. This viper has a reputation of being sluggish but if provoked can lunge at its tormentor with such force that the whole of the body leaps from the ground. It will hold on after striking. Rats are the chief prey. It is ovoviviparous and the young are vicious and cannibalistic.

Lacertilia   Lizards

The only group of lizards which contain species habitually associated with buildings is the family GECKONIDAE, the geckos, of which there are about 675 known species. Some are very small, being only 5 cm in length, and the largest are about 30 cm. They are nearly all nocturnal and most possess remarkable climbing ability using a series of friction pads under the digits. Some scales under the toes are highly specialised, forming a patch looking like a minute pin-cushion. Every tiny bristle is branched and at the top of each is a concave disc. It is by means of these intricate organs that geckos can walk not only up vertical panes of glass, but also run across ceilings. The scales of geckos are minute so that some species have a peculiar naked appearance especially when viewed on a glass pane against the light. Geckos are ignored by most people; to a few they are repugnant. Many of them make characteristic sounds – the only lizards with well-developed voices.

Chapter 8

Mites

ARACHNIDA – ACARI EXCLUDING METASTIGMATA

The ARACHNIDA is a class of the phylum ARTHROPODA, almost equal in importance to the INSECTA. It includes the spiders, scorpions, harvestmen, mites and ticks, and almost all are terrestrial. About 35,000 species have been described, which is one-twentieth of the number of known insects, but even so, arachnids often compete with insects on favourable terms, and in many ecological niches are dominant. The evolution of the arachnids is of interest in that the earliest forms, the eurypterids, were at first exclusively marine and some species attained a length of 2 m. Some later species, however, inhabited fresh water, but all eurypterids were extinct at the start of the Carboniferous Period. The two modern marine arthropod groups, the XIPHOSURA or king crabs which are allied to the eurypterids, and the PYCNOGONIDA or sea spiders, are closely related to the ARACHNIDA.

Arachnids differ from insects in having the body divided into two parts with four pairs of walking legs, instead of three parts with three pairs of legs. They do not possess the compound eyes found in insects, but have only simple eyes; there are no antennae, and their developmental biology is much more simple in that they have no pupal stage with complete metamorphosis.

The ARACHNIDA is classified into 10 subclasses (or orders), although nine-tenths of all species are included in only two: the ACARI or mites and ticks, and ARANEAE or spiders (Ch. 10). Three small subclasses are also mentioned in this book as having species which enter buildings (see Ch. 11).

Acari

The mites and ticks are all small in size and many are minute. Although only 17,000 species have so far been described and named, the rate at which this is presently progressing indicates that they may eventually surpass the INSECTA in speciation. In some respects they resemble the

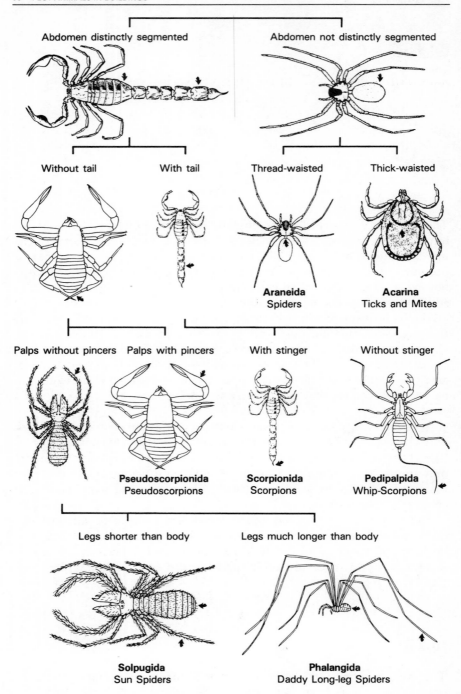

Fig. 8.1 Key to the common subclasses of the ARACHNIDA. (US Department of Health, Education and Welfare)

long-legged OPILIONES or harvestmen (but not in the length of their legs!) and although some ACARI may be fairly closely related to them, it would seem probable that the group is polyphyletic in origin. It appears certain, therefore, that it is not a natural grouping.

The mites and ticks are very little understood, and in spite of the work of specialists, it requires a great knowledge of them for identification to be attempted. This is in spite of the fact that in a number of environments they are the most abundant animals, and this is especially so in forest and grassland soils where they have attained such numerical proportions that they are to be counted in hundreds of thousands per square metre of surface. Apart from the part they play in the general dynamics of the environment, they are of immense importance in Man's social economy. Many species are injurious to his crops, but others control the numbers of other harmful animals. Some are important human and domestic animal ectoparasites, a number of which transmit disease organisms.

General anatomy
With few exceptions there is no obvious segmentation of the body as occurs in the insects. However, the acarine body may be divided into several parts, although these differ from other groups of the ARACHINIDA.

The simple mouthparts consist of a pair of jaw-like chelicerae, and a pair of leg-like sensory palps. These make up one region of the body known as the gnathosoma, which articulates with the remaining part of the body, known as the idiosoma. Sometimes the latter is entire, or it may be divided into two parts by a furrow between the second and third pairs of legs. The propodosoma bears the first two pairs of legs while the hysterosoma bears the last two pairs of legs, the anus and the genital aperture.

In a number of cases, inflexible, sclerotised plates or shields cover some part of the body surface and have a characteristic shape and distribution according to family. The degree of sclerotisation is much less in the immature stages.

Again, with few exceptions, there are three pairs of legs in the larval stage of mites, while in the later nymphal stages and in the adult stage there are four pairs. The number of developmental stages varies. A larva hatches from the egg and is followed by up to three nymphal stages before the sexually mature adult is reached. In some cases the larval and nymphal stages closely resemble the adult in appearance, but in others they differ widely. A resistant, non-feeding stage, or hypopus, may replace the second nymphal stage.

Classification
There are several hundred families of ACARI and these are grouped into seven orders. Three families constitute the METASTIGMATA or ticks: these are considered in the next chapter. Members of the other six orders are known as mites.

Notostigmata

These mites, which are up to 2.5 mm in length, are almost unique in showing body segmentation. They are found mostly in subtropical regions preying on small arthropods. They do not usually occur in buildings.

Tetrastigmata

These are large for mites, being up to 7 mm in length; the body is unsegmented and shows no subdivisions. They are probably carnivorous and as far as is known are not found in buildings.

Mesostigmata

Several thousand species are contained in the MESOSTIGMATA which are all small, mostly less than 2 mm in length. These mites have been placed with the ticks in the suborder PARASITIFORMES. The chelicerae are pincer-like in the free-living species but in the parasitic forms they are stylet-like; in both cases they can be withdrawn into a cavity of the body when not in use. Although the body shows neither segmentation nor regional divisions, it is sometimes covered in whole, or in part, with shield-like plates. Usually one plate almost covers the dorsal surface or it may be divided into two or more regions. Ventrally there is usually a series of three plates arranged in a longitudinal row. A single pair of spiracles is present which are located lateral to the base of either the second, third or fourth pair of legs. There is a similarity with the ticks here, but the spiracles are situated behind and above the fourth pair of legs. Almost always the legs are well developed, projecting well beyond the body. The long and slender first pair are covered with sensory hairs and are held forwards as feelers. In males, some of the legs may bear curved spurs for clasping the female.

The life-cycle of mesostigmatids consists of egg, larva, two nymphal stages (protonymph and deutonymph) and the adult. In some of the parasitic species, however, one or both nymphal stages are absent. A peculiar mating behaviour is shown by some: the chelicerae of the male are specially adapted for grasping a parcel of sperms (spermatophore), which is taken from his genital aperture and placed in the genital tract of the female.

A large group of the MESOSTIGMATA is known as the GAMASINA or gamasids. The greater number of these are free-living and usually predacious on other mites or small insects. They are most abundant in situations where their prey is most numerous, such as among organic debris and humus-rich soil and litter. Here their role can be looked upon as wholly beneficial as they restrict the number of species of small arthropods that might otherwise get out of hand. One example is the gamasid *Macrocheles muscaedomesticae* which feeds on eggs and larvae of the House fly, *Musca domestica*. When the female mite is mature, it clings to adult flies or dung beetles and thus gets carried to new breeding locations. This type of transport, where the carrier is not harmed in any way and which is commonly found among mites, is known as phoresy.

A number of gamasid mites are parasitic and some of these are important as reservoirs and transmitters of pathogenic organisms. The family DERMANYSSIDAE is of particular significance. The Red poultry mite, or Common bird mite *Dermanyssus gallinae*, is parasitic on chickens, turkeys, pigeons and many other birds, and carries the virus of St Louis encephalitis to Man. The Tropical rat mite, *Ornithonyssus bacoti*, is the intermediate host of a filarial parasite (a roundworm). In certain areas, usually in the vicinity of ports and in rural areas, Man sometimes acts as an intermediate host with a resulting painful skin condition. The mite also carries the rickettsial organisms causing murine typhus (see page 21).

Mites of the genera *Laelaps* and *Haemogamasus* are ectoparasites on mammals such as small rodents, and are very abundant. It is thought likely that these too may act as transmitters or as reservoirs of pathogenic organisms.

A small number of gamasids are endoparasitic and species of HALARACHNIDAE are known to infest respiratory passages of mammals. It is probable that in tropical areas human involvement may occur but as relatively little harm is occasioned to the host, such infections have gone undetected.

Cryptostigmata

The CRYPTOSTIGMATA are among the most abundant of all animals. They comprise a significant component of the fauna of humus-rich soil and organic debris of all kinds. Often they occur at the rate of 1 million per 10 square metres, a population density seldom achieved by any other animal. They are all very small, seldom being more than 1 mm in length – which is perhaps just as well!

They are usually globular in shape with a hard shell-like cuticle and often dark in colour with a division of the body into propodosoma and hysterosoma. These are generally immovable but in some, the two divisions articulate and enable the mite to close up like an armadillo. As they mostly feed on organic debris and fungal material the chelicerae are adapted for cutting and tearing decomposing plant remains. They play an important part in breaking down the humus component for soil formation.

A few species, however, feed on living plant tissue and burrow into twigs and leaves. The life-cycle is made up of six stages; egg, larva, three nymphal stages (protonymph, deutonymph, tritonymph) and adult.

As far as is known no member of the CRYPTOSTIGMATA transmits any disease-causing organisms except that some species found in grassland are intermediate hosts for sheep and cattle tapeworms. Infection takes place when herbage is eaten together with the mites. A few species enter buildings, but are of nuisance value only. *Phauloppia lucorum*, the Window-sill mite, for example, sometimes occurs in very large numbers in roof voids of houses in Europe, dropping into rooms below, and being found on and in all household accoutrements. As the mites are dark brown and very hard, they are readily noticed by the occupants and cause much alarm.

Astigmata

This order consists of an assemblage of small mites, almost always less than 1 mm in length; the body is mostly separated into propodosoma and hysterosoma and there is often a distinct separation between the anterior, forwardly directed two pairs of legs and the posterior two pairs which point backwards. Many species are abundant and distributed widely, many are of economic importance in the storage of food and a number are of importance as parasites of mammals and birds.

Most of the mite species harmful to stored grain and its products as well as other stored foods such as dried fruits are placed in the family ACARIDAE (TYROGLYPHIDAE) and are commonly called forage mites.

ACARUS SIRO   FLOUR MITE

This is one of the most important pests of grain and flour. Heavy infestations give off a strong odour known to bakers as 'mintiness' which taints the foodstuff making it unsuitable for human consumption. It is probable that only damaged grains are attacked, but the embryo is eaten and the moulted mite skins, faeces and minty odour cause serious losses. When heavy populations of *A. siro* are present, flour becomes a dirty grey colour, again causing it to be rejected for baking purposes.

Adult *Acarus* species are pearly white or colourless, with the legs yellowish, pinkish or brownish. Males are easily recognised, even with a ×10 hand lens, by the large spur projecting from the underside of the femur on the front legs. Up to 500 eggs are scattered over the food material over a period of about 40 days and they hatch in three to four days. The life-cycle is completed in 17 days at 18–20 °C, while at 10–16 °C, it takes 28 days. Relative humidity is an important factor:

Fig. 8.2 A heavy infestation of *Acarus siro*, the Flour mite. (Crown Copyright)

below 60 per cent RH, the population dies out; most mites will congregate in areas of 80–85 per cent RH when available. The optimum temperature for rapid reproduction is 25–32 °C.

The method of dispersal of this mite has been the subject of much conjecture as it always appears when conditions are appropriate for it. It occurs naturally in birds' nests and nests of rodents and probably occurs in small numbers among old straw and barn debris. When wheat and barley are harvested two mite species of the same genus, *A. farris* and *A. immobilis*, are often found on the grain but in storage they disappear. When the relative humidity is high enough, *A. siro* appears and flourishes when the moisture content of the grain exceeds 14 per cent. A hypopus stage seldom occurs in this species. *Acarus siro* is preyed upon by the predacious mite *Cheyletus eruditus*, especially in summer.

TYROLICHUS CASEI   CHEESE MITE

This common mite is 0.45 to 0.70 mm in length, the females being the larger and a hypopus stage does not occur. It is well known as feeding on cheese, where its presence is generally considered obnoxious. Some brands of cheese, however, are considered to be 'ripe' only after heavy infestation with *T. casei*. Heavy accumulations of dead mites, cast skins and faecal matter produce a grey or brownish dust which is brushed away from time to time. The Cheese mite also occurs on damp grain and flour. At 23 °C and 87 per cent RH the life-cycle takes 15–18 days.

TYROPHAGUS PUTRESCENTIAE

A cosmopolitan mite occurring on fatty foodstuffs and those of high protein content. Examples are groundnuts, sunflower seeds, dried egg, ham, copra, cheese (see note under *T. longior* below), different kinds of nuts, dried bananas, wheat and flour spillage.

TYROPHAGUS LONGIOR AND T. PALMARUM

These mites commonly occur in haystacks and in stored grain. The former is also found on poultry droppings in broiler houses. Both mites feed on cheeses, but there is a tendency for them to occupy different, though overlapping, geographical zones: *T. longior* is a cool to temperate form, *T. palmarum* mainly temperate and *T. putrescentiae* tends to be found in subtropical and tropical regions. The three species of *Tyrophagus* have all been implicated as the cause of dermatitis in those handling infested materials.

GLYCYPHAGUS DOMESTICUS   FURNITURE MITE

This widely distributed species of the family GLYCYPHAGIDAE, sometimes occurs in extraordinary numbers in buildings, especially dwellings. The nearly spherical body is covered with long setae from which small hairs arise. Legs are long and slender. The female bears a short tube-like organ, the bursa copulatrix, at the posterior end of the body. The male is from 0.32 to 0.40 mm and the female from 0.40 to 0.75 mm in length.

At 23–25 °C and 80–90 per cent RH the life-cycle without a hypopal stage takes about 22 days. In about five days the eggs hatch into wrinkled six-legged larvae which after feeding for two days become shiny and swollen. After resting for two days they moult into a protonymph. This feeds for four days then rests for a further two days before moulting into a deutonymph. This then feeds for five days, rests for another two then finally moults into an adult. Some of the protonymphs do not moult directly to the deutonymph stage but assume a cyst-like form devoid of appendages remaining within the protonymph cuticle. This hypopus is specially adapted for resisting adverse conditions; it may remain in this stage for as long as six months or even a few years before moulting to the normal deutonymph stage.

Glycyphagus infests a wide range of foodstuffs including flour, sugar and cheese, and in addition is often found on furniture in damp, badly ventilated rooms where it feeds on fungi growing on the surface of stuffing material of soft furnishings. When found in such situations, usually observed as dust-like particles slowly moving on dark-coloured woodwork, distress and anxiety is often caused, although no physical damage is done. Dermatitis sometimes results, from handling infested food and symptoms of asthma can occur.

LEPIDOGLYPHUS DESTRUCTOR

Similar to *G. domesticus*, but with a more pear-shaped body, this species is one of the most common of stored product mites and is frequently found with *Acarus siro*. It is cosmopolitan and can survive in granaries where the temperature may fall as low as $-18\ °C$.

CARPOGLYPHUS LACTIS   DRIED FRUIT MITE

Somewhat similar in appearance to *Acarus siro*, but with a more shiny cuticle. The legs are slightly pinkish in colour. In length this mite is from 0.38 to 0.42 mm, with males and females approximately equal in size. The adult mite lives for up to 50 days.

This species lives usually on foods containing lactic, acetic or succinic acids. It is common on dried fruits and stored Christmas puddings, rotting potatoes, cheese, old flour and groundnuts. It is another species known as the cause of dermatitis in persons handling infested material. The mite is also recorded from human urine, with the possibility that it inhabits the urinary tract.

DERMATOPHAGOIDES   HOUSE DUST MITES

Mites of the family PYROGLYPHIDAE are exceedingly common and widespread, probably throughout the world. They are often associated with Man, being found especially in and around beds; mattresses are usually well populated. Comparatively little is known about these mites, but they are attracting increasing interest as being the cause of certain asthmatic conditions in Man. Although undoubtedly important in this regard they do not, however, constitute the only allergen in house dust. The whole mites, fragments, excretory and secretory products are all implicated as producing human allergic responses. In addition, mites of

Fig. 8.3 *Acarus siro*, Flour mite. Female; magnification × 250. (Crown Copyright)

this genus have been associated with pruritic dermatosis principally in Japan, but also in the United States of America.

The chief species concerned are *D. pteronyssinus* and *D. farinae* but several others occasionally occur in house dust collections. Mature females are only about 0.4 mm in length, the males somewhat smaller, so that they are not readily observed. At 25 °C and 80 per cent RH, both species develop from egg to adult stage in about 30 days, passing through a larval and two nymphal stages. *Dermatophagoides farinae* lays one egg each day over a 30-day period and the adult lives for about 60 days. *Dermatophagoides pteronyssinus*, over a period of 45 days, lays up to 80 eggs and lives for 100–150 days.

Dermatophagoidid mites feed on human shed skin, of which about 5 g is sloughed per person per week. Each inhabitant of a house provides sufficient skin fragments to maintain large numbers of mites even though part of the day is spent away from home. Humidity appears to play an

Fig. 8.4 *Dermatophagoides farinae*, House dust mite. Front view of gnathosoma. (Crown Copyright)

important part in the population growth of the mites; thus at a RH of less than 60 per cent the numbers cease to increase, and then die out, but it has been pointed out that when a bed is occupied not only is the temperature adequate but the humidity is high.

Pyroglyphid mites hide in cracks and crevices, whereas acarids tend to browse in the open together with their cheyletid predators, so that the increasing use of the vacuum cleaner in recent years has favoured the pyroglyphids. It is believed that the latter are more abundant in dwellings today than in the past.

Dermatophagoides species are not only found in house dust. Other sources recorded are poultry and pig-rearing meal, dogmeal, wheat pollards, biscuit meal and tanned mammal skins.

Prostigmata

The PROSTIGMATA is perhaps the most varied of the main groups of mites. Not only is it a very large group in numbers of species but it is of great economic and medical importance to Man. In size these mites range from 0.1 to 1.5 mm in length and although most are pale in colour, a few are of the brightest red or green. The well developed mouthparts show a number of specialisations according to mode of life and there are often prehensile palps present. Sometimes the life-cycle includes three nymphal stages in addition to egg, larva and the final adult, but in some cases active nymphs are absent, their place being taken by a hypopus. Among the prostigmatids are many pests of agriculture, some of exceptional economic significance, but more relevant in this present work are the parasites of man and domestic animals.

SARCOPTES SCABIEI   ITCH MITE

Several species of the family SARCOPTIDAE are parasites of Man, other mammals and birds. The unpleasant human disease known as scabies is caused by the mite *Sarcoptes scabiei*. Closely related species infest the skin of a number of domestic and wild animals and sometimes transitory infestations of these are acquired by persons handling them. Permanent infestations of Man by such forms, however, do not occur.

Appearance
The body is oval and somewhat tortoise-shaped, rounded above but flat beneath. The male is only about 0.23 mm in length whereas the female

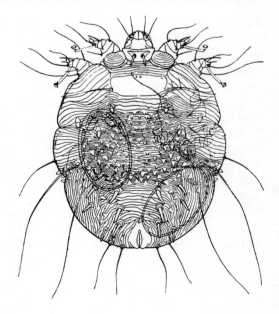

Fig. 8.5   *Sarcoptes scabiei*, Itch mite. Dorsal view of female. (Btitish Museum, Natural History)

may be twice this size. The whitish cuticle is covered with striations. Ventrally there are thickened rod-like structures which support the stumpy legs. The latter are capable of limited movement only, yet progression over flat surfaces is quite fast, an adult female being capable of about 25 mm per minute.

In both sexes the first two pairs of legs terminate in a long, stalked, sucker-like organ. The third and fourth pairs of legs of the female bear long whip-like setae whereas in the male these are borne on the third pair only, the fourth pair bear the sucker-like organs similar to the first and second pairs. The head bears a pair of toothed chelicerae somewhat similar in shape to the mandibles of insects and there is a pair of palps.

Breeding
The adult female searches for a crevice or furrow in the skin, then immediately commences to burrow. This is achieved by the use of the chelicerae and the front two pairs of legs. The suckers on the legs make a firm hold and the cutting edges on the last joint are used to tear at the skin. Purchase is afforded by the bristle-bearing legs, while the front end of the body digs into the skin. On the back of the mite are a number of short pointed projections and rather longer spines, all pointing backwards, so that as the mite pushes forwards into the newly formed burrow it is prevented from slipping back. From 15 minutes to an hour is taken by the mite to bury itself in the skin. Only short burrows are made by males and immature stages, but after fertilisation the females make long characteristic meandering burrows.

The active males leave their burrows from time to time to search for females. Copulation takes place in the burrow with the pair facing in opposite directions. The male lifts up the hinder part of his body and rests it on the top of the female's hinder body. The sexual organ of the male is ventral while that of the female is dorsal so that copulation can then be effected. The female possesses a separate oviduct which leads to a ventrally placed egg pore; she lays her eggs in the burrow.

Egg
Smooth, whitish and glossy, the egg is about 0.17×0.09 mm in size but increases slightly during development. This is relatively large compared to the adult female. Eggs are sometimes brought out adhering to the needle when burrows are probed in order to extract the female. Eggs hatch in from three to four days.

Larva
Similar in shape to the adult but possesses only three pairs of legs. The front two pairs bear suckers and the hind pair each terminate with a long bristle. The larvae are sometimes found inside a burrow and may attempt to dig into the floor but they quickly emerge and either excavate their own burrow or descend into a hair follicle. The larval stage lasts three days.

Nymph

The larva moults into a nymph with four pairs of legs. The two hind pairs terminate in long bristles. The nymphs either make a short burrow into the skin or find a hair follicle and retreat into it. From time to time, however, they emerge and after a brief sojourn on the skin surface find another follicle or make another burrow. This stage lasts from three to four days, after which a male adult or a second stage nymph (deutonymph) is produced. Little appears to be known about the second nymphal stage except that an adult female finally emerges, and that the stage also lasts three to four days.

The time from the egg being laid to the first eggs of the next generation is 14 to 17 days. Since the male does not undergo a second nymphal stage, its total length of life-cycle is only 9 to 11 days. The female lives for from one to two months and during this time lays on average slightly more than two eggs daily. Although a high proportion of the eggs hatch, the mortality is high among the immature stages, only about 10 per cent reaching maturity.

Scabies

The worldwide human disease known as scabies is caused by infesting mites of *Sarcoptes scabiei*. Only small populations are built up; large numbers are rarely encountered. Due to the great difficulty of finding the immature stages on the human body, but the comparative ease by which the adult females can be detected by trained persons, the severity of an infestation is usually assessed by the number of adult female mites present. The average number of female adult mites carried by a scabies sufferer is about 11, whereas about 50 per cent of scabies patients carry less than 6. Few infested persons harbour more than 50.

If this information is compared with the number of progeny estimated from life history data it is seen that the number of infesting mites is very small compared with the number expected under 'ideal conditions'. One hundred adult mites would be produced from one initial infesting female after one month and several thousands after two months.

If the course of a primary infestation by a single female is followed, the adult females of the second generation begin to appear after three or four weeks. The population builds up slowly at first but later more quickly for about two months. Until this time, the infested person experiences no symptoms, but now sensitisation occurs and a skin reaction gradually becomes more severe, accompanied by a mounting irritation. The skin reaction is shown by erythema and by the formation of translucent vesicular swellings. This is the 'scabies rash' which *does not correspond with the location of the burrowing mites*. The irritation increases until it becomes continuous and unbearable. Scratching is inevitable and although a few ovigerous female mites may be hooked out, it often leads to secondary bacterial infections such as impetigo.

Reinfestation results in a prompt reaction. Within 24 hours there is local irritation and redness. Experimental reinfestation, however, has been shown to be very difficult on account of this intense skin sensitisation. The infesting mite either gets scratched out or the colony multiplies very slowly. Such infestations often die out spontaneously.

The most common sites of infection are between the fingers and elsewhere on the hands and wrists; nearly three-quarters of all scabies patients showing this. To a lesser extent they invade the palms, elbows, feet and ankles, external genitalia, groin, breast, shoulder blades and navel. Other sites are comparatively rare.

It appears certain that the majority of new infections are caused by the fertilised adult female mite wandering from an infested person to an uninfested one during an intimate association. This is usually while in bed and is most likely to occur in the course of ordinary family life, such as children sleeping together or with their mother. Experimental transmission using larvae has invariably failed, whereas it has often proved successful when female adults have been employed. Furthermore, the immature stages of the mite have a high mortality rate, as have the adults when away from the host's body. This also makes transmission uncommon through the use of bedclothes, bedding and garments which have been used by an infected person, even only a short time previously.

The amount of infection brought about by transitory contract such as between children playing, hand-holding, caressing at dances, cinemas, etc., would be difficult to assess but it no doubt occurs.

Scabies is said to have a predilection for the poor, crowded and unwashed, and also to possess a venereal connotation. The former, unfortunately, may be true, but the second has little basis in fact.

SARCOPTIDAE MANGE MITES

A large number of mite species of this family are parasites on the skin of domestic animals and birds. While some live on the skin surface, others burrow into the skin or enter various glands. Running sores are sometimes caused and the condition is known as mange or scab. Some species exist in a number of different forms, each form only able to infest a particular animal species. Some forms are identical in appearance, but physiological differences are obviously present.

Sarcoptic mange in dogs is caused by a subspecies of *Sarcoptes scabiei*. Scabs and sores develop from the bare areas of skin scratched by the dog. It first appears round the eyes, outside the flaps of the ears, and on the elbows, hocks and abdomen. Closely related is *Notoedres cati* which causes mange of cats. It usually commences on the neck then travels over the head. Unpleasant greyish crusts are produced and the affected skin assumes a leathery appearance. Rats are attacked by *N. muris* which may be fatal to both tame and wild animals.

Psoroptes communis, a surface-living mite infests tame rabbits, horses, sheep, goats and cattle, forming characteristic scabs. 'Scaly leg' in poultry is caused by *Cnemidocoptes mutans* (= *Knemidokoptes*), and *C. laevis* var. *gallinae* is the cause of 'depluming itch' which also irritates the birds so that they continually pluck at the feathers. Canker of the ear of dogs is caused by *Otodectes cynotis* and that of cats by *O. cynotis* var. *felis*.

DEMODICIDAE

Mites of this family are very small; some are minute. They have a worm-like body with lateral striations, and the legs are short and stumpy. They occur in the sebaceous glands and hair follicles of various mammals including Man. *Demodex folliculorum* is distributed throughout the world

Fig. 8.6 *Demodex folliculorum*, Follicle mite. Ventral surface of female.

and is a common human parasite although apparently it does no appreciable harm to its host. As many as a dozen or more all with their anterior ends pointing inwards inhabit a single sebaceous gland or blackhead, usually on the face. The females measure from 0.27 to 0.44 mm and the males are smaller. Little is known of their biology, but it is probable that mites are much more common on the human body than is realised or appreciated. *Demodex canis* is the cause of mange of dogs. Bare skin appears in small patches and this is followed by pimples and pustules. A number of secondary infections may occur which cause suppurating sores often with a disgusting odour. It is not very infectious, although human infestations of a transitory nature sometimes occur.

TROMBICULIDAE

The TROMBICULIDAE family consists of a number of species which are free-living predators in the adult stage but which are ectoparasites of vertebrates and arthropods as larvae, when they are known as chiggers or red bugs. Some species attack Man. The adult mite is constricted in the centre giving it a fiddle shape and the cuticle has the appearance of velvet due to a covering of stiff serrated bristles. Many species in this family are highly coloured, either red or orange, and are conspicuous.

NEOTROMBICULA AUTUMNALIS   HARVEST MITE

This mite is found throughout northern Europe. The bite of the larva causes intense itching.

The eggs which are laid in the soil measure from 0.1 to 0.2 mm in length. The six-legged larvae which hatch, climb upwards and wait on the tips of leaves for the opportunity to attach themselves to a vertebrate host. Rodents and rabbits are most usually attacked, the mites attaching themselves to the ears or around the genitalia. On birds they are found on the thighs, under the wings or around the anus.

The larvae are red or yellowish in colour and measure from 0.2 to 0.3 mm in length. Numerous feathered setae are present. On Man they usually crawl upwards until a constriction in the clothing is met, such as a garter, waistband or belt. They then insert their mouthparts and the salivary glands secrete a substance which causes histolysis; a hardened tube is formed through which the host's lymph is sucked. Blood is not taken. The mites remain attached for several days. Finally the larva drops off and moults to an eight-legged predacious, but not parasitic, nymph which eventually moults to the adult.

The length of the complete life-cycle is subject to much variation, requiring from 2 months to upwards of 12 months and it is thought that one to three generations occur annually. Intense irritation caused by the parasitic larval stage, however, occurs only in the summer and autumn months.

EUTROMBICULA ALFREDDUGESI

A troublesome biter, in the larval stage, widely distributed throughout the eastern, mid-western and southern states of the United States. A number of closely related species occur in the southern states, of which *Acariscus masoni* and *Eutrombicula batatas* are better known. The life-cycle follows closely that of *Neotrombicula autumnalis*.

Scrub typhus

Leptotrombidium akamushi and *L. deliense* and other related species, occur widely in the Asiatic–Pacific area. *Leptotrombidium scutellaris* is known from certain Japanese islands. Again the life-cycle appears to be similar in many respects to that of *N. autumnalis*, the six-legged larva being the parasitic stage, and nymphs and adults being non-parasitic. The oriental species, however, are of extreme importance as the carriers of *Rickettsia tsutsugamushi*, the causative organisms of the disease, scrub typhus. Although this disease was known in China in the sixteenth century, it was not until around 1942 that it became of great importance to Western military medicine. From 1942–45 the cases of scrub typhus occurring in the Allied armies, incompletely documented, totalled 18,450. The United States Army suffered 243 fatal cases among its 6,685 men who contracted the disease.

Scrub typhus occurs mostly along the coastal regions of Asia, from Korea to India and among the islands of the western and southern Pacific from Japan to northern Australia. The mite larvae are red and between 0.15 and 0.4 mm in length. The rickettsiae are transmitted from the mite to the vertebrate host during the period of parasitism. On the other hand, if the mite is uninfested it may take in rickettsiae from an infected host. In which case, as only one vertebrate host takes part in the mite's life-cycle, the rickettsiae remain within the mite and, at least in some species, are transmitted transovarially to the next generation. Not only, then, do small mammals and birds act as reservoirs for the rickettsiae but the trombiculid mites act as reservoirs as well as vectors.

Although endemic areas of the mites and rickettsiae occur over a wide range of habitats, the most usual site is where fields have been allowed to become overgrown with scrub vegetation and this has given the common name for the disease. These areas are often sharply defined. The seasonal occurrence of the disease is related to the abundance of the mites. In Japan the disease occurs in summertime when the mites are most abundant, although in some Japanese islands where the vector is *Leptotrombidium scutellaris* it occurs in winter. In tropical or subtropical areas infections are more prevalent during the wet season.

The incubation period can be 6 to 18 days, although in most cases it is from 10 to 12 days, when a sudden headache, feverishness and intermittent chilliness is complained of. An eschar or scab may be observed at the site of attachment of the infected mite. The temperature reaches 40–40.6 °C either quickly or step-wise. A red, spotted rash appears on the trunk on the fifth day and may extend to the limbs. It usually persists for several days. In Asians the eschar and the rash are more often absent. During the first week, despite the fever, headache and apathy, the patients rarely appear dangerously ill. During the second week the temperature remains elevated and even the less severely affected begin to show the debilitating effects of sustained illness. In severe disease the systolic blood pressure drops to below 100 mmHg. The central nervous system may become involved, with delirium, stupor and muscular twitchings occurring. With the temperature falling after the fourteenth day, the fever is reduced and the pulse rate and blood pressure return to normal. For those who do not receive antimicrobial therapy, return to full health is generally delayed for several months.

In children the disease is rarely fatal, but the infected elderly usually die. There is great variation in mortality rates. The highest has been in Japan where about half the patients have died. In Malaya about 7 per cent of cases end fatally. During the Second World War the United States Army recorded fatality rates of from 0.6 to 35 per cent in the southwest Pacific.

PYEMOTIDAE

This family contains a large number of species, most are parasitic on insects, but some suck the juices of plants. Only one species occurs commonly in buildings.

PYEMOTES VENTRICOSUS   STRAW ITCH MITE

Widely distributed in buildings in North America and Europe and probably cosmopolitan, this mite is often extremely abundant. It is an important parasite of invertebrates and, when attacking the insect pests of stored foods (larvae of the Angoumois grain moth are preferred) and wood-boring beetles in buildings, it is performing a useful service to mankind.

When extremely numerous, however, and particularly in hay and straw, the mite will often turn its attention to Man, when it becomes a serious nuisance, its bite causing a severe itching. Weals and pustules accompanied by red blotches are caused by the bites. Many lesions are surmounted by a minute vesicle, although no puncture is visible. Fever and cold sweats are often produced which with the intense itching make sleep impossible. As far as it is known, no pathogenic organisms are carried by this species.

This mite, previously known as *Pediculoides ventricosus*, is now perhaps

Fig. 8.7 *Pyemotes ventricosus*, female in distended condition with attendant parasitic male offspring. (Partly after Vitzhum)

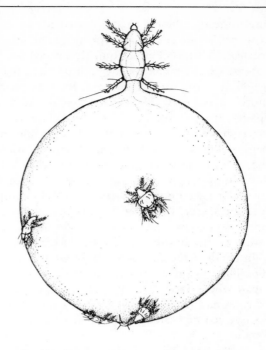

more correctly named *Pyemotes herfsi*. The greyish or yellowish female is about 0.22 mm in length before feeding, but is almost 10 times this size when fully fed. The young female attaches itself to an insect larva and there sucks its juices. One mite only is sufficient to kill its host, although quite commonly a number of mites may be present. Unfed, the mite is practically invisible, but when fully fed it is easily seen with the naked eye as a glistening yellow globule.

Life-cycle
This is peculiar in that the immature stages are passed within the parent's abdomen and emerge only when they are mature. The males wander over the distended body of the female and help emerging females by dragging them through the genital opening, before copulating with them. Parthenogenetic reproduction sometimes takes place, but only males are born. Mites can produce young as little as six days after emergence from the mother's body. One female has been recorded as bearing 270 young, 52 of which emerged in one day.

TETRANYCHIDAE

Many of these plant-feeding mites are of economic importance as agricultural pests. Some are pests of buildings, the most common being *Bryobia praetiosa*: this may be a single species, or it may be a complex of several species.

BRYOBIA PRAETIOSA   CLOVER MITE, RED SPIDER MITE

All stages of this mite are vegetarian only. They feed on a wide range of herbaceous plants and grasses, as well as a number of trees. Until about 1945, *B. praetiosa* was an occasional invader of dwellings both in North America and Europe. From that date onwards it became of much greater importance, entering buildings on a much wider scale and often in such numbers as to cause serious anxiety. Their crushed bodies often cause spotting on wallpapers, curtains and other surfaces. The reason for their great increase in numbers has been given as the housing boom and the growing of well-fertilised lawns adjacent to the walls of buildings.

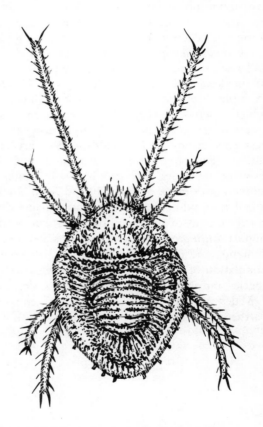

Fig. 8.8   *Bryobia praetiosa*, Clover mite.

Bryobia praetiosa is greenish or reddish and relatively large, being 0.8 mm in length. The front pair of legs are very long and there are a number of fan-shaped plates (setae) on the back.

This mite is parthenogenetic. The eggs, which are bright red, are laid in cracks around window frames, in mortar crevices, cracks in concrete foundations and under bark at the base of trees. Large accumulations of eggs, egg shells and moulted skins are often found in these and similar situations. The young larvae crawl downwards to feed on grass and other plants, and subsequently return close to the crevice where the egg was laid, there to moult to the first nymphal stage. This process is repeated for the second nymphal stage, then finally for the moult to the adult female. Males are extremely rare. Maximum activity takes place in late spring, while it is dormant during hot summer months; activity is renewed in autumn. Overwintering takes place as adults as well as eggs. There may be five generations annually.

CHEYLETIDAE

Free-living predatory mites, very common in leaf litter and topsoil, granaries, warehouses and barns. Most species feed on other mites, but some can penetrate the skin of mammals giving rise to a form of mange.

Cheyletus eruditus is a common predator of *Acarus* and *Glycyphagus* and always appears to be present wherever its prey occurs. Many species of cheyletids are found in buildings; *C. eruditus* is taken here as an example. It is easily recognised by the large and powerful pedipalps with which it seizes its prey. However, the mite is a solitary creature and seems never to be so effective in limiting the numbers of its prey that pest populations are not built up. As a predator it is most efficient in late summer and early autumn when moderately high temperatures prevail. Its value in biological control is least in winter and spring, for when temperatures range between 10 and 2 °C, *Acarus siro* can increase its population, while that of *Cheyletus* remains static. Much depends on the water content of the commodity; *A. siro* is dependent on this for population increase, whereas *C. eruditus* can do well at lower moisture contents, and under these conditions can control very effectively the population of the Flour mite. *Cheyletus eruditus* will feed on other mite species and in their absence will turn to cannibalism.

Males are very rare. Generally, parthenogenesis is the rule, with parthenogenetic eggs producing females only.

CHAPTER 9

Ticks

ARACHNIDA – ACARI METASTIGMATA

Three families of the subclass ACARI are usually grouped together to form the order METASTIGMATA or ticks. They are all ectoparasitic on mammals, birds, reptiles or amphibians and feed on the blood and tissue fluids of their host, usually becoming much increased in size (engorged). The three families are: the ARGASIDAE, with nearly 100 species and referred to as the soft ticks; the IXODIDAE or hard ticks with about 550 species; and the NUTTALLIELLIDAE with one species only which is of no consequence in this account.

Ticks are generally larger than mites, the other members of the ACARI, and are characterised by the presence of spiracles or breathing pores situated behind the third or fourth pair of legs. Located on the distal segment of each of the first pair of legs is a unique organ of sense – Haller's organ. The gnathosoma, besides the basal area, bears a pair of four-segmented palps and a pair of cutting chelicerae which penetrate the skin prior to insertion of the barbed hypostome for anchoring to the host.

When a tick has fixed itself on its host, feeding commences by means of a pumping action of the powerful pharynx. Blood and tissue liquids are sucked in and forced into the oesophagus and thence to the stomach. Some ticks produce a saliva containing blood anticoagulants, thus allowing feeding to take place over a prolonged period without clogging the mouthparts. The tick's cuticle is capable of extraordinary distension so that when engorged with a full blood meal, the tick may be many times larger than it was before feeding commenced.

There are four stages in the life-cycle of all ticks. The egg hatches into a six-legged larva which in turn changes to an eight-legged nymph before the sexually mature eight-legged adult stage is reached.

Ticks do not usually inhabit buildings, but the reason for their inclusion is because they are brought into buildings by their hosts – man and his domestic cats and dogs. It is in dwellings that such infestations are usually found and as a number of species transmit the pathogens of several serious diseases, it is obviously of importance that some account of them should be given here.

ARGASIDAE SOFT TICKS

Argasids are characterised by the lack of a sclerotised plate, the scutum, on the back, while the integument of the nymphal and adult stages is leathery and ornamented with wrinkles or protuberances of various sorts. The much thinner integument of the larva lacks wrinkles and protuberances. The gnathosoma is located ventrally in nymph and adult and the sexes are similar.

The adult argasids, which are usually nocturnal, are intermittent feeders found usually in and around the resting places of their hosts such as nests, dens or lairs. They are able to live for prolonged periods without feeding, a most advantageous character in an animal where chance plays a large part in finding a host. Adult ticks are known to have been kept alive for 40 years while only being fed at five-year intervals. Relatively few eggs are laid. When the larva has taken its first meal, which may last for as little as a few minutes to several days, it drops from the host and moults into a nymph. The nymph may feed several times, moulting after each feed to a more highly developed nymphal stage, then finally transforming into the adult.

A number of argasids are vectors of important and serious diseases of man and domestic animals. in addition, the bites of many species are venomous; that of *Ornithodoros coriaceus* occurring in California and Mexico is especially so. Secondary infections are common as a result of the hypostome remaining in the wound following the forcible removal of the tick. Tick removal is best carried out by dabbing it with chloroform or ether, pressing the tick inwards to release the barbs of the capitulum, then pulling to gently away.

ARGAS

Species of this genus are distributed throughout the world. *Argas persicus* is well known as a pest infesting poultry and may transmit a spirochaetal disease to fowls. The tick will sometimes bite Man and cause a severe wound. *Argas reflexus*, the Pigeon tick, often attacks humans in the vicinity of pigeon breeding sites. Other *Argas* species attack birds and bats and may then attempt to feed on anyone who ventures near the roosts.

ORNITHODOROS MOUBATA EYELESS TAMPAN

This tick occurs in dry situations in East and Central Africa and is principally a human parasite. It hides in native huts, rest houses and camp sites, emerging at night to feed. The tick transmits the bacterium *Spirochaeta duttoni*; the organism causing African tick-borne relapsing fever. The spirochaetes are passed from one generation of ticks to another through infected eggs. The organisms are passed to humans either by the tick bite or by the tick faecal fluid contaminating the skin.

Fig. 9.1 *Ornithodorus moubata*, Eyeless tampan. (British Museum)

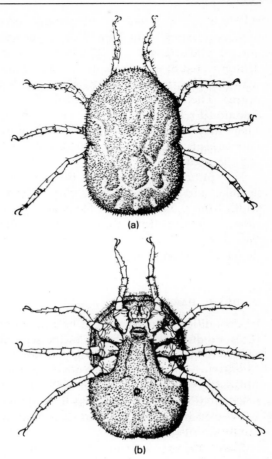

ORNITHODOROS SPP.

In the Americas *Ornithodoros turicata*, *O. hermsi*, *O. talaje* and *O. rudis* are important vectors of tick-borne relapsing fevers, whereas in Asia the species concerned is principally *O. tholozani*.

Relapsing fever

Relapsing fever is an acute infectious disease caused by a bacterium of the genus *Spirochaeta* (=*Borrelia*). The main characteristic of the disease is the occurrence of a fever with body temperatures up to 40–40.5 °C for two to eight days. Commencing with a chill, a high fever follows with intense headache, pains in muscles and joints, vomiting, photophobia and bronchitis. There may be psychic disturbances and delirium. During this stage the spirochaetes can be demonstrated in the peripheral blood. This is followed by an abrupt fall in temperature to normality, but in the elderly or weak, a state of collapse may occur. There is then a period of remission lasting from 3 to 10 days when the temperature remains normal, and strength and appetite return. Spirochaetes can no longer be demonstrated in the blood. If the case is untreated then relapse occurs, with the important symptoms recurring, but usually to a rather milder degree. The relapse does not usually last as long as the initial attack. Additional relapses may occur, but are progressively shorter and milder. There may be many side complications during convalescence. The fatality rate varies from 2 to 8 per cent, being highest among the undernourished and the aged.

The spirochaetes responsible for tick-borne relapsing fever are highly flexible, spiral organisms; they vary in length from 8 to 30 μm and in width from 0.3 to 0.7 μm. There are 5 to 10 loosely wound and irregular spirals. The organisms are very active. Although many species of *Spirochaeta* (=*Borellia*) have been described and named, there is some doubt as to their being separate species.

IXODIDAE   HARD TICKS

Ixodids differ from argasids in bearing a dorsal plate or scutum. The sexes are dissimilar: in the males the scutum covers the whole upper surface of the body while in the females it occupies only part of it. In addition, the gnathosoma extends forwards, and prominent spiracles are situated behind the hindlegs in both adult and nymph. Generally these ticks are distributed throughout the environment of their host although in a few species they do frequent their nests and resting places. Large numbers of eggs are laid.

Each stage requires to feed once only, and each blood meal takes several days. The female drops to the ground after she has fed and mated, rests for a while, then lays several thousand eggs. When these hatch the larvae wander around, seeking attachment to a suitable host by climbing to the tips of the herbage and waving the front legs like antennae. When successfully transferred to a host animal, a blood meal is taken and the larva then drops to the ground. It moults into a nymph which again seeks a host; if one is found it feeds again. Afterwards it falls to the ground and moults into an adult.

Although this three-host cycle is usual, it is not universal among all ixodids. There are a number of one-host ticks, exemplified by the genus *Boophilus*. In this genus, when the larva finds a host and attaches to it, it feeds and moults on the same host until, as an engorged adult female, it drops off to lay eggs. Yet again, a number of ixodid ticks, known as two-host ticks, spend the larval and nymphal periods on one host, but, when fully engorged as nymphs, fall from the host and then moult to the adult stage. The adult thereafter must find a second host.

Ixodid ticks are of the greatest importance to man as they are the carriers of the causative organisms of many serious diseases of humans and, in tropical and subtropical areas, of cattle. In addition, they can cause injurious disorders or even death through toxins contained in the saliva secreted by the tick in order to prevent coagulation of the blood. The most serious cases occur when the tick is sited at the back of the neck or base of the skull when 'tick paralysis' may result. Commencing with fever then ascending motor paralysis, it is possible for cardiac and respiratory failure to occur. It is most common in human females due to long hair hiding the tick. If the tick is removed in time, recovery takes place in from one to six days. In heavy infestations, especially of cattle, direct loss of blood can cause anaemia.

DERMACENTOR ANDERSONI   ROCKY MOUNTAIN FEVER TICK

Also known as *D. venustus*, this tick is abundant and widely distributed throughout western North America. It is a three-host tick, feeding during the larval and nymphal stages on small mammals such as rabbits, chipmunks, ground squirrels, pine squirrels and field mice. In the adult stage, large mammals such as horses, cattle, goats, sheep, deer and often Man, are the hosts. The tick is notorious as transmitting three serious diseases to Man.

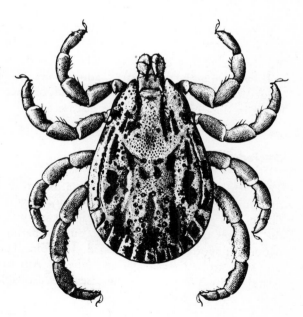

Fig. 9.2 *Dermacentor andersoni*, Rocky Mountain fever tick ('wood' tick). (British Museum)

Rocky Mountain spotted fever

This disease is caused by the organism *Rickettsia rickettsii* and is very much more widespread in the United States than the name implies, as it is found in every state with the exception of Maine and Vermont. The disease occurs also in Mexico, Brazil, Colombia and Canada. Although *Dermacentor andersoni* is the principal tick vector in the West, at least 10 other ixodid species have been implicated.

Rocky Mountain spotted fever is an acute specific infectious disease chiefly of the peripheral blood vessels. It commences with chills, then continues with fever for nearly two weeks. It is accompanied by headache and severe pains in the bones and muscles. An eruption of spots appears after about four days, first on the wrists, ankles and back, then spreading over the whole body. Before the initial chill there is sometimes a rise in temperature in the evenings but afterwards the temperature rises to 38.8–40 °C by the second day and remains at 40–40.5 °C until the end of the second week when it drops to normal or subnormal. In severe cases death takes place generally between the sixth and twelfth days of the disease. Recovery from the disease is followed by immunity for a number of years. Strains of the rickettsia vary in virulence and that of 'Bitter Root Valley', Montana is exceptionally virulent. Before 1947 the mortality rate in this area for non-vaccinated adults was 80 per cent and 37.5 per cent for children. Before the advent of the drugs chloramphenicol and the tetracyclines in 1947 the country-wide mortality was about 20 per cent, but deaths of treated patients are now rare and the mortality rate has fallen to about 3 per cent.

In the eastern states of the United States the principal vector is *Dermacentor variabilis*, the American dog tick. Women and children are mostly affected by the disease, whereas in the west where *Dermacentor andersoni* is mainly responsible, adult males are most affected. It has been argued that this variation shows the importance of the household dog as a factor in the distribution of the disease.

Colorado tick fever

This virus disease, in spite of its name, is known in all western states of the United States where *Dermacentor andersoni*, the Rocky Mountain wood tick occurs. In the State of Colorado it is by far the most important of the diseases transmitted by ticks, outnumbering the cases of Rocky Mountain spotted fever, it is said, 100-fold. The virus has been isolated, however, from *Dermacentor variabilis* the American dog tick in Long Island, New York so that the disease is probably much more widespread than is actually recorded.

Like other tick-borne diseases the onset is closely paralleled by the times of greater activity in the life-cycle of the tick, in the spring and early summer. Invariably the victim will have been in a tick-infested area between four and six days previously and often the presence of the tick or ticks goes unnoticed, as their bite is painless and they quite often fix themselves in the hair of the head or upper neck. The virus is transmitted from adult to egg, so that although no small mammals have so far been demonstrated as constituting a reservoir of the virus, the ticks themselves are an important component of continuity. Only one death has been reported and no one has been recorded as having had the disease twice.

Clinically the disease is indistinguishable from dengue except that a rash is absent. The symptoms are a sudden onset of chilly sensations with mild photophobia, quickly followed by generalised aching, deep ocular pain and backache lasting two days. This is followed by a remission of about two days and then a second attack lasting somewhat longer than the first.

Tularaemia

An infectious disease, caused by the bacterium *Pasteurella tularensis*, of wild mammals such as rodents, rabbits and hares. It is transmitted by blood-sucking flies, ticks and other arthropods. Man catches the disease either by being bitten by an infected tick or by handling diseased mammals such as rabbits when hunting. Tularaemia occurs widely in the northern hemisphere including practically the whole of the United States, Canada, Alaska, Mexico, Japan, central and south Russia, Norway, Sweden, Czechoslovakia, Austria, Poland and, since 1946, France, Belgium and Germany. From 1924 to 1959 more than 26,000 human cases were recorded. There has been no proved human to human transmission.

There are a number of different manifestations in human beings of *P. tularensis* infection. There is a typhoid type with an incubation period of 1 to 10 days, proceeding as a general disease with a sudden onset followed by severe headache, vomiting, chills and fever, with temperatures rising above 40 °C. This is followed by prostration, a weight loss and in severe cases there is delirium and stupor; the fever period lasts 10 to 15 days.

Other tick-borne diseases

Space has limited the number of diseases dealt with in any detail, but it should be made clear that numerous other tick-borne viruses and bacteria are found in various parts of the world. Viruses causing encephalitis and haemorrhagic fevers incapacitate Man, particularly in northern and temperate parts of Europe and Asia, Indo-Malaysia and North America. The ticks involved are of many genera and species. They do not all transmit diseases to Man, but may act as natural reservoirs, infecting alternate hosts.

The rickettsias are also spread by many species of ticks in most parts of the world; for example, boutouneuse fever, caused by *Rickettsia conori*, occurs throughout Africa, the Mediterranean, Middle East and South-East Asia. Other species infect Man in Siberia and Queensland, Australia.

Rhipicephalus sanguineus, the Brown dog tick, is widely distributed throughout the Old World and also in America. It is implicated as a vector of boutonneuse fever in Africa and also of Rocky Mountain spotted fever in the warmer parts of the Americas.

Spirochaete bacteria occur in three main geographic zones: central and western United States south to Argentina; central Africa from Kenya to Cape Province, and the Mediterranean from Portugal and Dakar to Arabia and Transcaucasia. As mentioned earlier, several species of *Spirochaeta* (=*Borellia*) cause human relapsing fever; all are carried by ticks of one genus – *Ornithodorus*.

Chapter 10

Spiders

ARACHNIDA – ARANEAE

Perhaps the best-known characteristic of spiders is their ability to spin a silken thread from organs situated at the rear of the underside of the abdomen, known as spinnerets. The spinning of silk has given rise to a large number of specialisations, the most primitive of which is a simple dragline; the thread of silk laid wherever the spider moves, which adheres to whatever substrate is present. From this, sheet webs have probably evolved and then funnels and later trap-doors. Webs have evolved from the dragline being produced when the spider jumps from the end of one twig to another.

All spiders are predacious, the vast majority preying on insects. Consequently they are to be found wherever there are insects, with the exception of Antarctica. Spiders are exceedingly numerous and in some situations 5 million per hectare have been calculated as being present. It has also been estimated that in Britain alone, spiders destroy annually a weight of insects greatly exceeding that of the human inhabitants.

Spiders share with the scorpions and harvestmen the anatomical feature of the combined head and thorax, known as the cephalothorax or prosoma. This is joined to the bulbous and generally larger abdomen by a more or less pronounced pedicel or stalk. A pair of palps, which in the male adult become modified as mating organs, arise from in front of the first pair of legs. A pair of chelicerae is situated in front of the mouth and each is composed of three segments ending in a single, sickle-shaped fang with which the spider injects poison into its prey. No spiders possess wings, so they are unable to pursue insects (their common prey) into the air.

Many spiders enter buildings to search for prey or to hang their webs in windows where flying insects are likely to be found. Most of these common inhabitants of our buildings are quite harmless to Man (although some people find them unpleasant) and are generally beneficial. It is impossible, in this book, to mention the many species involved; space is, therefore, allotted only to those spiders whose bites cause illness to Man.

LATRODECTUS   RED-BACK SPIDERS

The family of web-spinning spiders, the THERIDIIDAE contains the important genus *Latrodectus* of which *L. mactans* is the most widely known of all venomous spiders. As well as occurring throughout the whole of the North and South American continents it is found in the West Indies, Bahamas, Hawaiian Islands and the Pacific area generally, Australasia, India, eastern Arabia and southern Europe.

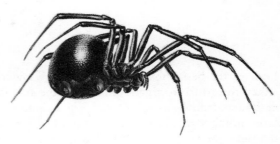

Fig. 10.1   *Latrodectus mactans*, Black widow or Red-back spider. (Shell Chemical)

The female's bite feels only like a pinprick at first, and a reddish patch appears with two red spots showing the actual site of the bite. Pain then occurs almost immediately in the region of the bite, and this becomes intense within one to three hours, and may last as long as 48 hours. The pain has been described as 'acute agony' and there is usually profuse perspiration. A dull pain spreads to all the body muscles and the diaphragm becomes rigid. There is nausea and a low fever, but little, if any, swelling. The symptoms disappear after three days.

The male Red-back spider is much smaller than the female, and is rarely observed.

The web consists of an irregular snare secured by lines of silk attached to objects in the immediate environment, and is usually located among old leaves, piles of rubbish or in rough boxes used to protect water, gas and electric meters and unsewered privies. In addition, the corners of brickwork in little-frequented basements and dusty garages are often used by the Red-back for its web. Several egg sacs are produced by each female, each containing about 100 eggs. The sacs are suspended among the threads of the snare.

The prey consists of insects, commonly beetles, woodlice (slaters), centipedes, and occasional feeding on small lizards has been recorded.

Latrodectus mactans is known by a variety of popular names. In North America it is generally called the Black widow, but in Australia it is the Red-back, and significantly in New Zealand it is known as the Katipo, Maori for 'night stinger'.

The female has a globular abdomen, about 6 mm in diameter, and the legs are comparatively long. The body, excluding the legs, is about 12 mm in length. There is great variation in colour marking. Generally the body is black or dark brown, and has a velvety sheen caused by the covering of fine, short hairs. There is a red or orange mark or stripe on the upper surface of the abdomen, with a smaller mark – either red or white – underneath. In North America, the mark on the upper surface is hour-glass shaped. All-black spiders often occur.

In the event of a bite from *Latrodectus*, a doctor should be called immediately; in some countries where the spiders occur, antivenens (antivenins) are available. There are other methods of treatment and the patient is advised to take hot baths at frequent intervals. Alternatively, intravenous injections of calcium salts (usually calcium gluconate) give rapid relief from the pain caused by the venom. Closely related to *L. mactans*, and often confused with it, is *L. curacaviensis*, which has also caused serious illness and death due to injecting venom when biting. This spider is of retiring habits, but will bite when held captive, or partly crushed in bedding or clothing, or when accidentally provoked in the dark. This often occurs in dwellings, outhouses, garages and privies.

In the United States and Australia where spider bites have attracted attention, a number of species, other than those previously mentioned, have been recorded as biting and causing pain. Thus, in the United States *Peucetia viridans* squirts venom into the eyes causing severe pain. *Chiracanthium inclusum* causes pain by biting, as does *C. diversum* in Hawaii. Slight pain and swelling is caused by the bite of *Phidippus audax*, *Lycosa carolinensis* and *L. helluo*. Other genera thought also to be implicated are *Misumenoides*, *Pamphobeteus* and *Ummidia*.

ATRAX FUNNEL-WEB SPIDERS

Three species of the family DIPLURIDAE or funnel-web spiders occur in New South Wales and are of importance on account of the venomous nature of their bites. One species *Atrax robustus*, the Sydney funnel-web spider has been the cause of a number of human deaths. Much less is known of the habits of the other two species, *A. formidabilis*, the Northern Rivers, (or North Coast) funnel-web spider, and *A. venematus*, but all three species are similar in appearance and generally require an expert to identify them.

The male of *Atrax robustus* is about 25 mm in length and black to reddish-brown in colour, depending on location and whether skin has recently been shed. The palps of the male are leg-like and terminate in a swollen area with a single spine at the end. The abdomen is small and the legs are longer than those of the female. In addition, a spur is present on the underside of the second pair of legs: as with many spiders, courtship is a hazardous procedure, at least for the male, and these spurs hold off the female from predatory intent.

The female is similar to the male in colour, but is slightly larger. The fangs of both sexes are usually folded away into two parallel grooves beneath the cephalothorax. They contain poison ducts connecting with venom sacs at their bases. The base of the fangs, as well as the general underside of the body, is covered with fine red hairs.

Atrax robustus is usually found in dampish areas such as suburban gardens which are watered and where there is litter. The burrow is either in the ground or in rock crevices and is lined with silk. A silk curtain sometimes hides the entrance. The life-cycle appears to be from three to five years.

LOXOSCELES RECLUSA   BROWN RECLUSE SPIDER

Several species of the genus *Loxosceles* are known to cause, or are suspected of causing, a severe gangrenous ulcer as a result of their bite. This is referred to as 'loxoscelism' or nectrotic spider bite. These spiders occur in North and South America but one, *Loxosceles reclusa*, the Brown recluse spider has attracted the most attention. It occurs in Oklahoma, Kansas, Missouri, Arkansas, Texas, Louisiana, Alabama and Tennessee.

Life-cycle
The eggs are contained in a whitish, silk-like globular case about 8 mm in diameter. It is deposited in the spider's dark retreat. During the summer from 40 to over 50 young spiders emerge from the case after 24 to 36 days, but as the abandoned egg case contains the cast skins of the first instar they have hatched from the egg some time previously. Development is slow and depends on weather conditions and availability of food, but in favourable circumstances is completed in seven to eight months. The immature spider resembles the adult except in size and often is slightly lighter in colour.

Fig. 10.2 *Loxosceles reclusa*, Brown recluse spider. (Shell Chemical)

Loxosceles reclusa is a soft-bodied spider varying in colour from yellow to dark brown, with the cephalothorax lighter than the abdomen. The spider measures 8 to 13 mm, averaging about 10 mm in length: females are usually slightly larger than the males. The legs are long and thin and covered with short dark hairs. When resting, the front two pairs of legs are held forwards while the two hinder pairs point backwards. There are three pairs of eyes on the forepart of the head arranged in a semicircle and immediately behind there is a violin-shaped mark. The flattened carapace has a distinct median groove.

Habits

The Brown recluse spider is usually found in buildings, but when in dwellings most often occurs in bathrooms, bedrooms, closets, garages, basements and cellars. The spider hides in old clothes, on the undersides of tables and chairs, behind baseboards and door facings or in any corner or crevice. A nondescript web is usually present, although this spider is a hunter rather than a trapper.

Loxosceles reclusa retreats into a crack or other cover when disturbed and shows little aggression. Most bites have occurred when old clothes have been put on that have been hanging in a garage and a hiding spider has been partly crushed. Similarly people have been bitten while sleeping in bed when rolling on the spider accidentally.

Bite

The following account of the effects of a bite from *Loxosceles reclusa* has been taken from a leaflet issued by the Entomology Department of Oklahoma State University.

The victim may not be aware of being bitten for two or three hours, or a painful reaction may occur immediately. A stinging sensation is usually followed by intense pain. A small blister usually rises and a large area around the bite becomes congested and swollen. The victim may become restless and feverish and have difficulty in sleeping. The local pain is frequently quite intense, and the area surrounding the bite remains congested and hard to the touch for some time. The tissue affected locally by the venom is killed and gradually sloughs away, exposing the underlying muscles. The edges of the wound thicken and are raised while the central area is filled by dense scar tissue. Healing takes place quite slowly and may take six to eight weeks. The end result is a sunken scar which has been described as resembling a 'hole punched or scooped from the body'. Scars ranging from the size of a penny to half-dollar have been reported.

In the case of a bite, the victim should immediately consult a physician and, if possible, bring along the spider which caused the bite for positive identification.

OTHER *LOXOSCELES* SPECIES

At least four species of *Loxosceles* occur in the United States, which in addition to *L. reclusa* may cause loxoscelism or necrotic spider bite.

Loxosceles unicolor is found in southern California, Nevada, Utah, Arizona and south into Mexico. It does not yet appear to have been found in buildings.

Loxosceles arizonica occurs in southern Arizona, southern New Mexico, south-western Texas and Mexico. The spiders are found in debris, under boards and under bark and probably in or under buildings.

Loxosceles rufescens has been introduced into a number of ports and cities and is well established along the east coast and Gulf states of the United States. A necrotic lesion has been produced in the laboratory on rabbits. The response, however, was less severe than that produced by the bite of *L. reclusa*.

Loxosceles laeta has been well known for many years as the cause of severe bites with resulting symptoms similar to or worse than those from the bite of *L. reclusa*. It occurs in Central and South America.

Chapter 11

False spiders, tail-less whip scorpions, false scorpions

ARACHNIDA – SOLIFUGAE, AMBLYPYGI and PSEUDOSCORPIONES

Solifugae

The SOLIFUGAE, SOLIFUGIDA or SOLFUGIDA are known as false spiders, sun spiders, camel spiders or wind scorpions. They bear only a superficial resemblance to true spiders: the body is hairy are divided into a prosoma or cephalothorax, and an opisthosoma or abdomen. These are joined broadly but not by a thin pedicel as in spiders; the abdomen is segmented and spinnerets are absent. The well-developed chelicerae have two, pincer-like claws with which their prey is caught and broken up. They are said to be, for their size, the most formidable jaws in the animal world. The six-segmented pedipalps are leg-like and end in a suctorial organ, not a claw.

The first pair of legs are long but feeble; They are carried stretched out in front as tactile organs, and not used for walking. Five special organs, known as malleoli or racquet organs, are borne on each of the fourth pair of legs. Their function is not known. In length, the body varies from 10 to 50 mm, but some of the larger species can span about 120 mm with their limbs and their hairiness gives them a formidable appearance. Solifugids are usually yellowish or brownish in colour but a few are black and some are patterned with longitudinal stripes on a yellow background.

False spiders almost always inhabit hot, dry areas. None are found in northern Europe but a few are found in the south. They are abundant in Africa and the Americas, but absent from Australia and New Zealand.

Most solifugids are nocturnal, hiding under stones or in the soil during daylight, but a few are active during the day. *Gluvia dorsalis*, for example, is a common sight in the streets of Madrid, and *Mummucia variegata* and *Pseudocleobis morsicans* are known as sun spiders.

All species are predacious, preying on insects, but in addition will kill and eat large spiders, scorpions and lizards, and even mice and small birds are known to be eaten by them. As many prey on harmful insects they can be considered as beneficial; the householder, influenced by their awesome appearance and ill-deserved reputation, would not, however, consider them so.

Whether the bites of solifugids are poisonous to human beings has been shrouded in some doubt, but it is now generally accepted that they are not poisonous. There appear to be no glands which could secrete venom but, on the other hand, a number of cases have been recorded of solifugid bites which have resulted in severe effects, and occasionally death. It appears likely that the fatal causes have been the result of secondary infection.

Amblypygi

The AMBLYPYGI or PHRYNICIDA, the tail-less whip scorpions, are of characteristic flat appearance but their most easily identifiable features are the extremely long, flexible tips of the first pair of legs, which are made up of many segments. Amblypygids do not possess a caudal appendage as do the UROPYGI or THELYPHONIDA (true whip scorpions). In size they vary from 8 mm to as much as 45 mm.

The cephalothorax bears a pair of median and three pairs of lateral eyes. A poison gland was thought to be situated in the end of the pedipalps, either in the fixed part of the pincer, or in both parts, but this is now in doubt. The basal segments, or coxae, of the pedipalps extend forwards to form masticatory plates or to hold the prey with the mouth located between them. The abdomen or opisthosoma is attached to the cephalothorax broadly, there being no stalk-like segment as in the ARANEAE. There are 12 segments comprising the abdomen, the last being very small and consisting only of a ring surrounding the anus. The genital orifice is situated between the first and second sternites.

A number of amblypygids are found in buildings in tropical regions where they hide themselves in damp corners. Thus, *Phrynichus ceylonicus* in India; *Paracharon caecus* in Guinea-Bissau; *Masicodamon allanteus* in Morocco; and *Damon medius* in Mauritania, Mali and Niger, commonly occur in dwellings. In addition, it has been stated that *Damon variegatus* occupies all buildings in Pietermaritzburg in South Africa, where it lives in the subfloor ventilation areas.

These arachnids are entirely predacious, feeding on insects and other arthropods. Their presence is not much relished in dwellings, however, as their strange crab-like walk and the unusual way in which the first pair of legs are waved like antennae, probably contribute to a feeling of distaste.

Pseudoscorpiones

Another order of the ARACHNIDA, the PSEUDOSCORPIONES or false scorpions, sometimes referred to as CHELONETHI or CHERNETES, have some significance in buildings. Their common name refers to the resemblance to scorpions in the shape of the pedipalps and body (but the hind part of the abdomen is not narrow and a caudal sting is absent). The dorsal surface of the prosoma or cephalothorax consists of a single large sclerite which bears the eyes when present and six pairs of legs. The small pre-oral chelicerae are composed of two segments, the tips of which are finger-like and bear complicated structures set with fine teeth; there is a flagellum formed of setae attached to the tip of the fixed part of the chelicera. Silk produced from these organs is used in nest building. The large, six-segmented pedipalps are highly developed and resemble the claws of scorpions. They are prehensile and pincer-like and capture and kill the prey.

Venom is produced in poison glands which open at the terminal tooth of either the fixed or the movable finger, or both. The normal food is insect and spider eggs, and small animals of all kinds. Some species live in groups, the individuals of which work together to hold and to kill large insects such as ants. The venom does not affect Man since the pedipalps are too weak to penetrate the skin.

Pseudoscorpions are often found attached to the legs of flies and harvestmen. They apparently do not harm their hosts, but simply use them as a means of transport.

There are about 1,500 species in the order and a number are habitually associated with Man in buildings throughout the world. *Chelifer cancroides, Allochernes italicus* and *Cheirisium museorum* are common examples. The first-named has sometimes been found on Man but as the phoretic habit is common in this group it seems probable that dispersion is the sole reason for this. The last named is well known as an inhabitant of old, slightly damp books and is known as the Book-scorpion. It is found also in warehouses and in mills.

Toxochernes panzeri often occurs in old buildings, especially in stables and barns.

Chapter 12

Scorpions

ARACHNIDA – SCORPIONES

Present-day scorpions differ little from their ancestral forms which were present during Silurian times 300 million years ago. Additionally, there is little diversity among the 600 or so species found today. They are of very characteristic shape. There is an anterior prosoma (cephalothorax) covered with a shield. The rest of the body (the opisthosoma) is segmented and consists of two distinct regions. The first seven segments make up a broad pre-abdomen (mesosoma) tapering posteriorly; the last five segments forming a narrow cylindrical post-abdomen (metasoma), or tail. The latter is terminated by the sting which is curved and ends in a sharp point. Usually the tail is bent forwards over the back of the pre-abdomen. From the mouth region extend a air of formidable crab-claw-like pedipalps which are held with the joints pointing backwards and the tips of the claws forwards.

A number of the larger species produce rasping sounds when they are menaced. These are made in various ways, such as by movements of the mouthparts or by the sting being rubbed over a groove in the pre-abdomen when the tail is arched over the back.

Scorpions are usually yellow, brown, black or dark-reddish, and occasionally bluish-green. Generally they are large, *Pandinus imperator* from West Africa, the largest, being 200 mm in length.

Prey is not usually obtained by hunting but by seizing it with the large claws. Arthropods are the scorpion's main food but on occasions lizards, mice and other vertebrates are taken. The prey is broken by the small jaws (chelicerae) and the soft tissue sucked out. Many months, or even a year, may elapse between meals.

Scorpions are generally and widely distributed in dry, tropical areas, especially in desert and semi-desert regions, although a few species occur in more temperate conditions such as southern Europe and British Columbia.

Breeding
Scorpions are remarkable for their elaborate courtship behaviour or 'dance'. Although there are a number of variations in procedure the usual

pattern is for the male to seize the claws of the female in his, while he moves backwards and forwards. The male extrudes a spermatophore from the genital aperture and cements it to the ground. He then pulls the female over it when she opens her genital aperture and the spermatophore is pushed inside. All scorpions produce living young, eggs not being laid. The brood stay with their mother, on her back, until after their first moult. They then disperse and find suitable hiding places under stones, logs or dark crevices, or by tunnelling according to the habit of the species.

Stings

Scorpions are dreaded wherever they occur on account of their poisonous sting. The latter is brought into use only as a defensive mechanism but, nevertheless, many human beings are stung through accidental contact when treading on them in the dark, or disturbing them in bedding. Children are often victims when they tease or play with them. The effect of the sting varies according to species from a mere pinprick to a few hours' pain, or to death.

Their importance as undesirable animals should not be underestimated. The symptoms arising from a scorpion sting of the virulent species are similar to those of strychnine poisoning. Speech fails or becomes difficult, frothing at the mouth and nose usually occurs while in the terminal stages there are convulsions. In serious cases death usually occurs within three hours. Scorpion-sting antivenins are maintained where dangerous scorpions occur; these are very effective if administered early.

The most well-known venomous scorpions are *Buthus occitanus* found in south-west Europe, the Middle East and North Africa; *Androctonus australis* from North Africa; *Lieurus quinquestriatus* in upper Egypt and the Sudan;

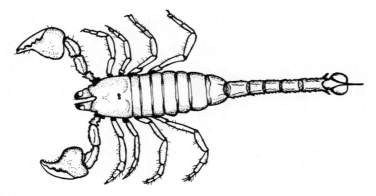

Fig. 12.1 Great Indian Scorpion.

Centruroides species from southern USA and Mexico and species of *Tityus* from Brazil.

Small yellow scorpions with delicate claws are generally poisonous but it would be unwise to treat any scorpion sting in any way other than seriously and medical help should be sought without delay.

Of all the countries in the world it is in Mexico that the effects of scorpion stings have the most serious consequences. During the periods 1940–49 and 1957–58, no fewer than 20,352 persons were killed by this means. During the same periods 2,068 persons died from snake bite, 274 by spider bite, and 1,933 from bites or stings caused by unidentified venomous animals. Many of the latter, however, perhaps the majority, may actually have been caused by scorpion stings, In many areas of Mexico scorpion stings are one of the main causes of death and although the rates have consistently shown a downward trend, the absolute number of deaths due to this cause has averaged more than 1,000 per year.

Three-quarters of the deaths occurred in children from birth to three years of age, and mortality predominated during the summer.

Chapter 13

Woodlice or sowbugs

CRUSTACEA – ISOPODA

The great majority of the CRUSTACEA are aquatic, indeed, most are marine but the order ISOPODA contains a number of species adapted to varying degrees for terrestrial life. These are known as Woodlice or slaters in the British Isles and sowbugs in the United States.

Most terrestrial isopods are between 5 and 20 mm in length, oval in shape, with the body arched. A pair of antennae are present as well as a pair of smaller antennae in front of them. The thorax, or perion, is composed of seven segments and these are usually wider than the following six that form the abdomen or pleon. A pair of walking legs are borne by each thoracic segment. There are also plate-like appendages on each abdominal segment with the exception of the last. These pleopods are variously modified and in the more developed forms bear tufts or invaginated tubules for respiration, called pseudotracheae or lung-trees.

Numerous species of ISOPODA inhabit the damper, cooler areas of buildings but few have been specifically recorded. However, in Michigan, USA, *Armadillidium nasatum* occurs in buildings and not outside, while *A. vulgare*, is chiefly found around human habitations. In Europe the latter species is often found in buildings in chalk or limestone areas, while *Oniscus asellus* is everywhere to be found in damp buildings. *Porcellio scaber*, also very common, may be recognised by the transverse rows of small tubercles which cover the back of the head and body giving it a mat appearance. This species is better adapted to live in drier conditions than are most other woodlice.

Most woodlice are nocturnal and are able to seek out damper areas if their habitat is too dry. Woodlice have little ability to prevent water loss as they do not possess a waterproof integument. They can, however. regain lost water by drinking and by absorption through the pleopods. They survive on land as a result of behaviour mechanisms that keep them in cool, moist places.

Thus woodlice occur in or under buildings where damp conditions prevail, especially in association with fungally decayed wood or other decomposing vegetable matter, on which they feed. They are generally found in kitchens where there are water leaks or badly ventilated

Fig. 13.1 *Oniscus asellus*, Wood louse.

cupboards. Woodlice do not cause damage to sound materials, but their dark faecal pellets are objectionable.

Breeding
Woodlice, like other CRUSTACEA, carry their eggs in a brood pouch or marsupium under the thorax, and the newly hatched young are found huddled together on the underside of the female. When first hatched, the head, eyes, segmented body and short, stumpy limbs are discernible. For the first few days the limbs are incapable of movement, but eventually they emerge from the brood pouch. This brooding stage is critical as the young may otherwise die from desiccation or become so wet that they are invaded by fungi. Woodlice seldom breed until they are two years old.

Chapter 14

Beetles

COLEOPTERA

The chief characteristic of the order COLEOPTERA or beetles is the modification of the first (mesothoracic) pair of wings into more or less hardened, non-folded, rigid elytra which meet edge to edge when at rest, and which partly, or wholly, cover the wings and the abdomen. The second pair of wings (metathoracic) are membranous, folded and are used alone for propulsion in flight; they are not always developed. The mouthparts are mandibulate and used for biting and chewing. The prothorax is well developed and free, forming with the head a distinct forebody contrasting with the hindbody formed by the elytra covering the meso- and metathorax and abdomen. The mesothorax is usually small and the abdominal sternites are more sclerotised than the tergites. In the larva, thoracic legs may be present or absent, but abdominal prolegs are rarely present. There is a distinct head capsule with the antennae and mouthparts.

The number of species of COLEOPTERA in the world is between 277,000 and 350,000. This is about one-third of all known animal species, but the reason for the obvious success of this group in nature is a matter of conjecture. Perhaps water conservation is an important factor: abdominal spiracles do not open directly to the exterior but into an enclosed space beneath the elytra, thus restricting water loss.

Beetles have colonised almost every habitat, so that it is not surprising that the largest part of this present work concerns them. However, it is as consumers of stored foods, wood and animal products such as wool, feathers and fur, that they concern us in buildings. There are no coleopterous ectoparasites of Man, nor do they carry any pathogenic organisms that affect him, with the possible exception of the DERMESTIDAE.

With a large number of beetles to describe, it was considered necessary to arrange the families in alphabetical order.

ANOBIIDAE

The family ANOBIIDAE is related to the families BOSTRYCHIDAE, LYCTIDAE and PTINIDAE. It contains small beetles (2–6 mm) with a hard, brittle integument (although it is soft in the species *Ernobius mollis*). The pronotum is shield- or hood-like, sometimes with lateral projections. The body surface is clothed with fine recumbent or semi-erect setae. The antennae which are 9- to 11-segmented, sometimes with the last three segments markedly elongated, are inserted in front of the eyes and are widely separated. The legs are capable of being retracted into the angular cavities between the head and prothorax, prothorax and mesothorax and behind the declivity of the hind coxae.

The larvae are soft and hook-shaped or 'C' shaped with small two-segmented antennae and 10 abdominal segments. Many are wood-borers, spending the whole of the larval stage within the wood. Several species are found in timber in buildings, especially in temperate regions. By far the most important of these is *Anobium punctatum* the Common furniture beetle or Houseborer but other species such as *Xestobium rufovillosum* and *Ptilinus pectinicornis* are of importance too. Some anobiids are pests of stored foods and other dry vegetable matter.

Many anobiids have been shown to possess mycetomes in the form of caeca in the mid-gut. Symbiotic micro-organisms from these bodies pass to the anus where, in the female, they are stored in special sacs. From this position they can be transferred to the usually pitted surface of the eggs, as they leave the oviduct. The micro-organisms are essential for nutrition. Without them the larvae would die.

STEGOBIUM PANICEUM   BISCUIT BEETLE, DRUG STORE BEETLE

The Biscuit beetle is an important household insect which attacks stored food products and breeds in dried vegetable matter of many kinds. In domestic larders it feeds readily on flour, bread, farinaceous foods of diverse kinds, meat, soup-powder and spices (especially red pepper). In chemists' shops it used to infest poisonous materials such as strychnine, belladona and aconite. *Stegobium* damages books and manuscripts and has been known to bore in a straight line through a whole shelf of books. Tinfoil and sheet lead are sometimes perforated. The beetle is also a common inhabitant of birds' nests.

Experiments on its nutrition show that the beetle can complete its life-cycle in foods containing very small quantities of carbohydrates, and its symbiotic organisms resemble yeasts which produce vitamins of the 'B' group. The larva is, therefore, independent of external sources for this important nutritional factor.

Stegobium paniceum is a small reddish-brown beetle, only 2.0–3.5 mm in length. The pronotum, which is rounded and is just as wide as the elytra, completely masks the head which is inserted almost vertically beneath it. Very fine hairs are arranged in longitudinal rows on the elytra. It may be distinguished from the Common furniture beetle by the profile of the pronotum being evenly rounded, rather than crested.

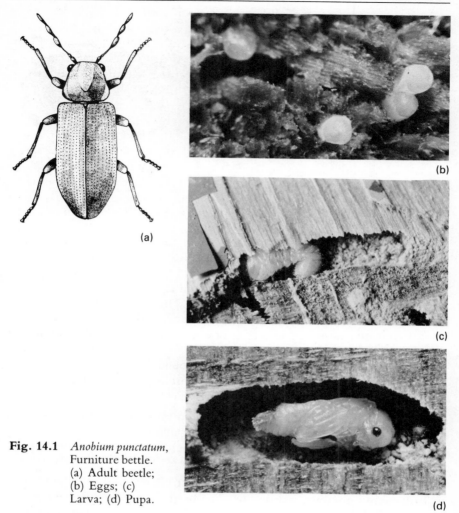

Fig. 14.1 *Anobium punctatum*, Furniture bettle. (a) Adult beetle; (b) Eggs; (c) Larva; (d) Pupa.

Breeding
The life-cycle is normally completed in about seven months, but increase in temperature diminishes this period considerably. About 100 eggs are laid singly in the foodstuff or in nearby crevices, over a period of about three weeks. The first-stage larva is extremely active and wanders about exploring its surroundings. It is small, measuring only 0.5 mm in length, and 0.125 mm in breadth, and can squeeze itself through very small crevices in packaged foodstuffs. At this stage it can survive starvation for about eight days.

The larva moults four times, and when fully grown, is 5 mm in length, white and typically anobiid in shape. This fully fed larva constructs a cocoon of food particles cemented together with a secretion from the mouth.

Fig. 14.2 *Stegobium paniceum*, Biscuit or Drug Store beetle.

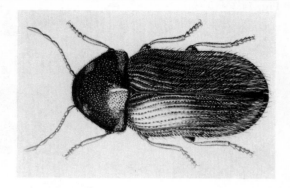

LASIODERMA SERRICORNE   CIGARETTE BEETLE

This species is probably subtropical in origin and does not survive winters in unheated premises in northern Europe. Nevertheless, it is generally cosmopolitan in distribution.

It damages tobacco, both in the leaf and in cigars and cigarettes; the greatest loss occurs in the finished products. The beetle and its larvae also attack spices and various seeds such as coriander and caraway. It occurs most frequently, however, on cocoa beans, oilcakes, especially cotton seed cake and locust beans (carobs).

This beetle is distinguished from other anobiids found in stored products by its antennae which are serrate throughout, by the absence of punctures on the elytra and by the strongly flexed head and pronotum. Its larvae differ from those of other anobiids found in stored products in having no rows of spinules on the abdominal segments.

ANOBIUM PUNCTATUM   COMMON FURNITURE BEETLE OR WOODWORM BEETLE

Anobium punctatum infestations of softwood structural timber in buildings in Britain are important and serious. The beetle occurs in about three-quarters of all such infestations and about £10 million are spent annually in its control. *Anobium punctatum* is injurious to timber indoors in a large part of the temperate world but it is most serious in areas with a cool, maritime climate. It remains a potential hazard to building timber and furniture in areas where, as yet, it has not been found. However, the beetle is not likely to injure timber in tropical areas except where altitude makes conditions suitable.

The Common furniture beetle occurs throughout Europe where it is indigenous, it appears to become less common towards the south but occurs in the east. Elsewhere in the world it has become established through being transported in infested woodwork, such as the furniture of migrants. The beetle is found throughout temperate Australia (it is called Common houseborer in New Zealand), and the timber of *Podocarpus dacrydioides* seems particularly susceptible in Tasmania, New Zealand and

Fig. 14.3 *Lasioderma sericorne*, Cigarette beetle and side view.

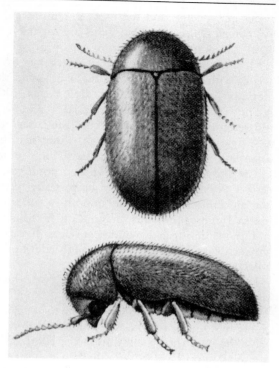

New South Wales. Blackwood, *Acacia melanoxylon*, is also attacked but *Eucalyptus* species are rarely infested. At one period every house in New Zealand that had been erected for more than 15 years contained an *Anobium* population, but control measures have been vigorous.

In North America the beetle appears to have a wide distribution on the eastern side but is rarely regarded as a pest. In south-eastern United States the anobiid, *Xyletinus peltatus*, is of more significance, and is found also on the western seaboard of North America, particularly in the Vancouver area. In South Africa, *Anobium* was not noticed until 1939 but is now widely distributed throughout the four provinces of the Union.

With regard to the areas where it has not been recorded, these include tropical Africa, tropical Asia, Central and South America.

Description

Adult beetles vary in length from 2.5 to 5.0 mm and are variable in colour also, from light reddish-yellow through dark chocolate brown to pitchy-red. Some variation in colour is due to the rubbing off of pubescence. The pronotum is almost as wide as the elytra, and, viewed from above, is pear-shaped with prominent, sharp angles at the base. Viewed from the side, the front margin is sinuate, rising to a prominent crest. A shallow depression occurs in the mid-line from the front of the pronotum to the crest. The wing-beat is rapid, the frequency being 9–130 cycles per second, and the flight is reminiscent of that of the Lesser house fly, *Fannia canicularis*. The wings are capable of complete retraction after flight, although often, the wing-tips protrude.

Mating takes place soon after emergence from the wood, and often occurs with the female in the flight hole and the male on the surface, or, they may retreat into the flight hole while still 'in copula'.

The egg is whitish and ellipsoidal, 0.35 mm wide and 0.55 mm long, and resembles an acorn in its cup, the latter being alveolate with minute pits. The alveolate end is slightly wider than the distal two-thirds which is smooth and terminates in a short, blunt point.

The site for egg-laying must incorporate a crack, groove or rough surface in which the eggs can be partially wedged or anchored. They are not laid indiscriminately on smooth or otherwise unsuitable surfaces. Normal oviposition sites indoors other than those given above, are rough end grain, unfinished wooden surfaces of various kinds, open joints between two pieces of wood, rough edges of plywood and old flight holes.

Eggs are laid singly or in rows of two, three or four. They generally take two to five weeks to hatch.

The emergence of the larva from the egg is of the greatest importance in the biology of this insect. The alveolate end of the egg is the last to leave the ovipositor, and during the contractions to force the egg down the tube, special yeast-like cells become attached to the alveolus. The embryo is orientated with the head into the alveolar portion of the chorion, so that, when the larva bites its way out, it takes into its gut a number of cells. The young larvae bite their way directly into the wood and the whole of the larval stage is spent in a gallery or tunnel within the timber.

The larval tunnels contain rather loose aggregations (known as frass) consisting partly of rejectimenta – small wood particles torn off but not taken into the mouth – and faecal pellets. When fully fed, the larva constructs a pupal chamber near the surface of the wood. The pupa is milky-white at first, gradually darkening to the adult colouring. The length of the pupal stage is given as from two to eight weeks. After rupture of the pupal cuticle, the beetle remains for a time in the pupal chamber while the exoskeleton hardens. It then bores out of the wood, making a circular flight hole or exit hole about 2 mm in diameter.

If the beetle is prevented from leaving the wood by the intervention of a covering such as glass, it will bore as much as 50 mm in a straight line in an endeavour to get round the obstacle.

The minimum length of the life-cycle is about three years in buildings in Britain. Depending on the timber and the conditions, it may be considerably longer, even 9–10 years. Thus an infestation can be present for many years in timber without flight holes being present.

Parasites and predators
Anobium punctatum larvae are preyed upon by a number of arthropod parasites and predators. The most frequently occuring parasite of the larval stage being mite *Pyemotes ventricosus*, and up to one-third of all larvae have been found to be infested. Other important parasites of *Anobium* are the ant-like wasp, *Theocolax formiciformis*, and the braconid wasp, *Spathius exarator*. In north Germany the latter appears to be particularly prevalent, on occasion parasitising all the *Anobium* larvae in the timber. The most important predator of *Anobium punctatum* in England is the steel-blue beetle *Korynetes caeruleus*, while in Germany the beetle *Opilo domesticus* takes its place. Both are members of the family CLERIDAE.

XESTOBIUM RUFOVILLOSUM   DEATH WATCH BEETLE

This beetle is found normally in old buildings in hardwood timber which shows some decay, either active or inactive, but secondary infestations in decayed softwood are not rare. Old, large-dimensional beams are, however, sometimes re-used and in this way new buildings become infested. It occurs throughout Europe, including Corsica. In Africa it is recorded from Algeria and in America from the New England states where it is almost certain to have been introduced from Europe, probably by an infested wooden-hulled ship. In England and Wales it is of great importance for the damage it does to historic buildings.

Xestobium rufovillosum is one of the largest of the ANOBIIDAE the length varying from 5 to 7 mm. Females are, in general, slightly larger than the males but this cannot be relied upon for separation of the sexes. The prothorax is widely flanged, with the head rather deeply sunk into it. The general colour of the insect is dark chocolate brown, tessellated with patches of yellowish pubescence which gives the upper surface a variegated appearance. The hairs are easily rubbed off, when the elytra become more reddish and shining.

Habits
The emergence of the adult is generally in the latter part of April but very occasionally in the autumn. The activity of the beetles is mainly influenced by temperature: at 14 °C, they move about slowly, but at 17–20 °C their activity is increased leading to tapping, pairing and egg-laying. *Xestobium rufovillosum* seldom flies indoors but under laboratory conditions it will fly at 22 °C.

The tapping noise is made by both sexes and is caused by knocking the front of the head against the wood on which the beetles are standing. They normally tap from 6 to 12 times: a set of 12 taps occupies 1.25 seconds, when active, and 1.25–2 seconds separate each set. The function is not known but is probably a sex call. Tapping may continue after pairing.

Breeding

Pairing takes place on the surface of the wood but it is probable that it occurs also in cavities inside large timbers. Eggs are laid singly, in pairs, or in small groups in cracks or crevices on roughened surfaces, among frass in old tunnels and in the vessels of oak. Decayed wood is not necessarily selected. When first laid the egg is pearly-white and somewhat lemon-shaped, slightly more pointed at one end than the other. It measures 0.6–0.7 mm in length and 0.4–0.5 mm in width. The egg becomes more opaque shortly before hatching which takes place after about five weeks out-of-doors.

On hatching, the larva walks actively over the surface before starting to tunnel into the wood. This contrasts with the larva of *Anobium punctatum* which bores straight into the wood through the base of the egg. The fully grown larva is typically anobiid – fleshy, soft, curved and creamy-white. It is up to 11 mm in length. The frass contains bun-shaped pellets.

Pupation takes place in the autumn, but the pupa remains in its chamber, just below the wood surface, until the following spring when the newly-formed adult tunnels out. The flight hole is circular, from 2 to 3 mm in diameter.

The length of the life-cycle is extraordinarily variable and is dependent on the degree to which the timber is fungally decayed. In heavily decayed sapwood the minimum length of life-cycle recorded is three years, under unheated conditions. But, on the other hand, if it is only slightly decayed, the development may take many years. Some fungal decay appears to be obligatory.

PTILINUS PECTINICORNIS

This is a common insect in Europe and Asia Minor where it damages the timbers of beech, maple, oak, plane, sycamore, elm, hornbeam and poplar. It is said to rank second only to *Anobium punctatum* as the cause of damage to dry wood in Europe. It is an elongate, cyclindrical anobiid and has an almost globular pronotum which is usually darker in colour than the reddish elytra. The latter are decorated with punctures arranged in

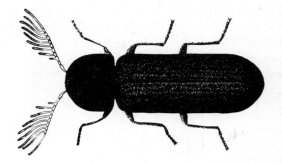

Fig. 14.4 *Ptilinus pectinicornis.*

regular, longitudinal rows. The male is easily recognised by its strongly pectinate antennae, while those of the female are serrate. It varies in length from 3.3 to 5.5 mm.

Breeding
During copulation, the female may pull her partner into a flight hole, giving the impression that it takes place entirely within the timber. Then seeking a crevice in the wood, she bores at right angles to the grain to construct a brood chamber. If the timber has already been infested, she enters a flight hole (very often her own) and makes a brood gallery leading off from the pupal chamber. The eggs are laid directly into the lumen of wood vessels and are extremely long and flexible, varying in size according to the lumen into which they are inserted. In poplar they have been measured as 1.5 mm in length and 0.075 mm in width. After completion of egg-laying. the female dies *in situ*, plugging the entrance to the brood chamber with her massive pronotum.

The larvae show extraordinary polymorphism. The first-stage larvae are thread-like, resembling nematode worms, but they gradually assume a normal shape through several instars.

ANOBIUM PERTINAX

This species is found in buildings, where the larvae are able to feed on softwood which has some fungal attack; wooden handles of old gardening implements are frequently attacked. *Anobium pertinax* occurs throughout Europe (except in the UK) and in Russia as far east as east Siberia. Its biology is similar to that of *Anobium punctatum*.

This anobiid is a dark, pitchy colour but is lighter around the front margin of the hemispherical pronotum. There are two depressions in the pronotum near the hind margin and there are a number of longitudinal rows of punctures on the elytra. The beetle is 4–5 mm in length.

PRIOBIUM CARPINI

Also found to be associated with softwood in which some fungal attack has taken place, the beetle occurs over the greater part of Europe, in Russia as far east as eastern Siberia and in Asia Minor and Cyprus. This species is 3–5 mm in length and medium brown in colour. The hind margin of the pronotum is angled.

ERNOBIUS MOLLIS   BARK BEETLE

This beetle appears to be indigenous to the north temperate region, but is now distributed almost worldwide. It is abundant in northern Europe, especially in Scandinavia. *Ernobius* was introduced into the North American continent during the last 100 years, where it now ranges from Novia Scotia to south central Ontario and southwards to Texas and

Florida. It is found everywhere in the Union of South Africa where unbarked, coniferous timber is used. In New Zealand the larvae tunnel throughout the wood of dead, standing pine trees having a diameter of up to 20 cm. In milled timber, the larvae are encountered at least 5 cm below the bark. In Australia, from time to time, it is present in great abundance especially with the timber of *Pinus radiata*. The adult beetles gnaw through floorboards and floor covering at the time of emergence.

In this species the pronotum is as wide as the elytra and the punctures on the disc are deep and circular. The eyes are large. The tibiae and tarsi are long and slender and the front tibiae are turned slightly outwards, more so in the male than in the female. The elytra are soft and the colour is a light yellowish-brown. The average length is about 4 mm but the recorded range in size is from 2.8 to 6.2 mm.

Breeding
Egg-laying takes place only at night. After examination of the bark by the female the egg is wedged into a crevice. Relatively few eggs are laid – usually 16 to 26 – but the potential egg number is 120. The length of the egg stage varies from 10 to 21 days according to conditions. On hatching the larva first eats the shell, thus transferring symbiotic organisms supplied by the female parent, then walks rapidly over the bark surface in order to find a suitable entry point. Once the larva has begun to bore, the body becomes curved. The total gallery length does not exceed 12 mm. Generally, the larval period is spent in the cambial layer, and the frass is characteristically bun-shaped and loosely packed in the galleries. When the larva has been eating sapwood, the pellets are yellowish-white, but when the larva is boring in the bark, they are dark brown. The larvae eat their cast skins at each instar.

The pupal stage lasts for rather less than 10 days and after a further 6 to 12 days spent in the pupal chamber, the adult bites its way out, making a flight hole about 2 mm in diameter. The adults fly strongly.

In Sweden there are two generations each year, but in England one only; indeed, the disparity in larval size suggests some individuals may take two years to complete their life-cycle. In South Africa adults may appear all year round in heated rooms.

OTHER SPECIES

Many other anobiids occur in timber in buildings, but their limited importance prevents their description here. However, most of these species are less widespread in their geographical range than those considered above.

ANTHRIBIDAE

The members of this family are characterised by the head being prolonged into a short broad snout. The antennae are thread-like, but with the last three segments enlarged to form a slender club. The legless larvae live *within* the infested produce.

This family of about 2,400 species is mainly tropical in distribution with particular emphasis on the Indo-Malayan region. The larvae of many species are associated with wood more or less fungally decayed but only seldom are they found infesting timber in use as the larvae usually die when the timber is dried.

ARAECERUS FASCICULATUS   NUTMEG OR COFFEE BEAN WEEVIL

Occurs in tropical produce imported into Europe, most commonly in nutmegs, more rarely in coffee and cocoa beans, but has been recorded in seeds, grain and dried fruit generally from warmer climates. In the United States it is very common in the south, and in Brazil it is of importance as damaging coffee. The larva is legless, and the complete life-cycle takes only 30 to 45 days, and from 8 to 10 generations annually have been recorded. It requires high temperatures and humidity for successful breeding. It is not established in Britain even in heated premises. *Poticus pestilano* is associated with dried apples in which the larvae develop.

BOSTRYCHIDAE

With but few exceptions all members of the BOSTRYCHIDAE possess wood-boring larvae. Usually only dead, dying or freshly-felled trees are attacked but many of the 550 or so species are of importance in the timber trade. Generally the damage is recognised but the timber is often used for low value timber usage such as packing-case manufacture. In this way a number of species have been distributed throughout the world. Although the family is of worldwide distribution the great majority of species are of tropical or subtropical origin.

Although the BOSTRYCHIDAE are joined with the LYCTIDAE as powder-post beetles they are easily separated. In bostrychids the club of the antennae is of three segments and the pronotum is subglobular and hoods over the head. In lyctids the club is of two segments only and the pronotum which is not subglobular does not hood over the head. In addition, in bostrychids, the front of the pronotum and the hind end of the elytra are often ornamented with spines and protuberances, and the hind end of the body is often truncated whereas in lyctids there is neither ornamentation nor truncation. Finally, bostrychids are cylindrical but lyctids are dorso-ventrally flattened.

Some species are very small, 2–3 mm in length but others are upwards of 20 mm.

Biology
Long after the eggs have hatched adults of both sexes remain in the oviposition galleries guarding the larvae from the attacks of predators.

Many adults die in the oviposition gallery and the heavily armoured front and rear end seal it effectively. As in many other bostrychid species the entire sapwood is often reduced to dust. The larvae are similar to those of ANOBIIDAE, except the thoracic folds are larger and they differ from larvae of LYCTIDAE in that the eighth abdominal spiracle is of similar size to the others. Only the cell contents of the new sapwood such as starch and sugars are of any nutritional value to the larvae as it does not possess cellulase. Mycetomes, however, are present at the anterior end of the gut, the micro-organisms being transmitted to the eggs among the sperm of the male.

SINOXYLON ANALE

This species is primarily a pest of trees of the family LEGUMINOSAE although it has been recorded from a large number of timber species. As a pest in timber-using industries it is of great economic importance. It is the commonest bostrychid in India and from there and other eastern countries it has been carried to many parts of the world including Australia and New Zealand. In Britain it sometimes emerges in great abundance from woodwork of various kinds originating in the East.

This squat-looking bostrychid is about 4 mm in length with a pair of conspicuous pointed processes emerging from the posterior truncated part of the elytra.

DINODERUS MINUTUS

Several species of the genus *Dinoderus* are important as damaging bamboo, mostly in Asia but they have been reported in many parts of the world to which bamboo articles have been exported. *Dinoderus minutus* is small, being only 2.5–3.5 mm in length, pitchy-red in colour but with the prothorax usually darker than the elytra. The larvae have occasionally been found in various timber species and sometimes even in other materials, but in the latter case they will have originated in bamboo.

Dinoderus ocellaris, imported into Europe from the East, is found chiefly in imported fruit. The raised hairs of the frontal region are short and few in *D. minutus*, but are long and thick in *D. ocellaris*. *Dinoderus bifoveolatus* is another species listed as imported into Europe from the East.

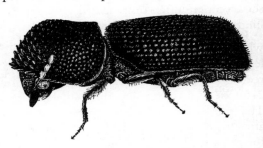

Fig. 14.5 *Dinoderus minutus*, Bamboo borer.

RHIZOPERTHA DOMINICA   LESSER GRAIN BORER

This is the only bostrychid found in grain. It originated in South America but is now found throughout the warmer countries of the world, more especially in Asia, Australia and the southern United States. In Australia it is the most serious pest of grain.

The life cycle at 30 °C and 30 per cent RH in wheat and maize takes between 30 and 40 days. The adult beetles have been maintained alive for 10 months. It can be readily identified by the rounded, rugose pronotum which completely conceals the head when viewed from above and the well-defined rows of punctures on the elytra. It is very small being only 2.5–3.0 mm in length.

This bostrychid is recorded as damaging wheat, barley, rice, maize, millet, sorghum, juar, dried potatoes, manioc roots and biscuits. It is also under suspicion of attacking a number of timbers in India.

BOSTRYCHUS CAPUCINUS

This is the most important European bostrychid. It is 8–14 mm in length and easily identified by the elytra and tarsi being red, while the pronotum is dull black. Although the larvae are usually to be found in the decaying wood of species of *Quercus*, oak, it nevertheless often emerges from new wooden structures in buildings where second-grade timber has been employed.

STEPHANAPACHYS RUGOSUS

Reported from Pennsylvania, USA, this species is found attacking bark-covered joists in dwellings. Members of this genus are unusual in that softwoods are infested. The length of the adult is from 9 to 14 mm. It is most usually known as a borer of the larger bamboo species but is recorded from a number of timbers in addition. The life-cycle is usually an annual one, but is sometimes prolonged to two or three or even six years.

Bostrychopsis jesuita may be up to 20 mm in length and its 6 mm diameter tunnels are commonly found in the timber of eucalypts in Australia.

SCOBICIA DECLIVIS   SHORT-CIRCUIT BEETLE, LEAD CABLE BORER

A peculiar type of damage is caused by certain species BOSTRYCHIDAE which cause great economic loss on account of their habit of boring into lead-covered cable. In California, the species *Scobicia declivis*, which is a reddish-brown, cylindrical beetle up to 6 mm in length, bores holes 2–5 mm in diameter into the lead sheathing of aerial telephone cables. The holes are bored adjacent to the cable-supporting rings (which may be attached to a building) where the insect can obtain the necessary leverages for gnawing the lead. Confusion often results from a large number of

Fig. 14.6 *Bostrychopsis parallela*

short-circuits occurring in telephone calls during the first rain after the emergence period of the beetle.

Other species of BOSTRYCHIDAE have similar lead-boring and cable-boring habits in other parts of the world. It is interesting to note that most records of insects attacking or boring into metals concern adults biting their way out from the pupal chamber, but in the case of these species they cause damage by biting their way in, in order to make the oviposition gallery.

Scobicia declivis also severely damages wine casks. The female bores directly into the wood where she lays her eggs. They hatch in 21 days and thereafter the larvae tunnel in the wood for about nine months. The pupal stage lasts for only about 14 days but after eclosion, the beetle remains in the pupal chamber beneath the surface of the wood for a month, before finally gnawing a flight hole to the exterior.

Fig. 14.7 *Bruchus pisorum* Kholhaas

BRENTHIDAE

The BRENTHIDAE contains about 1,700 species of elongated shape. The head is also elongated and includes a long rostrum but the antennae are not elbowed. The larvae have small thoracic legs. They are distributed exclusively in tropical, forested areas. Little is known of their biology, but the larvae of some are believed to be wood-borers. Others take over the workings of other wood-borers. They are only rarely encountered in timber of economic importance.

BRUCHIDAE

This family contains over 900 species. For the most part the larvae feed inside large seeds such as those of the LEGUMINOSAE, coconuts and palm-nuts.

Fig. 14.8 *Bruchus ervi.* (Rentokil)

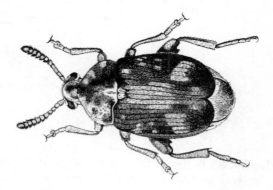

The beetles are small and characterised by the enlargement of the hind femora which also usually bear a tooth. The elytra do not completely cover the abdomen, and the exposed terminal segments are inclined steeply. The larvae of bruchids are ecruciform and apodous, and usually lie curled up in a small chamber within the seed. On emergence from the egg, however, slender legs are present but are lost when the larva has bored into the seed. Generally only one larva is to be found in each seed, except in *Acanthoscelides obtectus*, which may produce several larvae per seed.

Bruchid beetles may be divided into two groups, those in which the eyes are normal and those in which the eyes are emarginate, or

Fig. 14.9 *Acanthoscelides obtectus,* Common bean weevil. (Degesch)

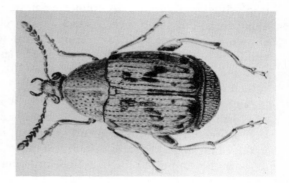

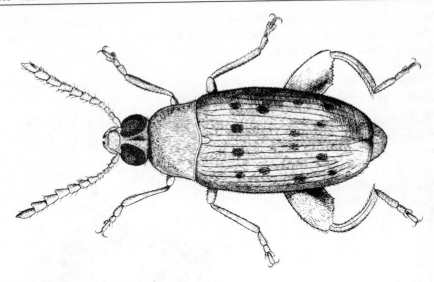

Fig. 14.10 *Caryedon gonagra*. (Rentokil)

U-shaped. In the first group the single genus and species *Caryedon serratus* damages groundnuts. In the second group three genera are important as harming pulses: *Acanthoscelides* from South America, East Africa and Portugal, *Bruchus* from the Mediterranean, Turkey, Iran and the USA, and *Callosobruchus* from the Middle and Far East, Africa and India. All these, however, can now be considered as cosmopolitan.

Bruchid beetles are of special significance as being destructive to the seeds of leguminous crops which are a vital source of protein for communities unable to obtain adequate supplies of meat or fish.

BUPRESTIDAE JEWEL BEETLES

This family of about 12,000 species is widely distributed throughout the world, although they are most strongly represented in the humid tropics. Buprestids are more usually found in forests where they attack weak and dying trees, but some species attack newly-felled timber. A number are known to complete their life history in structural or joinery timber in buildings, emerging as adult insects often many years after the wood has been put into service. Life-cycles of over 20 years have been authenticated.

Buprestis aurulenta is an example of this. It is native to the Pacific coast of North America, where it tunnels into the trunks of Douglas fir which bear a fire scar. It not infrequently turns up in countries as far away from its breeding place as Great Britain and Australia, when it has emerged from timber imported for house construction.

The length of the life-cycle is generally two years, the majority passing through the larval stage in the inner bark and sapwood before pupating deeper in the wood. The emerging beetle makes a semicircular flight hole. It is characteristic of the adults that they fly strongly in bright sunlight and are attracted to flowers. In addition, some species are known to fly into camp fires, to become very active during forest fires and even to fly in numbers around burning refuse and oil, and are thought also to be attracted to football crowds!

The beetles are elongate and depressed in shape, with the head retracted into the pronotum as far as the eyes. The pronotum possesses distinct lateral margins and fits closely along the posterior margin to the anterior end of the elytra. The latter converge posteriorly. Their most noteworthy feature, however, is the metallic lustre of a great number of species. This has earned them the name of jewel beetles in England and splendid beetles in Germany. Some species are said to be the most brilliantly coloured of all insects.

Larvae
The prothorax is flat, distinctly larger than the other segments of the body, which gives them the name of flat-headed borers in North America. There are plate-like areas on the pronotum and prosternum. On the former there is a V- or Y-shaped mark. Buprestid larvae are legless and the spiracles crescent-shaped. The larval galleries in the wood are 'flattened oval' in cross-section and are tightly packed with fine frass. The packing is carried out in a distinct manner, showing arc-like patterns described as 'cloud-like' in Germany.

CERAMBYCIDAE LONGHORN BEETLES

Members of family CERAMBYCIDAE, containing about 20,000 named species, are distributed throughout the world wherever trees or bushes grow and they may be found wherever timber is transported or used. They are, therefore, often of the greatest importance in buildings where wood is a constructional material. As the great majority of the larvae infest some part of the ligneous tissue of woody plants they are readily transported in timber and thus a few specialised species have caused serious damage in buildings in areas far removed from their place of origin. Most cerambycids, however, are of forestry significance. There are 64 British indigenous species and almost as many again are known as being regularly imported into Britain as larvae. From the Indian region 1,200 species are known and, of these, 200 were first described during the period from 1914 to 1941. No doubt the number of described species will grow for many years. Over 600 species are so far known from the African region but it is in the New World that the greatest multiplicity of species occurs.

Of all beetles, cerambycids show the greatest variation in size. Generally they are rather more than medium in size but some species such as *Macrotoma heros* and *Titanus giganteus* are among the largest known insects with a body size considerably larger than that of a small mammal. On the other hand some species are very small and a few are even minute. The general form of cerambycids is fairly well defined although there is some variation. The body is elongate with the elytra wider than the pronotum. The most characteristic feature, however, shared by all but a few species, is the extreme length of the antennae. They are usually as long as, or longer than, the body. In *Batocera kibleri* they reach a length of 22 cm.

The eyes of cerambycids are large and frequently bow-shaped, partially encircling the large tubercle from which the antenna arises. The antennae are most often thread-like but sometimes they are saw-shaped; occasionally they are comb-like and in a few species they are ornamented with tufts of hair. The legs are moderately long but in a few species the front legs are produced to an exaggerated degree, both the femur and the tibia being as long as, or even longer than, the body. Five tarsal segments are invariably present although usually the fourth segment is much reduced, only four segments being visible until the insect is carefully examined. The third tarsal segment has a definite bilobed appearance.

The larvae are generally whitish and fleshy, broadest in the thoracic region, or the head is partially invaginated into the prothorax. The intersegmental grooves are pronounced and legs are present although very small. They are of little value to the insect, their function having been taken over by ambulatory ampullae.

Many species attack only one timber species but some are polyphagous, attacking a large number. The tree, *Shorea robusta* (a member of the genus from which the timbers, meranti and serayo, are converted) is attacked by no fewer than 38 cerambycids. Whereas the species *Stromatium barbatum* is known to attack the wood of 311 tree species.

HYLOTRUPES BAJULUS

Perhaps the most important cerambycid species damaging timber in buildings is *Hylotrupes bajulus*. In Britain it is known as the 'House longhorn beetle', in Germany 'Hausbock', in Spain 'Cerambicido de la madere labrada', in France, 'Capricorne des Maisons' and in the United States, 'Old house borer'.

Description
Adult beetles are variable in size, the males being smaller than the females. The smallest males are about 7 mm in length while the largest females are up to 25 mm. The body is somewhat compressed dorsoventrally and the pronotum is almost circular. The colour is greyish-brown to black and covered with greyish pubescence. Two

shining black eye-like areas divided by a central longitudinal band are present on the pronotum and give it an owl-like appearance. On each elytron there are four greyish or whitish areas arranged in two transverse sloping bands which may coalesce. The males have longer antennae, while the females may be identified by a pronounced elongated pygidium.

Life-cycle
After a 90 second courtship in which legs and antennae are often bitten off, copulation takes place and usually lasts for a similar period. It may take place several times in the course of a single day, although this is not necessary. However, it appears to have a stimulating effect on egg-laying. After copulation the female searches for an oviposition site immediately. This is a crack generally caused by drying or shrinkage of the wood 0.25 to 0.6 mm wide, and extending to 2 to 3 cm into it. If the crack is wider than this the beetle will crawl under the piece of wood and lay in the extensions of the crack.

The average duration of oviposition is 12 days, when from two to eight clutches are laid, the later being smaller than the earlier. The smallest egg-number recorded is 86 and the largest 582, with the average lying between 140 and 200 eggs. They are dull-white and spindle-shaped, 1.2–2.00 mm long and 0.5 mm broad, and the clutch is usually fan-shaped.

The larva is typically cerambycid in form as already described. Although generally cylindrical it is somewhat flattened. The abdominal segments prior to the apical two enlarge somewhat before tapering again. In colour it is a shiny ivory white and the body has a sparse covering of long, thin, yellow hairs. There are three ocelli near the base of each antenna. The larva reaches a length of up to 24 mm and up to 7.5 mm in width at the prothorax.

Optimum conditions for larval growth lie between 28 and 30 °C and from 18 to nearly 20 per cent moisture content in the wood. The young larvae, which grow at a faster rate compared with older larvae, at first keep immediately beneath the outer skin of the sapwood. This is the area of highest nutritive value. The frass is rarely ejected from the wood by the larva but it can be detected under the surface as long blister-like protrusions and it sometimes happens that the greater volume of the frass causes the outer veneer of unconsumed wood to split, usually along the grain. The larvae can clearly be heard gnawing the wood especially in warm weather when they are most active. The length of larval life is usually from 3 to 6 years but 2 years and up to 10 years are not uncommon and there is one record of 32 years.

The pupa is white and including those that would produce diminutive males, varies from 14 to 25 mm in length. In Europe the length of the pupal stage is from two to three weeks. In South Africa it is given as from 29 to 44 days. In Germany it has been found to be 22.5 days at

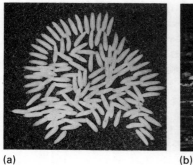

Fig. 14.11 *Hylotrupes bajulus*. (a) Eggs; (b) Larva; (c) Adult. (Rentokil)

21 °C, 17 days at 24 °C, 15 days at 25 °C, 14 days at 26 °C and 11 days at 28 °C.

In Europe the adults emerge from mid-June to September usually when the weather is hot. The female adults live on average 25 days.

The wood attacked is confined to the genera *Pinus, Picea* and *Abies* of the PINACEAE. In buildings the sapwood of the timber is infested and not before the major part of this is destroyed does the larva bore into the heartwood. Timbers in the roof space are usually at greater risk to attack. Various theories have been put forward to account for this but undoubtedly ease of access into roof spaces is a factor. The roof space is very hot in summer but perhaps the alternation of cold in winter is of equal importance. The great spread of this insect in Europe in recent years has been explained by the replacement of thatch by slates and tiles, the latter absorbing heat while the former had an insulating effect.

Hylotrupes bajulus is found from central Norway to North Africa and from Portugal to Siberia. In addition, it has become established in South Africa and the United States. At one time it had penetrated into Australia but vigorous control measures were instituted and it was eradicated. In central Europe and some south European countries it is the most important softwood-destroying insect. In many areas it is never found under natural conditions in the open, but its ability to live and breed in dry, barkless timber, under certain climatic conditions, is the reason for its extensive occurrence in buildings.

Distribution in Britain is singularly circumscribed, the majority of infestations occurring in south-west London and the adjoining area of the county of Surrey.

Larvae and adults are known to have bored through sheet lead and other metals.

A large number of cerambycid species are known to have emerged from timber in buildings but very few capable of producing a second generation in these conditions. However, some species occur so abundantly that a mass emergence of adults taking place a year or so after the installation in a building of high quality joinery may give rise to heavy economic loss.

PHYMATODES TESTACEUS   OAK LONGHORN

At one time, when low quality oak floors were laid containing a high proportion of sapwood, it was not uncommon for large numbers of the adults of this species to complete their development and emerge from a floor which had been put down only a few months before.

A most variable insect in colour from yellowish-brown through reddish to a deep blue-black. The eggs are laid under the bark of recently felled *Quercus* trunks or recently dead standing trees. The tightly frass-packed larval galleries are situated in the inner bark adjacent to the cambium. Larval development normally takes two years, the pupal stage about three weeks and the adults emerge from May to July in Britain.

CLYTUS ARIETIS   WASP BEETLE

This is one of the best known cerambycid beetles being generally common throughout Europe. Its transverse black and yellow banding and quick wasp-like movements make it easily observed. It has been recorded from a large number of hardwood timbers. A number of instances are known of the beetle emerging from beech furniture. It is unusual among cerambycids in producing round flight holes.

ERGATES SPICULATUS   SPINED PINE-BORER

The Spined pine-borer is very large reaching a length of 57 mm. It is found boring into conifers throughout the Pacific slope of North America. The full-grown larvae are from 50 to 75 mm long and are able to complete their development when the timber is converted and utilised in a building. The larvae are known to bore into lead cable.

PHORACANTHA SPP.

Phoracantha semipunctata and *P. recurva* are native to Australia where the larvae feed on the wood of *Eucalyptus* spp. Usually their attack is confined to newly felled timber from which the bark has not been removed. Once timber is dry it is never attacked but the larvae, once they have spent a period in the sappy wood, are able to complete their development after the timber has been dried and utilised in a building. Out-of-doors, however, the larvae are mainly cambium feeders for the greater part of their life but on becoming fully fed they bore straight into the heartwood and there pupate. This causes serious defects in the timber which are not always obvious when it is put into service and, moreover, it sometimes happens that numbers of the relatively large beetles suddenly make their appearance in a dwelling.

In addition to Australia, *Phoracantha* spp. are known from many parts of the world where *Eucalyptus* of various species is now grown or to which the timber is exported. Israel and South Africa are examples.

CLERIDAE CHEQUERED BEETLES

Most of the members of this family of about 3,000 species are tropical or subtropical and are predacious on wood-boring insects, although some are necrophorous, living on animal and plant remains or on mould fungi. (In Germany *Opilo domesticus* is said to be the most important predator on larvae of *Anobium punctatum*.) Members of the CLERIDAE family are brightly coloured, often banded, hairy beetles with cylindrical prothorax. The colours are metallic blues, greens and bronze, while some species show white and red markings on a black ground colour. The antennae are usually 11-segmented with an enlarged 3-segmented club. The tarsi are 5-segmented but the first and fourth segments may be so short as to be overlooked. (For this reason they are sometimes referred to in keys as 'tarsi', never 5-segmented.)

The species *Thanasimus formicarius, Thaneroclerus buqueti, Paratillus carus, Tarsostenus univittatus* and *Korynetes caeruleus* prey on anobiid beetles such as *Stegobium, Lasioderma* and *Anobium punctatum*. They are elongate beetles with the prothorax narrowing towards its junction with the abdomen. *Necrobia* spp. feed on copra, cacao and animal products such as bones.

Nearly all the larvae of clerid beetles are coloured pinkish or reddish, or reddish-brown with a brown or black horny plate on the last abdominal segment, bearing two horn-like urogomphi. The pronotum is also horny.

KORYNETES

One species, *Korynetes caeruleus*, is metallic blue, 3.5–6.00 mm long with sparse, black hairs. While it attacks hams and skins, it is also predacious

Fig. 14.12 *Thanasimus formicarius*. (Rentokil)

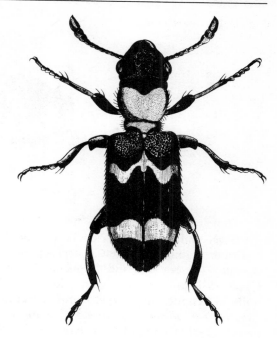

and necrophagous, feeding on the larvae of the Window fly, *Scenopinus*, on dead moths and commonly found associated with infestations of the wood-boring anobiids, *Anobium punctatum* and *Xestobium rufovillosum*.

NECROBIA

The species in this genus are less elongate than the predacious members of the genus *Tarsostenus*. Two are common in food storage warehouses and cargoes, *Necrobia ruficollis* and *N. rufipes*. The former is part-coloured, the prothorax and shoulders of the elytra being bronze, the rest of the body metallic blue or green; length 4–6 mm. *Necrobia rufipes* is inky-blue in colour, 4–5 mm long, with the basal segments of the antennae and the legs, red. Both species feed on hams, cheeses and copra. Whereas *N. ruficollis* is virtually restricted to animal products, *N. rufipes* also attacks oilseeds and oilcakes. Some years ago *N. rufipes* caused serious damage in warehouses storing copra. It was first reported in 1933 in Britain on a cargo of copra in Victoria Dock, London.

CRYPTOPHAGIDAE FUNGUS BEETLES, PLASTER BEETLES

A number of species are known as plaster beetles. This is because they are usually present in a house shortly after its erection when the plaster is new and still very damp. In such conditions surface-growing moulds and mildews occur on plaster, paper and wood and it is on these fungi that

Fig. 14.13 *Necrobia rufipes.* (Degesch)

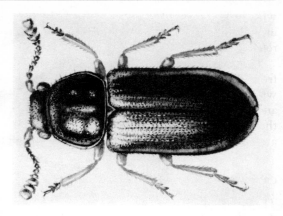

the larvae and the adult plaster beetles subsist. The most common species in Britain in buildings are *Cryptophagus cellaris, C. distinguendus, C. saginatus, C. scutellatus* and *C. pilosus.* Of the 80 species found in Britain *C. cellaris* is probably the most widely distributed, but it is found in mills, warehouses and large, damp cellars more often than in dwellings. *Cryptophagus acutangulus,* however, is most common in the latter type of premises. The larvae of *C. scanicus* feed on the moulds growing on damp, dried fruits in warehouses. Some seven species are common in damp stores and in warehouses storing dried fruits where moulds are present on

Fig. 14.14 *Lathridius minutus,* Plaster beetle.

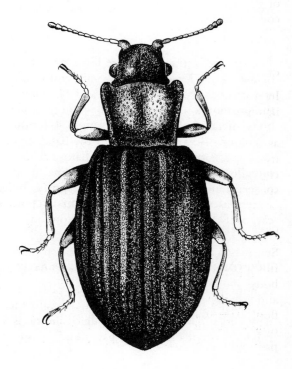

which the larvae can feed. Records of species of this family being found in buildings in countries other than Britain are sparse, but see closely related family LATHRIDIIDAE.

The CRYPTOPHAGIDAE are very small beetles, usually ranging in size from 1.0 to 3.8 mm. Some are unicolourous reddish-brown to black, while others are parti-coloured. The colour, however, changes after death and cannot be relied on to distinguish species or groups of species. All the cryptophagids are covered with a downy pubescence. The antennae are clubbed; the club consisting of three segments.

CRYPTOPHAGUS

These are small beetles with a somewhat square-shaped prothorax with the apical margins upturned or, more rarely, with a recurved tooth. All the storage species have a small tooth on the margin of the prothorax at or near the middle. They are not easily separated by external characters alone, the genitalia only affording a guide to their identification. The larvae are campodeiform with the last abdominal segment forked.

HENOTICUS

One species, *Henoticus californicus*, differs from *Cryptophagus* spp. found in stores, in having a series of small, but distinct, teeth on the side margins of the prothorax. It is not uncommon in stores housing dried fruits, cocoa and spices.

CUCUJIDAE

In recent years changes in harvesting and storage of grain have given increased importance to these beetles.

The species of this family, both adult and larva, can only be described as diverse. In size the adults vary from 1.5 to 25 mm, the antennae are usually long and usually more or less thread-like but they may be clubbed and there is much variation in the number of tarsal segments. All species, however, possess a flattened body.

CRYPTOLESTES (LAEMOPHLOEUS)

Species of *Cryptolestes* found in stored foods have elongate elytra and long filiform antennae. When seen in a sample of grain, the small size, flat body and curious swaying gait make them easily recognisable. On farm and storage premises *C. ferrugineus* is the predominant species, while in flour mills *C. turcicus* predominates. The former species is the major pest of grain in the prairie provinces of Canada. The larvae are elongate with pronounced tail-horns (urogomphi).

Fig. 14.15 *Cryptolestes ferrugineus*, Rust-red grain beetle.

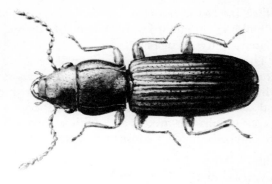

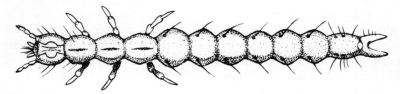

Fig. 14.16 Larva of *Cryprolestes ferrugineus*, Rust-red grain beetle.

CURCULIONIDAE WEEVILS

This is one of the largest families not only of the *Coleoptera* but of the whole of the animal kingdom, containing some 35,000 species. Many species damage growing crops of various kinds throughout the world, but only a few are harmful to crops after harvesting. Some of these are, however, of outstanding importance, namely the Grain and Rice weevils. Some species damage woodwork.

Most weevils are recognised by the beak-like prolongation of the head known as the rostrum, by the elbowed, clubbed antennae and the rigid palpi. Generally the tarsi are 4-4-4 segmented, the true fourth segment being minute and situated on the emargination of a bilobed thrid segment. The rostrum of the female, in many cases, is used for boring a hole before insertion of an egg. Often the rostrum is larger in the female, and most larvae live within the food material and are legless.

SITOPHILUS (FORMERLY CALANDRA) GRAIN AND RICE WEEVILS

These weevils are well known to all who have experience with stored grain. They are small (3–5 mm), brown to dark brown to almost black, with a long rostrum. The antennae terminate in a distinct club. The larvae are legless, scarabaeiform with a yellow or pale brown head and whitish body, and live wholly within the grain. Until recently two species were commonly recognised, *S. granarius*, the Grain weevil and *S. oryzae*, the Rice weevil. These two differ mainly in the punctures on the

Fig. 14.17 *Sitophilus granarius*, Grain weevil.

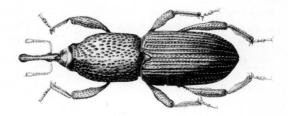

Fig. 14.18 *Sitophilus oryzae*, Rice weevil.

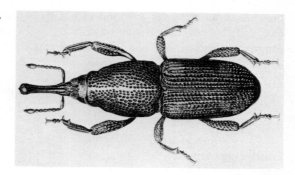

prothorax and in the fact that in *S. granarius* the hindwings are atrophied. In the latter species the prothoracic punctures are distinctly oval and the lines of punctures on the elytra are widely separate. In *S. oryzae* the punctures on the prothorax are round and the lines of punctures on the elytra are narrowly separate.

For some time two forms of *S. oryzae* have been recognised, and recently it has been confirmed that these two forms are, in fact, separate species, *S. oryzae* and *S. zeamais*. They are so alike in general appearance that they can be separated only by examination of the genitalia. *Sitophilus oryzae* and *S. zeamais* are very important pests of all types of cereals produced in the tropics and subtropics. The former tends to be more common on grains smaller than maize and the latter in maize, but both will occur on all kinds of cereals. *Sitophilus granarius* is essentially temperate in distribution.

These weevils still cause serious damage to grain in India, Africa, China and South America, where *Sorghum vulgare* is the staple food of many people. Millet infested by *S. oryzae* in India is *just* palatable to regular consumers of this grain after three months of storage but by which time, too, its food value is seriously diminished. The spontaneous heating of grain by insects was first noticed in connection with infestation by these species.

WOOD-DESTROYING WEEVILS

A number of species of CURCULIONIDAE are known to infest wood in buildings, particularly in damp situations where some fungal attack is present. Except for the four species *Euophryum confine, E. rufum, Pentarthrum huttoni* and *Caulotrupis aeneopiceus*, however, the amount of damage is not significant.

EUOPHRYUM CONFINE

First found in Britain in 1937 it had become widespread throughout London by 1946 and is now found throughout England although it is most common in London and the surrounding counties. It is a native of New Zealand where it is found in old tree-stumps and is not harmful. In England, however, it is often associated with damp softwood where the fungus *Coniophora cerebella* is causing some damage. The borings of the adult weevils and the larvae allow water to be absorbed more readily and probably accelerate the rate of decay.

The beetle is typically weevil-like with a rather broad snout, a deep constriction immediately behind the eyes and the tips of the elytra are dilated.

EUOPHRYUM RUFUM

Also originating in New Zealand this weevil was first observed in Britain in 1934. Its association with damp wood is similar to that of *E. confine* and it is widespread in Britain.

PENTARTHRUM HUTTONI

This weevil, widely distributed in England and Ireland, is brownish or pitchy-black in colour, cylindrical in shape and varies from 3 to 5 mm in length. The adults and larvae are found within softwoods and hardwoods including plywood, in ill-ventilated situations where fungal decay occurs. Most often the species concerned is *Coniophora cerebella*.

Life-cycle

At 25 °C and 95–100 per cent RH egg-laying commences about four days after mating. Using her mandibles holes are bitten in the wooden surface and single eggs laid in such sites or in cracks and crevices. They are then sealed over with a semi-transparent whitish secretion from the ovipositor. Two females were found to lay 50 eggs in 80 days. The shining white eggs, which are somewhat flexible and flattened at one end, hatch after 16 days. The young larva about 0.5 mm in length consumes the whole of the shell. After making a hole in the exterior of the wood the larva bores in at an angle of about 45 ° and then constructs a tunnel parallel to the surface. Five instars take place and then pupation occurs in from six to eight months after hatching. The pupal cell becomes lined with fungal hyphae. The fully-grown larva is about 3.3 mm in length and is widest at the third thoracic segment. The cuticle is covered with fine but conspicuous spines. The pupal stage lasts about 16 days and the adults live for about 16 months.

Habit
The adults readily feign death and although the males may be easily removed from the wood the females resist removal by clasping the surface with the tibial spines. The flight holes which are irregular and ragged are placed at an angle of 45 ° to the surface.

Other wood-damaging weevils found in buildings

Cossonus parallelepipedus and *C. linearis* are found in buried or damp wood, often in buildings in Europe. *Codiosoma spadix* occurs in damp softwood throughout Europe. *Eremotis elongatus* and *E. orcatus* are also found in damp wood; European. *Rhyncolus culinaris* is found in central Europe in pit-props and in damp wood in buildings but is absent from Britain. A number of other species have on occasion emerged in buildings from imported timber in a number of countries of the world but appear to be of little significance. In weevils which are without wings and are slow moving, it is difficult to explain how they spread from building to building.

PLATYPODINAE AND SCOLYTINAE

Included in the CURCULIONIDAE are the above two subfamilies each of which, according to some authorities, should be granted family rank. Many species, in both groups are known as bark beetles and are small and cylindrical with a superficial resemblance to bostrychids but possess compact antennal *clubs*. Some burrow between the bark and the wood of trees where the eggs are deposited. The pattern of the adult and larval tunnels is generally characteristic of the species. Some, however, known as ambrosia beetles bore into the wood of various tree species and the larvae feed on fungal mycelium growing in the tunnels. The wood surrounding the tunnels in many cases becomes stained so that when such timber is utilised in buildings the distinction between such damage and that caused by anobiids can be made with ease. (The chief diagnostic character, however, concerning the damaged timber is that the tunnels are dust-free) Ambrosia beetle larvae die off when the timber is dried and converted whereas anobiids infest such wood.

DERMESTIDAE

High place must be given to DERMESTIDAE as enemies of commerce. It has been calculated that the annual loss to furs, hides and skins caused by one species alone, *Dermestes maculatus*, was over half a million pounds sterling in 1945, in only a few countries in respect of only a few of the commodities which it regularly attacks. The main damage to furs, hides and skins is reported on those imported from West Africa and the Middle East, and nearly always to untreated hides. This loss is probably much less now, due to control measures.

The most important beetles in the family DERMESTIDAE from the economic standpoint are members of the genera *Dermestes, Anthrenus, Trogoderma* and *Attagenus*.

DERMESTIDAE is a family of small to moderate sized beetles varying from 1 to 12 mm and invested with fine hair or scales which may be in distinct patterns. The pronotum is strongly narrowed anteriorly and the hind angles are acute. In most, the antennae are clubbed, but the number and shape of the segments composing the club vary with species and sex.

DERMESTIDAE are readily distinguished in the larval stages. These are compodeiform, with numerous hairs, some of them remarkable in form. The ninth abdominal segment sometimes bears a pair of large horns. The legs are five-segmented. Adults of some species can draw the antennae and legs into grooves in the head and thorax, so that the beetles appear to be without legs and antennae and, as in the species *Anthrenus*, closely resemble plant seeds. The great majority of the DERMESTIDAE feed on dried animal matter: fur; wool; feathers; hides; skins and dried flesh on bones or skeletons; but some feed on fresh meat and cheese. Species of the genus *Anthrenus* are notoriously destructive to poorly maintained museum collections of insect and other animal specimens. The genus *Trogoderma* is vegetarian, and includes the notorious *Trogoderma granarium*.

ATTAGENUS FUR AND CARPET BEETLES

The members of this genus resemble in form small species of *Dermestes* (see later), but can be distinguished by the presence of a simple eye, or ocellus, on the forehead. Two species are common, both showing preference for animal foods such as hair and wool. A third, *Attagenus fasciatus*, is also common in many parts of the world, but not in Europe. Their hairy larvae are extremely destructive to hair and woollen fabrics, particularly rugs and carpets, in which they may feed unobserved, causing considerable damage before their presence is suspected. While they occur on food commodities as a result of 'cross-infestation' from woollen goods, they are scarcely to be regarded as pests of food. Two species are commonly found, *A. megatoma* which is wholly dark brown or black in colour, and *A. pellio* in which the base of the thorax has a patch of white hairs on either side and the elytra each have a small patch of white hairs near the middle. These two closely related species, the Fur beetle, *A. pellio*, and the Black carpet beetle, *A. megatoma* are described together in the following account.

The adult beetles are intermediate in size between *Dermestes* and *anthrenus*, being 3.6 to 5.7 mm in length and in breadth 1.8 to 3.0 mm. The antennae terminate in a three-segmented club. In the male, the last antennal segment is much enlarged, being about as long as the remainder of the antenna, but in the case of the female the terminal segment is only about as long as the combined length of the two adjacent segments.

After mating on flowers out-of-doors, the female flies into dwellings and lays her 50 to 100 eggs on materials suitable for larval food, such as woollen carpets, birds' nests, and any other dried substance of animal origin.

The larvae are characterised by the possession of a tuft of long hairs arising from the terminal segment of the abdomen, and have a banded appearance; the cast skins showing this to a much greater degree and often a feature of an infestation. The larvae avoid the light, and when disturbed they often remain immobile for a time in a slightly curved posture. The number of moults varies according to temperature and availability of foodstuff, but 6 to 20 is normal.

The pupa is never visible, as pupation occurs within the last larval skin. When the adult stage has been assumed the beetle remains within the larval skin for a period of 3 to 20 days before it emerges to seek the light and to feed on the pollen and nectar of flowers.

In Britain *A. pellio* is much the commoner of the two species, but in America and Asia *A. megatoma* and *A. fasciatus* are the dominant species

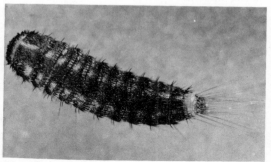

Fig. 14.19 (a) Larva of *Attagenus pellio*, Fur beetle. (b) Larva of *Anthrenus verbasci*, Varied carpet beetle.

of the genus. There is reason to believe that *A. pellio* has increased in importance as a pest in Britain since the war. It is a common inhabitant of birds' nests out-of-doors and from such sites the larvae probably invade domestic premises. Most commonly the larvae are found in homes infesting carpets, stored woollen garments, skins and furs, but in addition they have been recorded as infesting a large number of materials from those of high protein content such as museum specimens of insects, bones, smoked fish and meat, dried yolk of eggs, casein and silk in various forms, to products of high carbohydrate content such as grain, cereal products, flour, semolina, maize meal, rye bran, sugar and similar materials. It is not known to what extent in these cases the *A. pellio* larvae were feeding on dead insect remains of previous infestations.

The larvae of this beetle have been shown to be capable of carrying the bacilli of *Anthrax* both on their body and in their faeces. In addition, various species of nematode worms parasitic in Man have been shown to be carried by the larvae of dermestids.

TROGODERMA KHAPRA BEETLE

This insect attacks stored products, particularly grain. It can give rise to very large populations in a short space of time, but can also persist in small numbers for long periods, even for years, and then break out again if circumstances favour it. The genus is tropical, or subtropical in origin, but several species, including *T. granarium* (Khapra beetle), *T. parabile* and *T. inclusum*, are now cosmopolitan, but *T. granarium* is by far the most important.

The Khapra beetle is a small dermestid 2–3 mm long, elongate oval in shape, brown to brownish-black in colour, with the body densely clothed in fine hairs. The elytra show two or three indistinct transverse bands of sparse, yellowish and occasionally white hairs. The larvae are very hairy, some of the hairs being spear-headed. *Trogoderma granarium* survives in Britain in malting houses where it can do important damage; in warmer climates it is feared as one of the most destructive insects of grain. The reason is that it can enter 'faculative diapause', which enables it to survive unfavourable conditions, whether of low or high temperature, water shortage, or food shortage. In addition, it is the habit of the larva to penetrate deep into cracks in the structure of buildings or ships, making it difficult to kill by contact insecticides or fumigation. During diapause the larva may rouse itself to wander and seek food or better shelter and then resume its diapause again.

The males usually mate with several females and the eggs are laid singly and loosely among grain, except that occasionally they are laid in the groove of a wheat grain when several may be found together.

The small size of the beetles and larvae and their brownish or cryptic colouring make detection difficult. *Trogoderma granarium* has probably

Fig. 14.20 *Trogoderma granarium*, Khapra beetle. (a) Larva; (b) Adult. (Degesch)

(a)

(b)

been voyaging unobserved in ships' holds for many years. The Khapra beetle is thought to have been first found in Britain as long ago as 1860, but no certain identification was made until 1917. Just after the Second World War *Trogoderma granarium* was found infesting groundnuts in Nigeria, and rapidly became a serious problem. In 1952 it was found in California and then in Arizona, USA, and its occurrence caused widespread concern. It was eradicated from the United States at a cost of several million dollars because of the potential threat to the large grain stocks. In Nigeria, it is still endemic but is kept in check by routine fumigation of groundnuts.

All countries whose climate is favourable to the beetle are concerned to exclude imports likely to introduce it. Unlike other dermestid beetles, it is a vegetarian by choice. *Trogoderma inclusum* will feed and thrive on cereal food, but it more usually feeds on fur and dead insects.

Trogoderma granarium is an outstanding example of an insect which attacks stored products, having all the attributes we associate with the success of the insects as a class: small size; obscure colouring; adaptability to temporary changes in climate; and capacity to survive them. It has been recorded as living for eight years in the larval stage even when food was available.

Fig. 14.21 *Trogoderma inclusum.* (Degesch)

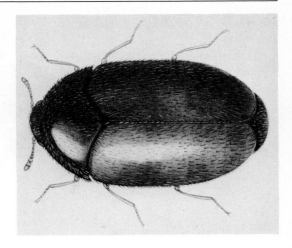

DERMESTES HIDE OR SKIN, AND LARDER OR BACON BEETLES

The members of this genus differ from all other dermestids in having no median ocellus on the forehead. They are the largest of the dermestid beetles occurring in stored products. The basic colour is black, or dark brown, with sprinklings or patches of whitish or yellowish hairs. The colour pattern formed by these hairs serves to distinguish some of the species.

Four species of *Dermestes* are common and almost cosmopolitan: *D. lardarius* and *D. ater*, in which the pronotum is entirely black, and *D. maculatus* and *D. frischii*, in which the margin of the prothorax is covered with whitish and pale-yellowish hairs. Meanwhile the following characters are useful for recognising the species of the genus *Dermestes* in the field (assuming that any dust on them has been blown or lightly brushed off).

Fig. 14.22 *Dermestes lardarius*, Larder beetle.

Dermestes lardarius – basal half of elytra pale grey-brown with patches of black hairs in the middle.
Dermestes ater, D. peruvianus and *D. haemorrhoidalis* – wholly black or dark brown.
Dermestes maculatus – apices of elytra produced at the suture to form a minute spine.
Dermestes frischii and *D. carnivorus* have blunt apices to the elytra.
Dermestes frischii also has a patch of black hairs on the first ventral segment of the abdomen which is absent in *D. carnivorus*.

The species of *Dermestes* are basically carrion beetles in habit and are most abundant in factories making bonemeal and fishmeal, and in warehouses where skins and hides are stored. They do considerable damage in the leather trade. They have been used for cleaning skeletons of small mammals and birds for museum collections. Infestations of *Dermestes* spp. are particularly important on account of the damage the larvae may do to other goods when seeking pupation sites. Because of this, cross-contamination often occurs on board ship. *Dermestes carnivorus* is an indigenous American species found in North and South America, but occurs also in India and reaches Europe from time to time. *Dermestes ater* differs from the other species in being associated with copra and by cross-infestation from copra, cocoa beans. It is seldom found on animal products.

ANTHRENUS MUSEUM AND CARPET BEETLES

A median ocellus is present on the forehead. The genus *Dermestes* is without one. In addition there are grooves in the prothorax which house the club of the antennae when the beetle shams death. This habit is characteristic of *Anthrenus* beetles and when it occurs the legs and antennae are withdrawn into grooves or cavities in the thorax, into which they fit so closely that even with a pocket-lens, the outline of the appendages can be seen only with difficulty. *Anthrenus* beetles are fairly small, measuring only 1.5 to 4.0 mm in length, and although oval in shape they are much rounder than other dermestids. The body is strongly convex, rather like the ladybird beetles (COCCINELLIDAE), but also convex below, unlike ladybirds, and they are characterised by the pattern of yellow, black, brown and whitish scales densely covering the entire body except the legs.

The larvae of *Anthrenus*, commonly known as 'woolly bears', are remarkable for the long spearhead-shaped hairs on the abdomen which form tail-tufts. Anthrenid larvae are scavengers, feeding mainly on fur and woollen materials (such as carpets), and on the bodies of dead insects. As stored product pests they are of little importance, but are among the most serious pests of fur and woollens. The adult beetles feed on pollen and nectar of a wide variety of flowering plants. Six common species are

Fig. 14.23 Head of *Anthrenus* sp.

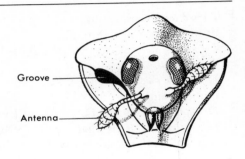

Anthrenus fuscus, A. verbasci, A. museorum, A. scrophulariae, A. flavipes and *A. pimpinellae*.

Life-cycle
Between 20 and 100 eggs are laid by the female during spring or early summer. The eggs are cemented to the foodstuff sufficiently strongly so that they are not dislodged by shaking. The female begins laying four days or so after fertilisation, and continues for one to two weeks.

The larva is hairy and brown with a bunch of special golden hairs on each side of the rear abdominal segments. These hairs are shaped like spears, the details of the heads of which are used for identifying the different species. If a larva is disturbed it rolls up and in so doing fans out these hairs, giving the appearance of a small golden ball. The hairs cause itching to human skin when the larva is touched. The larva usually moults six or eight times, but under adverse conditions such as in semi-starvation, the number of moults may be as many as 30. If a larva has been feeding well, it can resist starvation to a remarkable degree. One 10-month starvation period has been recorded. *Anthrenus* larvae avoid the light and they often pupate inside their food material, this being commonly the case if a dead insect is being consumed. The larvae reach a length of 4 to 5 mm. The last larval skin is not shed but remains as a complete cover to the pupa. The adult too remains within the last larval skin before emergence for a period varying from four days or so to as long as a month.

Anthrenus verbasci has been recorded from a host of different materials and although it is considered to be of greatest importance as attacking woollen materials, yet it has often been reported as infesting grain products, seeds of various sorts, cacao and other products of vegetable origin. Most of the older records concern the destruction of insect collections, but from the 1930s onwards it has become of increasing importance in its attacks on woollen carpets and garments. The manner of the insect's distribution is of interest and importance. On emergence, the adult beetles seek the light and fly to light-coloured flowers where they feed on nectar and pollen. After mating, the females enter houses and lay their eggs in birds' nests in roof voids and in other suitable places. The larvae feed on feathers and wool soiled with excrement, dead

Fig. 14.24 *Anthrenus verbasci.*

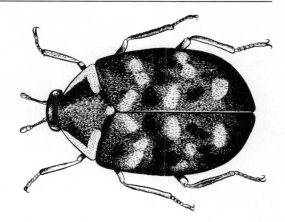

fledglings, etc. From this site they wander downwards, probably aware of the heat gradient, until they reach airing cupboards, clothes or wardrobes. They continue feeding in warm, dry conditions which are found in such situations and may wander further to carpets.

In Britain *A. verbasci* is abundant only in suburban areas in the southeast part of the country. The development of the larvae appears to be independent of temperature but there is a rhythm which synchronises the emergence of adults at a time when suitable flowers, especially those of *Heraclium sphondylium*, Hogweed, are in flower.

Anthrenocerus australis, Australian carpet beetle, was introduced into Britain in 1933 from Australia or New Zealand. It is, at present, only of minor significance as a household pest but has the potential to become more serious.

LATHRIDIIDAE PLASTER BEETLES

Plaster beetles and their larvae feed exclusively on the hyphae and spores of moulds, mildews and other fungi. They can, therefore, only exist under the same conditions of dampness as required by fungal growth. As a result, they are found in damp cellars, granaries and warehouses, as well as in new or newly converted houses where moulds may have developed on walls and unpainted woodwork, due to incomplete drying out of plaster and the associated high humidity of the air. Joinery with a high moisture content will often support surface-growing fungi on which plaster beetles and their larvae will feed. Under such conditions, if an infestation occurs, it will not continue for more than three or four months if the house is heated with good ventilation. Although plaster beetles occur in warehouses, they cannot be considered as important in damaging food, although they sometimes contaminate foodstuff with their faeces if the foodstuff is slightly damp, thus encouraging mould growth. It is thought that plaster beetles may be responsible for transmitting moulds from one commodity to another. *Lathridius minutus, Aridius nodifer* and *Cartodere constrictus* have a worldwide distribution.

All members of this family are small and from pale brown to black in colour. The tarsi are three-segmented with the exception of some males where the front legs have only two segments. The species commonly found in buildings have the prothorax and elytra rugosely punctured. The pronotum is generally much narrower than the elytra and the antennae are 8–11 segmented, including a 2–3 segmented club. Several are often found in large numbers in damp buildings.

Life-cycle of Lathridius minutus
The egg is laid singly and is 0.47 mm in length and 0.18 mm in width. It is oblong with one side slightly concave. In colour, the egg is shiny, whitish and opalescent. The larva grows to a length of 2.2 mm and each thoracic and abdominal segment bears a number of backward-curling setae which in the species *Aridius nodifer* are exceptionally long. When about to pupate the larva of *Lathridius minutus* secretes a sticky substance from the anus which cements it to the substrate. The length of the pupa of *L. minutus* is 1.53 mm; it is whitish in colour and after a few days the eyes are reddish. When viewed from above the head is concealed by the pronotum.

LYCTIDAE

LYCTIDAE together with BOSTRYCHIDAE are known as powder-post beetles on account of the larval habit of burrowing into the sapwood of certain hardwoods until virtually nothing is left but a fine powder. This may take several generations, but a thin skin of sound wood on the outside is left unsupported from the inside. This uneaten veneer may burst and become detached. The outer skin shows the flight holes made by the emerging, mature beetles. Over 60 species of LYCTIDAE are known, classified in 12 genera. Each of the world faunal regions possesses its own indigenous species in addition to species which have been introduced through commerce in timber, and many have become established.

LYCTIDAE are long, narrow, cylindrical beetles of small size. Although related to the families ANOBIIDAE and PTINIDAE, the head is not hooded by the pronotum but is clearly seen from above. The antennae are clubbed, the club consisting of two segments only. There is a median fovea or depression in the centre of the pronotum. The larvae are soft, crescent-shaped, enlarged at the anterior end and the posterior segments are larger than those in the centre. The larvae of LYCTIDAE may be separated from those of the related families by the possession of large spiracles on the eighth abdominal segment.

Life-cycle
A knowledge of the biology of the LYCTIDAE is not only of great

Fig. 14.25 *Lyctus brunneus*, Powder-post beetle. (Rentokil)

Fig. 14.26 Larva of *Lyctus* sp.

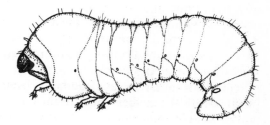

importance in the control of this insect in buildings, but is of exceptional interest.

Lyctus brunneus flies well to light. Beetles seen crawling round windows are often the first signs of infestation in a building. Females are more abundant than males, and the latter fertilise several females. The females live for about six weeks, but the males only for two or three weeks. *Lyctus brunneus* is a common pest species on which much research has taken place, so that most of the account which follows refers to this insect.

The ovipositor is of great length being almost as long as the length of the body. It is very flexible and at the tip are two two-segmented pygidial palps. Before the ovipositor is inserted the actual position is explored by the palps before it is inserted directly into the lumen of a vessel. A depth of 7.75 mm into a vessel has been reached and as many as eight eggs have been laid in a single vessel although more usually one, two or three eggs are deposited. Obviously the diameter of the lumen of

Fig. 14.27 *Lyctus* spp., laying eggs – diagrammatic.

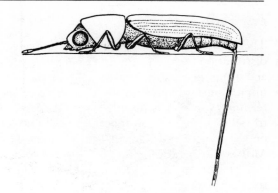

the vessel is critical as the ovipositor which is about 0.08 mm must be able to penetrate it. Indeed only the timbers known as the 'wide-pored' hardwoods are capable of being attacked by *Lyctus* spp.

The eggs are long and cylindrical, slightly tapering to the posterior end and there is a long thread at the anterior end. The average length is about 1 mm. Incubation periods show little difference between the species and are 19–20 days at 15 °C, 14–15 days at 20 °C and 7–8 days at 26 °C.

The first meal of the larva is a residual yolk mass left in the egg and this increases the size of the larva to the extent that it is able to make its way along the lumen of the vessel. In the early stages the larva excavates its tunnel with the grain but later its direction appears to be haphazard cutting through the tunnels of other larvae. In badly infested wood the larvae tunnel through mostly faecal matter in order to find wood tissue which has not been eaten.

The larvae of *Lyctus* spp. subsist only on cell contents, starch being the main foodstuff together with certain sugars, disaccharides, a polysaccharide and some protein as well as an unidentifiable water-soluble substance. Cellulose and hemicellulose are not digested. Faecal matter and rejected wood tissue are extremely finely divided. In freshly-sawn hardwoods, attacks can develop from eggs laid in surface-dry layers and proceed inwards with air drying. Thus, although the piece of timber as a whole may have a moisture content of 30 per cent, the region supporting the larvae has a much lower moisture content. For *Lyctus brunneus* the optimum moisture content of the wood is 16 per cent.

When fully grown the larva reaches a length of about 5 mm and apart from the giant spiracle on the eighth abdominal segment the larvae may be differentiated from those of anobiids by the three-segmented legs, the distal segment being paddle-shaped. The pupal stage is recorded as lasting from 12 to 30 days but it is certain that under favourable conditions it is considerably less than this.

The flight hole of *Lyctus brunneus* is circular with a clean-cut edge and averages 1.4 mm in diameter. The emerging adult often penetrates materials overlying the wood surface such as paper, leather, softwood, heartwood, asbestos, plaster and even lead and silver.

LYMEXYLIDAE

The family LYMEXYLIDAE consists of about 50 species of worldwide distribution. Some of the larvae are of curious and bizarre appearance, and a number cause a certain amount of damage to economically important hardwood timber. Some larvae are exceptionally long and the terminal segment is heavily sclerotised and functions as a tool with which to remove frass. They appear to feed also on fungal mycelium growing on the tunnel walls (ambrosia). Well known genera are *Hylecoetus*, *Melitomma* and *Lymexylon*.

HYLECOETUS DERMESTOIDES

The adult beetles are soft-bodied and elongate with five, six or seven visible sternites and with short antennae. The elytra are either long and narrow and when at rest covering the wings and abdomen except for the terminal abdominal tergite or very short not reaching the abdomen. The larvae are cylindrical with short but well-developed legs. The pronotum is enlarged and forms a hood over the head. In the male the head and thorax are black; the elytra are yellow but black at the apex. The female is entirely yellow. The larvae tunnel into both softwood and hardwood timber. The bore dust is regularly cleared from the gallery by the larva pushing out the dust with its long, serrated tail appendage, turning around in the tunnel in order to do so. The food of the larva is made up entirely of the hyphae of the fungus species *Endomycetes hylecoeti* growing in the tunnel introduced by the larva from its egg shell.

It is sometimes very common in central and northern Europe and destroys much valuable hardwood timber. In Finland it attacks birch, but in central Europe in addition, beech, oak, maple, alder, fir, pine, larch, spruce and Douglas fir, may be infested. The emergence period may take place at any time between the beginning of April and the first few days of July when the female lays her eggs in batches in wood crevices or in rough bark and occasionally in bore-holes. The clutch size varies from 4 to 91 and the largest number of eggs laid by a single female observed has been 146. The young larvae hatch from 7 to 14 days after laying, and they first eat part of the egg shell to which is attached a symbiotic fungus.

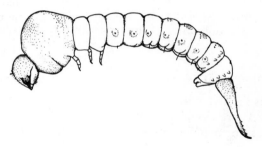

Fig. 14.28 *Hylecoetus dermestoides*, larva, length 10 mm.

LYMEXYLON NAVALE

The larvae of this widely distributed beetle bore into *Quercus* spp. including the heartwood. In Germany the larva is known as Haarwurm (Hairworm).

NITIDULIDAE

This is a large family of over 2,000 species of sap and ripe fruit feeders. They may vary greatly in form, but all have clubbed antennae and five-segmented tarsi. A number of species are found in stored products and in these at least two of the abdominal segments remain uncovered. The genus *Carpophilus* is of economic importance.

Fig. 14.29 *Carpophilus hemipterus*.

CARPOPHILUS

The members of this genus are flat, oval beetles. Most are pale to dark brown but one species, *C. hemipterus*, has a patch of yellow on the shoulders and tips of the elytra. The beetles vary in size from 2 to 4 mm. The whitish or pale yellow larvae are campodeiform with short weak legs; they bear a pair of small horns at the tip of the abdomen, and a pair of smaller horns or teeth just above these.

Carpophilid beetles and their larvae occur mainly in dried fruits, particularly currants, raisins and figs. In large numbers they soil the fruit badly, but as they occur in large numbers only in mouldy fruit, their true status can be assessed only with difficulty. Four species are cosmopolitan and common in dried-fruit warehouses.

Carpophilus hemipterus, *C. dimidiatus*, *C. obsoletus* and *C. ligneus* can be separated only after examination of the genitalia.

OEDEMERIDAE

This family, with wood-boring larvae is widely distributed. The adult insects are elongate, rather soft-bodied, and usually characterised by bright or metallic coloration. The importance of this family as causing

Fig. 14.30 Terminal segments of larva of *Carpophilus dimidiatus*.

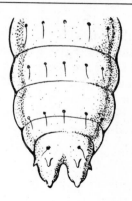

infestations in buildings is mainly due to one species, *Nacerdes melanura*, commonly known as the Wharf-borer. This is on account of the habit of the larvae boring into wharf timbers, piling, harbour and dock works, and other woodwork, at high tide mark. It is, however, unusual in its family as it is drab in colour. It has occurred widely in buildings such as in the City of London during the years following the 1939–45 war, when due to bombing, much structural woodwork was open to the elements and plagues of the adult beetles occurred during the summer on the streets of central London.

Fig. 14.31 *Nacerdes melanura*, Wharf-borer, larva. (British Museum)

Nacerdes melanura has a wide distribution in the coastal and estuarine areas in the North Temperate Zone and, in addition, is recorded from Gibraltar, Greece, Sardinia, Tangier, New Zealand, Costa Rica, California, and from much of the United States and Canada. There seems no doubt that its distribution has been brought about by transport in wooden-hulled ships in which the larvae were boring in the bilges.

The adult is from 7 to 12 mm in length and is soft in texture. It is yellowish-brown with black apices to the elytra. There are 12 segments in the antenna of the male but in the female 11 only. The eyes, sides of prothorax, legs and ventral parts, generally are blackish and the whole body is covered with dense yellow pubescence.

Several beetles of widely-differing families are often confused with the present species but *Nacerdes melanura* can be differentiated from them by the possession of three longitudinal ridges on each elytron.

The larva is from 12 to 30 mm in length and is of greyish-white colour. The intersegmental grooves are distinct. The large head is yellowish with black mouthparts. The large pronotum and the four following segments carry dome-like protrusions covered with spinules. The legs are moderately well developed, and two pairs of pseudopods are borne on the third and fourth abdominal segments.

Infestations mainly take place in softwood although occasionally oak and other hardwood species are attacked. The larvae do not bore definite sized or shaped galleries, but work in indeterminate spaces which are plugged here and there with long, torn fibres of the wood. These boring spaces are often adjacent to sound wood. This species is not a primary pest of sound timber, but bores into decayed wood which is in a damp condition and it often occurs in buildings where these conditions prevail. Some species of OEDEMERIDAE are known to be toxic and are presumed to contain cantharidin.

PTINIDAE SPIDER BEETLES

This family of about 500 species is related to the ANOBIIDAE. None is wood-boring. They possess a hood-like pronotum as in that family, but the antennae are thread-like, the legs long and the prothorax, being constricted at its base, is globular. These characters give the insects a spider-like appearance. The larvae differ from anobiids in not possessing the patches of spinules on the back of the abdominal segments. They are generally associated with the debris of cacao beans, dried fruits, various grains, seeds, droppings of rats and mice, dead insects and birds' nests, and are sometimes found in considerable numbers in dwellings.

PTINUS TECTUS AUSTRALIAN SPIDER BEETLE

Supposed to have originated in Tasmania, this insect is now to be found throughout the world although its resistance to low temperatures causes it to be of particular significance in the colder temperate regions. Although it is not able to survive in temperatures above 28 °C, at the other end of the scale activity does not altogether cease until it is as low as 2.0 °C.

The list of materials which this species has been recorded as damaging is a long one, including almost every type of stored food such as flour (often of consequence in cereal products in Canada and the northern United States) cayenne pepper, chocolate powder, desiccated soup, cacao, nutmegs, almonds, figs, ginger, sultanas, dried pears, dried apricots, beans, rye, maize, casein, stored hops, paprika pepper and dried crab meat. It was a major pest of flour stored for long periods during the Second World War. It also caused spectacular infestations of dried yeast and fishmeal. Bagged stacks collapse because of damage done to the bags where cocoons are formed. In addition, damage to carpets and furs has been reported.

While it is most commonly to be found in warehouses and mills, it often appears in dwellings, originating in birds' nests or infested food, and, seeming to subsist on wool dust, crumbs and general debris to be found in the crevices between floorboards, the numbers may be considerable.

The adult beetle varies from 3.5 to 4.0 mm in length, and is dull reddish-brown in colour. It is constricted at the junction of the prothorax and that part of the body covered by the wing case, to give a spider-like appearance. The wing cases are densely clothed with short, brown or golden-brown hairs which hide the series of longitudinal rows of small pits which are only visible when the insect becomes rubbed and worn. The legs are fairly long, a feature which adds to its spider-like appearance. Adult beetles readily feign death when disturbed and avoid the light.

Life-cycle
About 100 eggs are laid, either singly or in small groups. They are sticky when first laid and particles of food and debris adhere to them. They measure from 0.47 to 0.55 mm in length, and from 0.29 to 0.40 mm in breadth and are opalescent. Oviposition lasts about three to four weeks at normal temperature.

The larva is a whitish fleshy grub usually strongly curved, which rolls up into a tight ball when disturbed. The legs are small but bear strong claws and the whole body is covered with fine hairs. When the larva becomes fully fed it leaves the foodstuff and wanders about searching for a site for the cocoon. At this time the larva will bite its way through comparatively tough materials such as sacking, cellophane and cardboard. It will often hollow out chambers in adjacent woodwork. The spherical thin-walled but tough cocoon is constructed from an oral secretion applied by the mouthparts. The extremely delicate pupa is white at first, turning golden-brown later. The adult remains within the cocoon several days before biting its way out.

PTINUS FUR   WHITE-MARKED SPIDER BEETLE

Showing a well-marked sexual dimorphism, the reddish-brown elytra of the female bears two irregular, white patches which are absent in the male. The female is larger and more globular in shape than the male, which is long and slender. The larva feeds in the centre of an aggregate of the food material. The life-cycle has been observed to be completed in 32 days on fishmeal at an optimum temperature of 23 °C and 70 per cent RH. In Washington DC, the life-cycle takes $3\frac{1}{2}$ months during the summer. In centrally heated buildings two generations may develop annually, but the adult beetles have been observed crawling up the walls of cellars in unheated buildings in mid-winter.

In Britain *P. fur* tends to occur more widely in premises than does *P. tectus* which appears to be more confined to towns. It is to be found in warehouses, granaries, museums, libraries and dwellings and is reported as damaging feathers, animal skins, stuffed birds, herbarium specimens, stored seeds, ginger, cacao, dates, paprika, rye bread, flour, stored cereals and insect specimens. Generally this species is of slight economic importance in the United States, which must summarise the position of this cosmopolitan insect for many countries of the world, but which is, nevertheless, likely to occur wherever a wide spectrum of foods is stored.

PTINUS VILLIGER   HAIRY SPIDER BEETLE

Reddish-brown and usually with two white patches on each elytron, it occurs in many flour mills in Europe, Asia and North America, and it is said to be important in Canada. In the latter country the adult lays up to 40 eggs in spring on flour sacks, and the larval stage lasts about three months at 28 °C. A tough, silken cocoon is secreted, to which particles of flour adhere, and sometimes the cocoons are produced in the general debris at the base of the sack. Larval diapause often occurs within the cocoon, pupation not taking place until the following spring.

It was introduced into Britain during the Second World War on Canadian flour but it did not establish itself. The adult beetles require an exposure to low temperature before they are able to lay eggs.

PTINUS CLAVIPES (P. HIRTELLUS)   BROWN SPIDER BEETLE

This is another cosmopolitan ptinid. It resembles *P. fur*, and often occurs with it. It is uniformly brown in colour and in feeding can be considered a scavenger in premises not so well maintained as they should be.

NIPTUS HOLOLEUCUS   GOLDEN SPIDER BEETLE

Wingless and usually a little larger than the foregoing, *N. hololeucus* is

Fig. 14.32 *Niptus hololeucus*, Golden spider beetle.

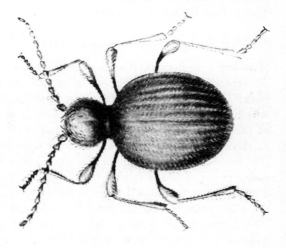

more globular and is covered with long, golden-yellow, silky hair. Relatively few eggs are laid (25–30), and the larva moults only twice. The complete life-cycle has been reported as taking six to seven months with two generations annually, but a further study has shown that at 25 °C only 11 to 15 days were taken from egg to adult stage.

Although not found in the tropics it is cosmopolitan in distribution. Mallis says that as a rule it is a pest because of its presence, although at times it may definitely be injurious. This species is found in dwellings, bakeries, flour mills, warehouses and granaries, especially if not subject to regular and periodic cleaning. The adults and larvae feed on woollens and silk, especially if soiled, as well as dead animal matter. They have been recorded as being present in sponges, bones, feathers, casein, brushes, leather goods, cacao, spices, dead insects, rat and mouse droppings, books, paper, bran, grain, flour, seeds and bread.

GIBBIUM PSYLLOIDES

This humped and shining beetle is brownish-red to almost blood-red in colour and varies from 1.7 to 3.2 mm in length, and is slow moving. It has been recorded in dwellings, warehouses, mills, bakeries and similar situations where it feeds on a variety of stored materials, from pepper to dog biscuits. It has been found breeding in rat droppings. It can maintain itself well in the residues in ships' holds.

MEZIUM AMERICANUM

This ptinid is dark reddish-brown to almost black in colour and is 1.5 to 3.5 mm in length. The elytra are highly convex, polished and bare. It has been reported from dwellings, bakeries and warehouses feeding on stored seeds, tobacco seed, cayenne pepper, opium and grain, as well as in rats' nests. It is found in North America and Hawaii and is common in Australia.

MESIUM AFFINE

Originating in the Mediterranean region, it occurs in Europe and also Australia, in flour, dry vegetable refuse and warehouse debris. The Australian record, however, may be a misidentification for *M. americanum*. Other species of PTINIDAE are known to damage foodstuffs and other stored material in buildings on occasion.

Ptinus raptor, found throughout the northern hemisphere, has been found in flour cargoes from Canada but it normally occurs in beehives. *Trigonogenius globulus* which occurs in Europe and North America, where its appearance in chocolate manufacturing establishments has sometimes caused serious problems, and *Tipnus unicolor* which occurs in Canada, Europe and Transcaucasia. *Pseudeurostus hilleri* occurs in Great Britain, Canada and Japan, and is said to breed in rat and mouse droppings.

SILVANIDAE

These beetles are so closely related to the CUCUJIDAE that many authors include them in that family. In so far as the pests of stored products are concerned, the SILVANIDAE can be distinguished by the antennae which end in a compact club. Two genera are represented in warehouses, granaries and food manufacturing premises: *Oryzaephilus* and *Ahasverus*.

Although there is some overlap in habits, *O. surinamensis* and *O. mercator* are fairly distinct in their nutritional requirements. The first tends to be a cereal and dried fruit pest. The second is more a pest of oilseeds and oilcakes which the former will not readily attack although it will feed on some nuts. *Oryzaephilus mercator* will, however, infest old stocks of dried fruit. While *O. surinamensis* cannot attack perfect grains, it can penetrate those damaged as a result of normal harvesting by combine and movement by augers and conveyors.

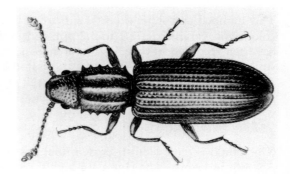

Fig. 14.33 *Oryzaephilus surinamensis*, Saw-toothed grain beetle.

ORYZAEPHILUS GRAIN BEETLES

In this genus the margins of the prothorax possess six teeth on each side, and there are three ridges on its dorsal surface. The two species, *O. surinamensis* (Saw-toothed grain beetle) and *O. mercator* (Merchant grain beetle), are distinguished by the length of the head behind the eye, which is greater in *O. surinamensis* than in *O. mercator*. That these are distinct species is well established as they will not interbreed. *Oryzaephilus* spp. feed on part animal and part vegetable debris, but sometimes the beetles are predacious, and often follow the more directly destructive insects such as grain weevils and the phycitid and oecophorid moths.

They are commonly called 'grain beetles'. The larvae are elongate, without tail-horns. Their small size often enables them to hide in crevices making control difficult. *Oryzaephilus surinamensis* is regarded as the most seriously damaging pest of grain stored in bulk in farms and in the stores of agricultural merchants and maltsters in Britain. It is favoured by the storage of grain which has been improperly dried and in which local heating occurs.

Fig. 14.34 Head of: (a) *Oryzaephilus mercator*, Merchant grain beetle; (b) *Oryzaephilus surinamensis*, Saw-toothed grain beetle.

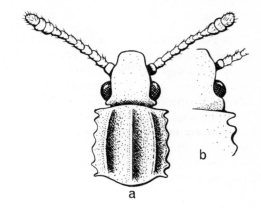

Fig. 14.35 *Ahasverus advena*.

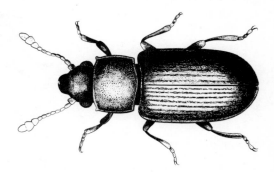

AHASVERUS

One species, *Ahasverus advena* is now cosmopolitan. It differs from *Oryzaephilus* in its squarer prothorax which has one prominent tooth at each apex. It is a scavenger living on plant and animal debris and moulds, and is not infrequent in cargoes of grain, cocoa and other commodities from West Africa, Burma and Malaya. It also occurs on farms in Britain, especially in grain which is mouldy.

TENEBRIONIDAE

This family, one of the largest in the COLEOPTERA and containing over 10,000 species, is represented by a number of species in the granary and warehouse fauna. Except for those of the genus *Tenebrio* itself, which are the largest of stored product beetles (14–18 mm), they are small beetles from 2.5 to 5 mm long. All are reddish-brown to pitchy-black in colour. The chief characters of TENEBRIONIDAE are the tarsi which in the front and middle legs are five-segmented, and in the hindlegs are four-segmented. It is estimated that more than 100 species are associated with stored products. The Churchyard beetle, *Blaps mucronata*, is an inhabitant of damp cellars in Europe.

TENEBRIO

This genus of large beetles contains two species, the Yellow mealworm, *Tenebrio molitor* and the Dark mealworm, *T. obscurus*. To the general public the mealworms are probably better known as food for birds. The damage done by the larger tenebrionids is not very considerable as they are seldom numerous. On the other hand, their presence in a consignment of grain or bran may well result in its rejection by a would-be purchaser. The life-cycle is normally an annual one and the adults live for about three months.

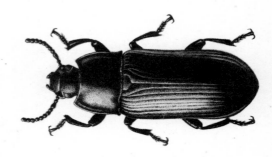

Fig. 14.36 *Tenebrio molitor*, Yellow mealworm. (Kohlhaas)

TENEBRIONIDAE

Tenebrio molitor varies from 12 to 16 mm in length and is dark reddish-black in colour and shining. The wing cases bear longitudinal rows of small pits. *Tenebrio obscurus*, which is less common that *T. molitor*, has dull wing cases but its biology is similar.

Life-cycle
Up to 576 eggs may be laid, either singly or in groups, by one female. They are bean-shaped and sticky and soon become covered with meal and debris. The larvae (known as 'mealworms') are long and cylindrical, the last two segments being conical. They are bright yellow, with each segment shading to yellowish-brown, and have a shiny, waxy appearance. The legs are small but well developed. The number of moults varies widely, from 9 to 20, and the larva reaches a length of up to 28 mm.

Mealworms not only consume farinaceous material but will eat animal matter such as dead insects. When fully fed, the larva passes through a prepupal stage when it lies on its side and assumes a curved position. The resulting pupa is also curved and lies among the foodstuff and debris it has been infesting.

TRIBOLIUM FLOUR BEETLES

In the beetles of this genus we find that small size is important. Although

Fig. 14.37 *Tenebrio molitor*, larva. (Kohlhaas)

Fig. 14.38 *Tribolium castaneum*, Rust-red flour beetle.

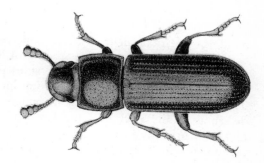

little more than one-third of the size of *Tenebrio*, *Tribolium* spp. are among the most numerous insects in stored products.

Tribolium castaneum is the most numerous insect found in cargoes imported into Britain. The two commonest species are *Tribolium confusum* and *T. castaneum*, and while they are best known as 'flour beetles' they occur on a wide range of commodities. The most heavily infested commodities are oilseeds, oilcakes, wheat by-products (e.g. bran), rice, maize and cocoa beans.

Tribolium confusum is distinguished from *T. castaneum* by the last segments of the antennae, which widen gradually towards the tip, instead of forming a three-segmented club. It occurs more frequently in milling machinery than *T. castaneum*, although both species occur on wheat in bulk. *Tribolium confusum* is more resistant to cold than *T. castaneum* and may have five generations in a year in heated premises. In unheated

Fig. 14.39 *Tribolium confusum*, Flour beetle. (Wellcome Foundation)

Fig. 14.40 Terminal segments of larva of *Tribolium castaneum*.

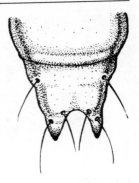

buildings in Britain neither species survives the winter. Their populations persist only by continued importation of infested commodities.

Flour beetles do not do well on 'short patent flours', thought to be due to the lack of vitamins of the B group in these processed flours. *Tribolium*-infested flour has a sour pungent smell and fails to rise in a dough as readily as clean flour. The reddish-brown colour of the beetles makes them easily identified by the consumer. Larvae of *Tribolium* are compodeiform with a horny bar on the last but one segment on the abdomen and two stout horns on the last abdominal segment.

GNATOCERUS

Two species in this genus, distinguished from *Tribolium* by the greater development of the mandibles and by the horny projections on the head in the male *Gnatocerus*, occur along with *Tribolium* but are much less common in cereals and still less common on other products. They are, however, of regular occurrence in flour-mills. The larvae are compodeiform but have no tail-horns.

Three other genera of tenebrionid beetles occur in stored grain – all small, *Palorus*, *Latheticus*, and *Alphitobius*. *Alphitobius*, represented by the species *diaperinus* is the largest of these (5.5–7 mm), and has recently attracted attention as a nuisance in the 'deep-litter houses' used in poultry farming.

Latheticus oryzae (2.5–3 mm) was originally regarded as damaging rice flour, but may occur on other cereals. It is much more common on rice bran (i.e. the highly nutritious layer removed during milling of rice) than on polished rice. It is also found in flour-mills in countries of high temperature and low humidity. It can easily be distinguished from *Tribolium* by the five-segmented club on the antennae and its long head. *Latheticus* depends for its occurrence in Britain on successive re-importation in tropical cargoes.

Fig. 14.41 *Tenebroides mauritanicus*, Cadelle. (Degesch)

TROGOSITIDAE

The members of this family of over 500, mostly tropical, species vary greatly in size, in form and in habits. Two genera are frequently associated with stored products, *Tenebroides* and *Lophocateres*.

TENEBROIDES

The most important member of this genus is *T. mauritanicus*, known as the Cadelle beetle. It is one of the larger beetles found in flour-mills (7–11 mm), it is dark-brownish to black in colour, and except that its antennae end in a club and that the prothorax narrows at its junction with the elytra to form a collar, it could be confused with a member of the CARABIDAE.

The larva is campodeiform, white with a sclerotised black head, a pair of large sclerotised black patches on the prothorax, and two prominent horns on a black terminal abdominal segment. The Cadelle is mainly harmful in flour-mills, where it feeds on flour but it preys also on other insects. It bites holes in the bolting or sieve cloths and is often more destructive in that way than as a consumer of flour. The larvae when full grown leave the flour in which they feed and may bore into surrounding woodwork.

LOPHOCATERES

One species of this genus, *L. pusillus*, is found in grain in tropical areas. However, it has been recorded in a wide range of commodities but seems to be a major pest of rice. It is a small flat beetle, 2.5–3 mm long, very different in appearance from the Cadelle. It is recognised by the upturned margins of the sides of the prothorax and elytra. Its larva, except that it is very much smaller, resembles that of the Cadelle, but it has no black patch on the first thoracic segment. Infestations of grain do not persist in temperate regions.

Chapter 15

Woodwasps, ants, bees and wasps

HYMENOPTERA

The HYMENOPTERA is one of the most abundant orders of insects, rivalling the LEPIDOPTERA and the DIPTERA in number of species with over 100,000 already having been described. They are highly specialised and show a broad spectrum of adaptive radiations, probably unequalled elsewhere in the INSECTA.

HYMENOPTERA members are characterised by the possession of two pairs of membranous wings, usually with greatly reduced venation. The forewings are larger than the hindwings and the latter bear a row of small sclerotised hooks which catch on to a fold or ridge on the hind margin of the forewings, thus holding them together in flight. The mouthparts are adapted for biting but may also be modified for lapping or sucking. The first segment of the abdomen is fused with the metathorax while there is a marked constriction of the abdomen between abdominal segments one and two except for the suborder SYMPHYTA.

The ovipositor is often highly developed and adapted for diverse functions other than egg-laying, such as sawing or boring a hole into wood, or for stinging. Although the larva is caterpillar-like in the SYMPHYTA, it is generally apodous and maggot-like.

There are two distinct suborders representing divergent evolutionary lines as follows:

SYMPHYTA is the most primitive and contains the woodwasps and sawflies. They are identified by the absence of the constriction at the base of the abdomen.

APOCRITA contains the most highly developed and specialised members of the order. The constriction at the base of the abdomen is always evident (wasp-waisted).

SYMPHYTA

SIRICIDE WOODWASPS

It is in the suborder SYMPHYTA that the family SIRICIDAE or Woodwasps is classified and in which they are joined with the sawflies. The SIRICIDAE is a fairly small family having 8 genera and about 70 species and subspecies. They are indigenous to the Holarctic and oriental regions, but they have been imported, in timber, into many countries of the world, in some of which they are viewed as causing serious harm to forests.

The biology of the SIRICIDAE shows a number of remarkable features. All the larvae are wood-borers but, except for the genus *Tremex*, attack only coniferous trees (PINACEAE) and as this group provides the world's softwood timber, their economic importance is substantial. They often occur in new buildings but are unable to continue to the second generation in dry wood. Their large, hornet-like, and fearsome appearance, however, often causes concern but they are unable to sting or bite.

The genus *Tremex* infests broad-leaved, deciduous trees, both in Europe and North America.

Siricids are usually introduced into a building in the larval stage, included in timber of low quality. The adult emerges subsequently and this may be up to two or three years afterwards. The large and frightening insect, up to 50 mm in length, often causes anxiety. Not only does it fly with a loud whirring sound or buzzing noise (called in Germany 'Schwirren') but the long, sharp ovipositor and its case resemble a sting. In North America siricids are referred to as 'horntails'. The female of *Urocerus gigas* is banded with yellow, thus mimicking wasp or hornet.

The circular emergence holes are up to 6 mm in diameter, so that where a number occur they may deface the appearance of flooring or other woodwork. Additionally, the adults will bite through a wide variety of materials in order to effect their escape from the wood. Hardwood, leather, linoleum and impermeable floor coverings of various sorts, as well as lead sheeting are perforated in this way.

Life-cycle
Mating usually takes place on the tops of trees and probably females outnumber males by two to one. The female *Sirex* examines the bark with its antennae, then after a few trial borings oviposition commences. This is accomplished by the insect lifting itself by forcing the ovipositor sheaths against the bark which thus exposes the elastic ovipositor which is then forced into the bark and moved rapidly up and down in a sawing movement. The abdomen moves rhythmically also. At the bottom of the shaft an egg is laid and then the ovipositor is slowly withdrawn a certain amount when another egg is laid. Coating the eggs and the tunnel wall is a glandular secretion which is white, glistening and glue-like. This is apparently a lubricant to facilitate the egg passage down the ovipositor tube. This secretion may contain the oidia of symbiotic fungi. Sometimes the female dies with the ovipositor stuck fast in the wood.

The number of eggs laid per tunnel is from 3 to 7, and the total number of eggs counted in ovaries is from 300 to 1,000.

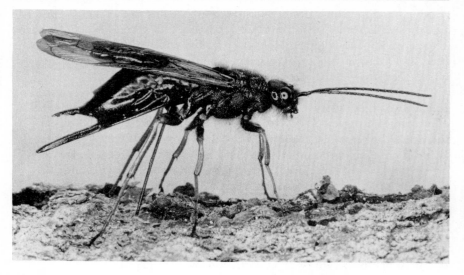

Fig. 15.1 Female *Sirex noctilio* ovipositing. (CSIRO, Australia)

Symbiosis

In a group exhibiting several biological phenomena of great interest it is perhaps the symbiotic relationship existing between siricids and wood-rotting fungi which is most remarkable. It is an obligate association. Neither can exist without the other. We may start with the initial injection of fungal spores into the wood substance of the tree when the female inserts her eggs in a mucous-like substance. The tree reacts immediately. The needles discolour and starch accumulates within them. The water stress which follows brings on conditions suitable for the spores to germinate and for the mycelium to proliferate.

The *Sirex* eggs can hatch only after the fungus has successfully started to grow in the wood and the larvae feed, not on wood substance but on fungal material. The fungus cannot reach new host trees without the aid of the *Sirex*, and the larvae of the latter are completely dependent on the fungus for food. The insects' physiology and egg-laying behaviour are extremely well adapted to the needs of the fungus, which can only grow in wood with a low water content. But, finally, it is the fungus which kills the tree, not the *Sirex*.

The ovipositor is made up three separate segments, a dorsal valve and two ventral valves. The former is thinner and extends around almost three-quarters of the ovipositor tube. The latter fit into longitudinal grooves at the lateral extremities of the dorsal valve. When an egg passes down the ovipositor the ventral valves are extended and the egg is visible through the slit.

In *Urocerus gigas* and *Sirex cyanea* the eggs are cigar-shaped with one end more pointed than the other, and in length are between 1.25 and 1.50 mm. The young larvae can be seen through the translucent chorion, the mandibles and terminal spine being dark in colour. The eggs hatch in about three to four weeks.

The young larva bores at right angles to the egg shaft keeping within the outer sapwood until about 9 mm in length. It then turns inwards to the heartwood, generally in a vertical direction. The tunnel becomes tightly packed with frass in which the moulted skins of the larva can be found.

After a time the larva directs its gallery to the outside of the wood and the pupal chamber is usually constructed 12 to 18 mm from the surface. The direction of the gallery is influenced by the position of knots which the larva tends to avoid. The gallery of *S. juvencus* is between 600 and 750 mm in length. A tunnel of *U. gigas* in silver fir was 375 mm in length. Galleries of *S. cyaneus* in larch were up to 255 mm long.

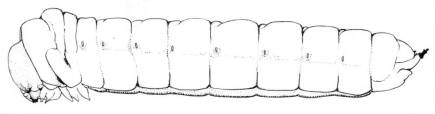

Fig. 15.2 Larva of *Urocerus gigas*.

Siricid larvae are cylindrical and whitish in colour. The head is yellowish-brown, directed downwards and the antennae are three-jointed. The mandibles are asymmetrical, the right lies above the left and bears four teeth. The left bears only three teeth. The pronotum is large and hood-like and there are three pairs of mamma-like thoracic legs.

The hypopleural organs are of great interest. These consist of a pair of deep folds between the first and second abdominal segments in which are situated epidermal structures overlying columnar epithelial tissue. In the pits between the folds tangled aggregates of fungal oidia occur. These organs occur only in the female larvae and are a means of identifying species. On the dorsal surface of the last abdominal segment of the larva there is a deep longitudinal groove and two small tubercles, below which there is an upwards-projecting terminal spine of characteristic shape and heavily sclerotised. It serves to pack the frass tightly in the gallery and also to act as a support as it is often driven into the tunnel side.

In the case of *Sirex cyaneus*, the minimum length of the life-cycle is two years, but many overwinter as adults to give a three-year period.

Trees attacked
Sirex usually attacks the suppressed trees within a forest. These are the trees that have had to give way to those more dominant. Slight damage to an otherwise healthy tree will often render it attractive to the

egg-laying females. This can be done by shooting off the top twigs. The most vigorous trees can cope with this by converting food reserves to polyphenols in the damaged area. Careful thinning of pine forests in order to increase the vigour of those remaining reduces the hazards of attack.

A damaged pine releases pinenes, associated with a water shortage or water stress which is attractive to siricids.

Fig. 15.3 *Sirex noctilio*, female, length 29 mm.

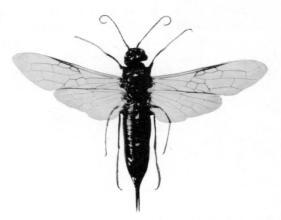

Parasites
Another remarkable feature of siricid biology is the phenomenon of parasitism. The two more important species concerned are parasitic HYMENOPTERA, *Rhyssa persuasoria* and *Ibalia leucospoides*. The former is a member of the PIMPLINAE group of the ICHNEUMONIDAE, while the latter is a cynipid. The female *R. persuasoria* explores the tree bark with her antennae which are continually palpated and curved downwards so that the dorsal surfaces are in contact with the bark. When boring the insect elevates itself to the fullest extent, appearing to stand on its head while the ovipositor is held downwards and guided between the hind coxae. When the ovipositor has cut a shaft near to its fullest extent an egg may be deposited – passing down the ovipositor canal which is so fine that a human hair cannot pass through. Before the egg is laid the woodwasp larva or pupa is injected with a paralysing fluid.

A large number of borings are made to no purpose as they are nowhere near a *Sirex* larva. Smell influences the final selection of the boring site, but the precise details are still obscure.

The egg of *Rhyssa* is at least 12 mm in length, but a narrow pedicel takes up 9 mm of this. The egg substance flows into the pedicel during oviposition then extrudes back again when it emerges from the ovipositor. The larva lies on top of the *Sirex* larva as an ectoparasite. On eclosion it bites its way through the remaining wood and bark making a

Fig. 15.4 *Rhyssa persuasoria*, female.

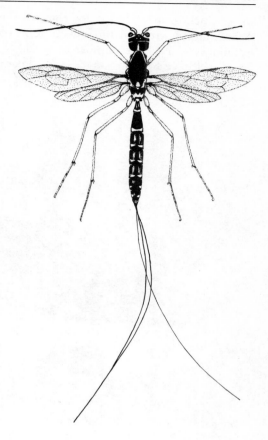

circular hole smaller than that of *Sirex*. The life-cycle occupies a year although the feeding stages occupy only about five weeks. Only nearly fully fed *Sirex* larvae are successfully parasitised, because at this time they are up near the wood surface and, therefore, within reach of *Rhyssa*'s ovipositor.

Ibalia leucospoides is most active immediately after the egg-laying period of *Sirex*. The female makes a detailed search of the trunks where *Sirex* has been inserting eggs and finds the actual egg shafts. She then uncurls the ovipositor and passes it down until the *Sirex* eggs are touched, when an egg is laid in each. The egg is similar to that of *Rhyssa* but whereas the latter is laid pedicel first, that of *Ibalia* is laid egg body first.

Distribution through commerce

During this century, siricids have considerably increased their range. In 1922 *Sirex noctilio* reached New Zealand, where it caused considerable damage on account of its parasites being absent. Probably in 1947 it entered Tasmania in a load of timber from New Zealand, although it was not discovered until 1952. In 1961 it was detected in Victoria, Australia, since which time a large-scale project for its control has been underway. 'Search and destroy' methods were followed by biological control techniques and a large number of siricid parasites from various parts of the world have been introduced. A nematode worm, *Deladenus siricidicola*, which causes sterility, is showing great promise.

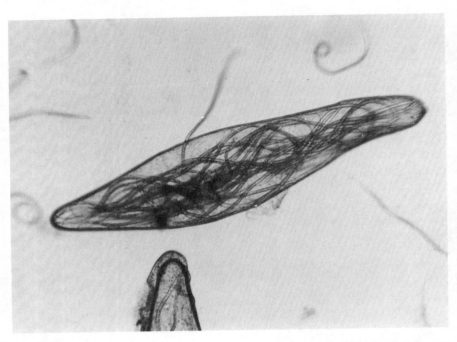

Fig. 15.5 Inside a *Sirex* egg parasitised by nematode worms. Such eggs are sterile. (CSIRO, Australia)

Species of economic importance
The following species are those most likely to be found emerging from wooden structures in buildings. Softwood timber is such a widespread material of international commerce that it is possible for siricids to occur almost anywhere, so that the country of origin cannot be given in every case.

> *Sirex areolatus* from Rocky Mountains and Pacific coast of North America.
> *Sirex noctilio*, Holarctic, established alien in Britain.
> *Sirex cyaneus*, Holarctic, but American in origin.
> *Sirex juvencus*, Holarctic, exists in two forms; established in Britain.
> *Urocerus gigas*, endemic in Europe but found throughout Africa. The subspecies *U. gigas taiganus* is native to northern coniferous forests of Eurasia, and the subspecies *U. gigas flavicornis* is native to northern

conferous forests of North America.

Urocerus fantoma (previously known as *U. augur*) is native to central and south-east Europe.

Urocerus tardigradus is native to central Europe and western Siberia.

Urocerus californicus is native to the Pacific coast of North America.

APOCRITA

The suborder APOCRITA is characterised by all the species being deeply constricted behind the first abdominal segment which is fused to the thorax. This is known as the propodeum. The APOCRITA is divided into two main groups; firstly, the TEREBRANTIA in which the great majority of species are parasitic and the ovipositor is adapted for piercing the cuticle of the host and for ovipositing in it, and, secondly, the ACULEATA. This includes the social or colonial HYMENOPTERA in addition to the fossorial or digger wasps. In all the aculeate families the ovipositor does not function as an egg-laying organ but for either paralysing prey for provisioning larvae or is used in defence of the nest site.

The VESPOIDEA consists of three families, EUMENIDAE, MASARIDAE and VESPIDAE. Only this last family contains social species.

Wasps and hornets make up the family VESPIDAE in the ACULEATA group. In the broad sense, wasps (both social and solitary) are important in regulating the populations of a wide range of insect and other arthropod species. Many of these are injurious to agricultural crops or to timber trees. The wasps are, therefore, of the greatest significance in benefiting Man. In the English-speaking world, other hymenopterous groups are referred to as wasps, often with a prefix, but the VESPOIDEA are the true wasps and generally recognised as such by the layman. In the United States they are known as 'yellow-jackets' or hornets, but in Britain the last name is now used exclusively for members of the genus *vespa*.

A number of VESPIDAE species occur in North America of which some of the more common stingers are *Paravespula vulgaris*, which is distributed throughout the United States and Canada, *Vespula maculifrons* of eastern North America, and *Vespula pensylvanica* of western North America. *Vespula maculata*, known as the Bald-faced hornet is found throughout Canada and the United States, and *Vespula arenaria* is distributed similarly. The last-named species nests close to the ground but *V. maculata* often locates its nest under the eaves of buildings.

EUROPEAN SOCIAL WASPS

Colony foundation and biology
New queens produced during the summer seek out sites for hibernation

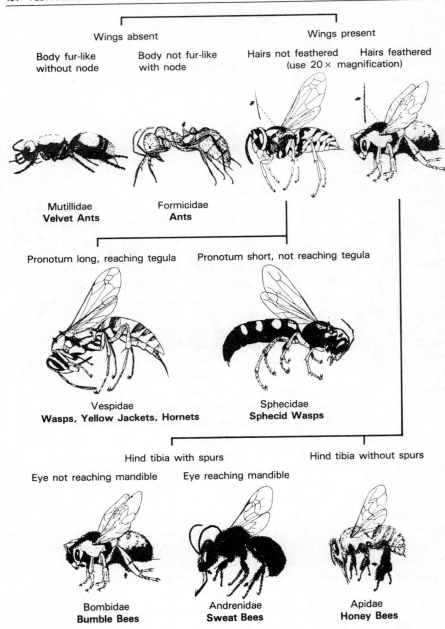

Fig. 15.6 Stinging HYMENOPTERA. Pictorial key to some common families in the United States. (US Department of Health, Education and Welfare)

in the autumn. Often this is in buildings (such as in the folds of curtains) or among rubbish in roof voids). Hibernation lasts for about six months, the metabolic rate being very low, and the new queens have ample fat

Fig. 15.7 *Vespa crabro* queen, European hornet. (Rentokil)

reserves. Sometimes many wasps hibernate together in the original nest. During the first warm days of April the queen wakes and commences to search for a site for her nest. At this time also, the queen will visit flowers for nectar and feed on tree-sap, as by now the food reserves have been much depleted. A hole in a bank, such as the disused burrow of a small mammal, is frequently selected by ground-nesting species, while situations well above ground level, such as a cavity in a tree, are preferred by the species of *Dolichovespula*.

When the queen has decided on the nest site she makes a study of its locality by flying to and fro, gradually increasing the distance and, presumably, local landmarks become fixed in the insect's memory. Wood-pulp is then scraped off fenceposts and sound, bark-less wood, taken back to the nest site where it is applied to the undersurface of the top of the chamber. Subsequently, wads of wood-pulp fashion the central pillar from which the whole nest is suspended. When the spindle has been extended, the first two cells are constructed before the outer paper envelope is formed, and the nest eventually attains the size of a football.

The first comb consists of about 35 cells of typical hexagonal cross-section and the larvae resulting from the eggs laid therein are progressively provisioned by the queen. The foot consists of protein in

Fig. 15.8 *Vespula vulgaris*, Common wasp. Nest in shed. (Rentokil)

various forms, principally of insects but also carrion and meat, and fish taken from shops is also fed to the larvae when available. The insect prey usually consists of flies, caterpillars, and sometimes bees, generally killed by a bite in the neck. Head, legs and wings are cut off before the bodies are carried to the nest.

When the first wasp larvae pupate and subsequently produce the first brood of sterile daughter wasps, the workers, the queen's foraging expeditions diminish until finally she is confined to the nest. Here her sole function is egg-laying while the workers take over all the duties of finding food, feeding the larvae, enlarging the nest and defending it.

Although only protein-rich food is given to the developing larvae the workers are attracted to sugary materials such as nectar. Many flowers are specially adapted for wasp·pollination. Honey-dew, the sweet, sticky secretion of aphids is also devoured as well as jam and other sweet confections. These simple carbohydrates are stored in the crop and on return to the colony are regurgitated and fed to other adults.

Much of the workers energy and time is taken up in enlarging the nest by manufacturing wasp paper from wood fibres and saliva and layering it on in thin strips and allowing it to dry. More horizontal combs are suspended from the original comb built by the queen by means of strong paper columns. Up to eight or nine combs are finally constructed which may contain upwards of 10,000 cells. The queen lays an egg in each cell

Fig. 15.9 *Vespula vulgaris*, Common wasp. Comb of nest. (Rentokil)

as it is completed or vacated by an emerging wasp. At the height of her powers, in mid-season, the queen lays about 300 eggs each day.

A feature of the wasp nest is its constant temperature. Heat is engendered by the activity of adults and larvae; the nest is insulated by the thick layers of wasp paper of which the envelope is constructed. If the inside of the nest gets too hot it is cooled by the wasps vibrating their wings at the nest entrance. In hot weather the wasps will introduce drops of water into the nest and allow it to evaporate.

At the end of August the number of adults present may reach 5,000 which is the maximum. Nearly three-quarters of all cells are used twice.

When the worker wasp feeds the larva the latter generally secretes from the mouth a droplet of clear liquid which is lapped up by the worker. It is also solicited by the workers by giving a gentle nip to the larva. This 'saliva' is of importance to the colony as it appears to serve several functions. It contains sugars derived from the protein diet of the larvae and thus the latter provide an energy-producing food for the active adults. In addition, there are protein-digesting enzymes present. This food exchange, found in other groups of social insects, is known as trophallaxis.

A number of insects are associated with the wasp colony, including syrphid flies whose black and yellow coloration mimics wasps, and the beetle *Metoecus paradoxus* whose larvae feed on the larvae of the wasp.

Wasp stings

The wasp is held in universal respect on account of the painful sting it can inflict. Late in the season wasps regularly invade buildings in large numbers where they are attracted by sugary materials such as jam and confectionery as well as ripe fruit. In chocolate factories production has often had to be halted until intruding wasps were removed.

Intense pain apart, stings (especially when multiple) have often led to death. In North America unaccountable road accidents have been attributed to the entry of a wasp into a motor vehicle and the subsequent loss of concentration by the driver on the emerging road traffic hazards. Queen wasps hibernating in buildings are also a stinging risk.

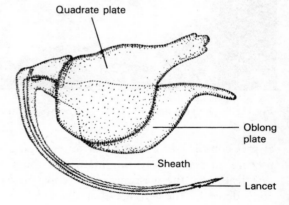

Fig. 15.10 Sting of wasp.

The sting of the wasp, like that of the honey bee is a specially modified ovipositor. The popular distinction between the two is that the sting of the wasp can be withdrawn while that of the bee cannot, due to the harpoon-like barbs. After stinging, the bee tries to pull the sting out of the wound but fails and the poison sacs are torn out of its body and the bee dies.

In the social wasps the workers use their stings only for defence of the nest while in the solitary species they are used for immobilising or killing prey.

Among other toxic substances in the venom, histamine is present as well as an enzyme hyaluronidase. The latter destroys the bonding cement between the body cells thus aiding the dispersion of the venom while the histamine is the cause of the inflammation and the weal as well as the itching. In contrast, the venom of the honey bee, while not containing histamine, directly releases an enzyme which has the effect of producing histamine from the tissues of the victim.

PAPER WASPS

The genus *Polistes*, the paper wasps, together with a number of other genera, constitute the subfamily POLISTINAE contained within the VESPIDAE. Species of *Polistes* occur throughout the warmer regions of the world and often extend into temperate areas such as France, Japan and Canada. Some species are fairly large, the wings are frequently brown or purple with a metallic sheen, and the 'waist' is usually extraordinarily narrow. The nest, which is generally composed of dead but undecayed wood, shows a wide range of form and complexity, although all hang from a narrow pedicel. The single-tiered comb possesses no outer envelope, and the number of cells composing the comb is very much smaller than that of *Vespula* and *Paravespula*. The colony is founded by a small number of queens, although after a time one only remains or becomes dominant.

In temperate regions the *Polistes* queen hibernates, sometimes for as long as six months or more, but hibernation does not occur in the tropics. The life-cycle from egg to adult is from 35 to 50 days in temperate regions, but in the tropical species *P. canadensis* it is unexpectedly about 65 days.

Generally *Polistes* wasps and other genera in the POLISTINAE play a beneficial role in their relationship with Man's economy, as they are carnivorous, provisioning their larvae mainly on lepidopterous larvae, many of which are injurious to crops. They are, however, often of considerable importance as a nuisance value when they construct their nest under house eaves, in porches or in roof voids. In defending the nest site they will sting viciously, causing considerable pain. In North America, *Polister fuscatus* readily attacks human beings when they approach too close to the nest, and the queen often hibernates in attics. In India, *P. hebraeus*, the Yellow house-wasp, is injurious on account of its habit of destroying paper in buildings and making its nest with it. Labels, notices, file-covers and books are often destroyed. Only the paper is removed; ink, paint or glaze remain untouched. *Polistes stigma* has similar habits but is not so abundant and widespread.

APOIDEA   SOCIAL AND SOLITARY BEES

The truly social species, those that possess a worker caste, are contained in the families BOMBIDAE and APIDAE, but the great majority of species of APOIDEA are solitary in habit. The principal diagnostic features of the adults are the presence of feathery or branched hairs on the head and thorax, and the bind tarsi are dilated or thickened. Their food consists of pollen and nectar and the mouthparts show a wide range of adaptions for visiting flowers of diverse shapes for the purpose of lapping up nectar, and the larvae are fed likewise except that the nectar is regurgitated from the crop as honey. The female bees possess pollen-gathering hairs, located on the abdominal sterna, on the posterior tibiae and tarsi, or on the femora. These are known as the corbiculae, but they are absent in those genera which are inquilines in nests of other species. Queens of social species do not posses pollen-gathering hairs.

The fertile female or 'queen' is fed a special protein-rich diet secreted from special glands in the sterile females, and the diet of the young bee larvae varies according to whether they will become queens or workers.

APIS MELLIFERA   HONEY BEE

One of the best known of all insects in *Apis mellifera*, the Honey bee, which has been introduced into almost every country of the world. It is, perhaps, also one of the most important as far as Man is concerned. Not only does it provide an important foodstuff, honey, which would otherwise be served to the bee larvae or be used by the adult bees during times when they must have recourse to stored foods, but their role as pollinators of a wide range of crop plants and trees is of the highest importance. Another characteristic of social bees is their secretion of wax in glands under the abdominal plates. This is used for building cells in which the bee larvae will be reared or for honey storage.

Colonies of *Apis mellifera* are ruled by a single queen. She usually lives for a period of several years, during which time she lays something like $1\frac{1}{2}$ million eggs. She can be recognised by her elongated abdomen. Most of the eggs develop into workers which are sterile females which lack the long abdomen. Only they are able to construct the combs of cells, keep them clean, feed the larvae (with nectar they have gathered) and also the queen, and do many other duties besides. In the summer the workers live for only six or seven weeks, but autumn-hatched workers may live over the winter in the hive and recommence operations the following spring.

In a strong colony the number of workers may range between 50,000 and 80,000 individuals. The males, called drones, are identified by their larger size and highly developed eyes. They number only a few hundred and perform only one task – that of fertilising the queen, although they may play some part in temperature control wing-fanning. New colonies are produced during the summer, when the old colony has increased in size beyond the capacity of the hive to contain it. The old queen then flies off, accompanied by a number of workers, to seek a situation for the new colony. The old colony is then taken over by a new queen which the workers have prepared for the task. The old queen kills all developing queens unless the workers consider the time propitious to send forth a

new swarm, when they prevent her from doing so. Fertilisation of the
virgin queen takes place in the air, after which she returns to the hive. At
the end of the year the drones are ejected by the workers and they die
from exposure or from starvation.

From what has so far been stated, *Apis mellifera* is wholly beneficial but
its undesirability in buildings is on account of its possession of, what is to
most people, a vicious sting.

APID DORSATA   ROCK BEE

In India this large bee is of importance to Man as it possesses a dangerous
sting and it is not unusual for deaths from stinging to be reported in the
newspapers. Elephants have been known to die after being stung by this
species. *Apis dorsata* is a particularly fierce and irritable bee, attacking all
those who approach its nest. Intruders are frequently pursued for miles.
The worker is 20 mm in length, about the same size as the female *A.
mellifera*. A large, fully-exposed comb is built about 1 m in width, but as
much as 1.6 m in length. It is usually suspended from the face of a cliff,
from the larger branch of a tall tree or sometimes from the walls or eaves
of a tall building. Often a number of nests are found together in a
suitable place and such sites are used year after year. *Apis dorsata* migrates
from the plains to the hills in the monsoon and returns in mid-winter. It
is especially attracted to the flowers of *Strobilanthes*. *Apis florea*, the Little
bee, rarely stings. Only a slight swelling takes place, which is not nearly
as painful as that resulting from the sting of the larger bees.

APIS INDICA   INDIAN BEE

Nests of this species often occur inside buildings, either in the roof or in
disused rooms, and even in boxes and cupboards. The nest consists of
several parallel combs about 30 cm in width. This bee shows some
tendency to migrate. The form occurring in the hills is slightly larger and
darker than that of the plains and less liable to sting.

No indigenous species of *Apis* occurs in Australia, but a closely related
genus, *Trigona*, is found in the northern half of the continent. These are
known as 'native bees', but do not sting.

The bee sting

The bee sting is situated in a cavity at the end of the abdomen and is
protected by the terminal dorsal and ventral plates which are separated
only by a narrow, curved aperture, projecting from which the tip of the
sting can often be seen. If the lower plates of the abdomen are removed
the sting is seen to be made up of a sting sheath which is grooved below
and bulbous at its base, with two long stylets which lie below and run in
grooves under the sheath. The stylet-tips project a little beyond the sheath

and are sharply-pointed and barbed.

The sting-sheath base diverges into two arms which curve outwards and the basal parts of the stylets follow them outwards. On internal abdominal pressure being applied the sting is pushed out and muscular action, anchored by the abdominal plates, drives it into the victim. Thereafter venom, secreted by two glands in the abdomen and stored in the poison sac, runs down the bulb-like enlargement into the stylets and thence into the wound, caused by the barbed, stylet-tips. Usually the sting apparatus with the poison sac remains attached to the wound after the body of the bee has been roughly removed, owing to the barbs on the stylet-tips holding firm.

Usually about 0.3 mg of bee venom is injected into the wound, which includes two recognisable elements. About 2 mg of histamine accounts for the local, cutaneous reaction, but more important is the presence of a dialysable protein called apitoxin which has a molecular weight of less than 10,000.

The immediate effect of a bee sting is of intense pain. What follows depends very much on the degree to which the victim has been exposed to previous been stings. Many stings may have serious consequences, especially in young children. These may be fatal. Anaphylactic shock may occur if a person becomes sensitised to the venom. Violent, burning pain erupts over the body, accompanied by itching and 'pins and needles'. Areas of anaesthesia may suddenly occur, as well as skin eruptions such as oedema, erythema or intense, giant urticaria. The sting site often remains swollen and may become septic or even gangrenous, but the timing of the onset of these various symptoms is variable. In some cases, the swellings may remain for as long as 14 days and then may be slow to heal. In intensely sensitised persons generalised symptoms, such as headache, muscular weakness or spasms with perhaps difficulty in breathing may terminate with collapse, with fatal consequences. It is obviously unwise to minimise the effect of any bee sting, especially if there is a history of hypersensitisation, and medical help should be sought.

XYLOCOPIDAE CARPENTER BEES

Xylocopid bees are very large, hairy, robust and usually black in colour, often with a metallic blue or green sheen, although in some species some hairs may be of a light colour. The wings are usually dark and smoky, and sometimes iridescent. The females bore holes in dry wood, sound or rotten, with their strong mandibles, often making a tunnel about 25 cm in length. Exposed woodwork in buildings is often destroyed when several of these large bees bore into one structural member, such as a rafter end. Sometimes a bee will utilise an existing gallery and so lengthen it still further. Lengths of 3 m are known.

When the tunnel has been constructed, it is divided into a number of chambers with fragments of wood stuck together. An egg is laid in each cell on the surface of an amount of pollen with some regurgitated nectar. In the case of United States species, the life-cycle is an annual one, the adults overwintering in the galleries to emerge in spring.

Fig. 15.11 *Xylocopa virginica*, Carpenter bee. (Shell Chemical)

Distribution
Although widely distributed throughout the world they are most abundant in the tropics.

Xylocopa orpifex, in which both sexes are black and which varies from 12 to 17 mm in length, causes significant damage throughout the western United States and lower California. The larval period is recorded as being from 37 to 47 days, and the pupal period 15 days.

Xylocopa varipuncta is 18–20 mm in length and while the female is black with metallic iridescence, the male is reddish-brown in colour. It occurs at lower altitudes in Arizona, California and lower California where it is known as the Valley carpenter bee. Another species, *X. virginica* is found in the eastern United States. In southern Europe, *X. violacea*, which is distributed northwards as far as Paris, causes a certain amount of damage to wooden structures.

In Africa, two species, *X. nigrita* and *X. incoustans*, are well known as being mimicked by asilid flies. In India a number of species are known and some are extremely large. *Xylocopa latipes* is about 44 mm in length, possesses a wing span of almost 80 mm, and the diameter of its tunnels is 40 mm. There are sometimes branch tunnels in this and the following species, and the life-cycle is completed in six weeks.

In *X. leucothorax*, which is found also in North Africa, the tunnel is bored by both sexes. One species, *X. rufescens*, occurring in Burma, is nocturnal, its loud buzzing being a feature of moonlight nights.

Xylocopa nasalis and a number of its subspecies occur widely throughout India, Ceylon, Malaya, Upper Burma and China. It is a large, black bee 32 mm in length, and its purple and metallic-green wings extend up to 70 mm. It often bores into constructional timber in buildings. *Xylocopa tenuiscapa* is also black and similar in size to the last-named, Its tunnels are 25 mm in diameter and about 12 cm in length, with short branches. After the eggs are laid and provisioned, the gallery entrance is plugged with pieces of leaf. It damages wooden buildings and is active throughout the year in south India.

Xylocopa verticalis occurs in India and Malaya and bores into constructional woodwork of a number of species. In Australia the closely related genus *Lestia* displays the same habits as *Xylocopa*.

Female xylocopids possess a sting, but it is rarely used. Males are stingless.

FORMICIDAE ANTS

Ants constitute the family FORMICIDAE of the HYMENOPTERA and their general facies are so characteristic that they are usually easily recognised. The non-entomologist, however, may refer erroneously to mutillid and thynnine wasps as ants. About 10,000 species are known and all of them live only in social organisations. This latter characteristic is also so well known that another non-related order, the ISOPTERA, has been universally known as 'white ants'. As in all HYMENOPTERA, the first segment of the abdomen is fused with the thorax and is termed the propodeum. In all ants, however, this is followed by an extremely accentuated waist which consists of either the second abdominal segment or the second and third. This is known as the pedicel or petiole, and the remainder of the abdomen, usually bulbous, is the gaster. Almost all ants possess large metapleural glands which open on each side at the base of the hind thorax. Ants are usually small in size, the largest being only about 25 mm in length and a number of species are minute.

The ant colony consists of a reproductive female, or queen, a reproductive male, and a number of sterile females which are known as workers, or in some species, workers and soldiers. Wings and well-developed associated flight muscles are possessed by the virgin queen as well as large compound eyes and three ocelli. She is usually larger than the workers her ovary-containing gaster being especially rotund. After

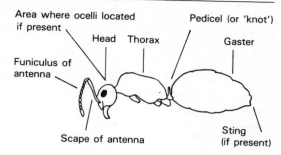

Fig. 15.12 Diagrammatic side view of ant. (NPCA)

pairing, the wings of the queen are shed. The male bears wings but they are not shed. His head is small although bearing large compound eyes and ocelli, similar to those of the female, but the mandibles which are functional in the queen, are usually modified or reduced in the male. Usually the size of the male is intermediate between that of the queen and that of the worker.

The worker ant is smaller than the reproductive castes and is sometimes present in two or more 'phases' usually called worker minors and worker majors. The worker is always wingless and the thoracic structure is generally modified or reduced. The compound eyes are much smaller and may be absent in some species and the ocelli likewise. The antennae and mouthparts are usually similar to those of the queen. Sometimes intermediates occur between queens and workers known as ergatogynes which accompany the queens and may, indeed, replace them.

A nuptial flight occurs from the old colony when the pairs mate. The queen then sheds her wings and enters a crevice or similar situation for the founding of the colony where she commences to lay eggs. The male dies.

The soldier caste, when present, is larger than the worker and is usually more substantial. It has a disproportionately larger head which is sometimes larger than that of the queen. The only task carried out by the soldiers is to guard and defend the colony, especially from the attentions of other ant species. When the nest is disturbed the soldiers rush to the perimeter and take up aggressive attitudes towards the intruder. They are fearless, whatever the size or numbers of their attackers. The antennae of ants in the more primitive species consist of 13 segments in the male while in the female there are 12, but often the number is much reduced. Except in some males, the antennae are elbowed, the long basal segment being known as the scape. Characteristics of the antennae are often of importance in identification.

Life-cycle

The eggs of ants are usually very small. The larvae are white and legless and more or less bottle-shaped. The head is small and in some species very small. The body may be covered with tubercles or hairs and the latter are sometimes twisted, enabling the larva to be 'fastened' to the wall of the nest or to be fastened together with others so that a number may be transplanted at once when danger is threatened. There are probably three larval instars. The first brood of larvae are fed only by the queen from her own stored food reserves, but after the first workers are produced the larvae are fed by them. In some groups, a thin paper-like cocoon is spun but in the remainder the pupae are naked.

Most countries of the world possess a number of ant species which either live permanently in buildings or invade them from time to time in search of food or shelter. The immediate environs of dwellings are often attractive to a number of species due to the presence of concrete or stone slabs under which the species will form a colony. Lawns are also often colonised. In both cases foraging ants will usually enter the building.

According to the degree of the colony age and specialisation, the ant colonies may contain as few as 10 adults or as many as several millions, but usually it is less than 2,000. In a number of regions of the world, in spite of their small individual size, ants play a most important part in the natural environment. They play a beneficial role in their control of a number of arthropod groups, especially termites, sawflies and predatory beetles. They are of importance also in soil formation due to the vertical mixing of soil particles and to aeration. On the other hand they conserve aphids and coccids which are harmful to agriculture but which the ants milk for their secretions.

As far as pests in buildings are concerned, in some countries such as the southern states of the United States, they rank high in importance. Not only do a number of species enter buildings foraging for food which may be carbohydrate in character such as sugary materials, or protein such as meat, but some species can sting viciously, others bite, a few will inject formic acid into the wound caused by the bite, and a few will eject a fine spray of poison up to 50 cm.

Beneficial and pest species
The ant fauna of many countries, however, consists of two components, the first of which contains the indigenous species where buildings are rather stony extensions of their normal environment. The second, however, consists of accidentally imported species, probably brought in with plants and timber or other agricultural material from tropical or subtropical regions, but which have become adapted to a protected life in heated buildings. Some pest species now have an almost cosmopolitan distribution.

IRIDOMYRMEX HUMILIS   ARGENTINE ANT

This ant, native to Argentina and Brazil, is an aggressive species now found in many parts of the world. The workers vary from about 1.5 to 3.0 mm in length and are light brown to dark brown in colour. There

Fig. 15.13 *Iridomyrmex humilis*, Argentine ant. (NPCA)

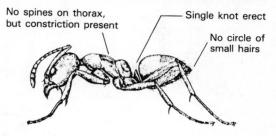

are no spines on the thorax, but a transverse constriction is present. The pedicel consists only of a single 'knot' or node which points upwards. There is no circlet of small hairs around the tip of the abdomen.

Iridomyrmex humilis generally occurs in very large colonies, and a number of queens, which move about freely, are usually present. The aggresive workers eliminate other ants of different species in their immediate environment, but tolerate their own species of different colonies. Colonies move frequently and the workers travel in well-defined columns. Nests occur in a wide variety of localities, from exposed soil to cavities in plants or rotting wood, or dark, moist situations generally. In temperate or cold temperature regions, however, the nest is indoors in walls or under floors. Swarming only rarely occurs, mating taking place within the nest. While preferring sweet materials such as various plant secretions and honeydew from aphids, meat, insects and seeds are also eaten.

Other species of *Iridomyrmex* are pests in buildings in many parts of the world. In Australia they are known as 'meat ants' and create problems in meat-processing plants as well as in dwellings. In Tasmania, the only ant species listed as pests in buildings are three species of this genus.

MONOMORIUM PHARAONIS   PHAROAH'S ANT

This ant is a very common pest in hospitals, restaurants and other establishments where food is handled on a large scale, in large apartment houses and hotels. It will travel considerable distances along established trails, and although it will eat a wide variety of substances, it appears to prefer grease and vegetable oils. It requires moisture and may commonly be seen getting water in kitchens and bathrooms. In hospitals it sometimes infests bedding, and may play a role in the transmission of pathogenic organisms.

In hot climates it nests out of doors, but indoors the location is generally inaccessible such as wall voids and foundations, between floors and commonly in warm places near hot water heating systems, hot air ducts, etc.

Native to Africa, this small ant is yellowish to light brown in colour with a dark tip to the abdomen, or may often be reddish, and it is one of the most widely distributed of all ants. The size of the workers is only

Fig. 15.14 *Monomorium pharaonis*, Pharoah's ant, 2.5 mm. (Wellcome Foundation)

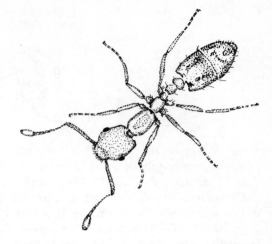

1.5 mm, but the size of the colonies is extremely large, reaching up to 300,000 in number. There are many queens, and new colonies are formed when a number of queens and workers split off from the mother colony. Mating occurs in the nest and may be at any time of the year, but swarms are never observed. *Monomorium pharaonis* may be identified by its size, the pedicel consisting of two nodes, the relatively small size of the gaster, antennal club of three enlarged segments and thorax without spines.

MONOMORIUM MINIMUM LITTLE BLACK ANT

This species of *Monomorium* is dark brown to black in colour and has a worker size of about 1.5 mm. The antenna has a three-segmented club; there are no spines on the thorax and the pedicel consists of two nodes. *Monomorium minimum* is abundant in northern and eastern states of the United States, and probably occurs in every other state. It often nests in the decaying timber and in the masonry of buildings. It feeds on insects, dead or alive, on honeydew, sweet materials of many kinds, meat, bread, grease, oils, vegetables and fruit. The colonies are often very large with many queens being present.

Fig. 15.15 *Monomorium minimum*, Little black ant. (NPCA)

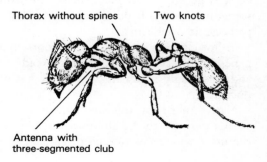

PHEIDOLE BIG-HEADED ANTS

These ants are abundant in a number of tropical and subtropical regions of the world. They are mainly characterised by the head being very large in proportion to the rest of the body, by there being two nodes in the pedicel and one pair of spines on the hinder part of the thorax. The antennae are clubbed, the club consisting of three segments.

Although they never appear to nest indoors, they often occur around buildings, either in lawns where they are exposed or sometimes in rotting wood. Although they seem to prefer meat and other foods of high protein content, they also feed on honeydew and infest grease and bread in buildings. The colonies are always small. In the southern two-thirds of the United States several species, including *P. hyatti*, occur. *Pheidole megacephala* is a house-invading ant in South Africa and tropical Australia.

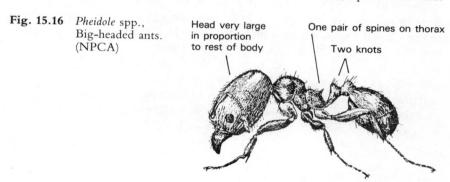

Fig. 15.16 *Pheidole* spp., Big-headed ants. (NPCA)

WASMANIA AUROPUNCTATA LITTLE FIRE ANT

It feeds on dead insects and honeydew, but in buildings prefers meat, fats, oils and dairy products. The colonies are generally large and more than one queen is usually present. It is said to be attracted to sweaty clothing and bedding.

This stinging species, now established in Florida and California, occasionally nests in houses. The workers are about 1.5 mm in length

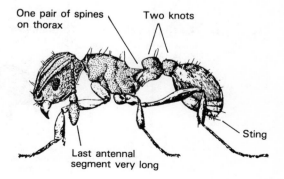

Fig. 15.17 *Wasmania auropunctata*, Little fire ant. (NPCA)

and are light brown to golden brown in colour. The head is heavily ridged; there is a pair of spines on the thorax, and the pedicel consists of two nodes.

TETRAMORIUM CAESPITUM PAVEMENT ANT

This ant has a worker size of about 3 mm and is light brown to blackish in colour. It is fairly easy to identify on account of the fine, distinct parallel lines which cover the head and most of the thorax, the small thoracic spines, two nodes in the pedicel and a small sting.

This European ant was taken to the United States by early colonists, and it is now common in urban situations in the Atlantic coast states. In addition, it is found in Chicago, Cincinnati and the Sacramento Valley.

Tetramorium caespitum nests most usually beneath stones and cement slabs in pavements, but appears to be extending this adaption to buildings in the United States where it works its way into cracks in concrete slabs into buildings. It will also build its nest in heat insulation materials. It also occurs in upper floors of multiple storey buildings. The large colonies often exterminate other ant species from the immediate locality. It feeds on insects, seeds and substances of high protein or high carbohydrate content but appears to prefer meat and grease. This ant bites and stings but it is not to be compared, in this respect, with *Solenopsis*, fire ants.

SOLENOPSIS MOLESTA THIEF ANT

Slightly smaller in size than *Monomorium pharaonis* at 1.5 mm, *S. Molesta* is yellowish, light brown or dark brown in colour, and is shiny. The thorax is without spines, there are two nodes on the pedicel and there is an antennal club consisting of two enlarged segments. A sting is present at the tip of the abdomen. This ant is distributed throughout most of the United States, being especially abundant in the eastern half.

Nests are formed in soil out-of-doors but also in buildings in decaying wood and in the masonry. Swarming takes place in later summer to early autumn, and the colonies grow to a very large size. It feeds on ant larvae

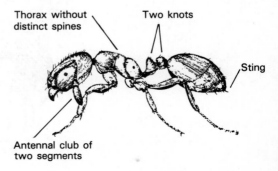

Fig. 15.18 *Solenopsis molesta*, Thief ant. (NPCA)

of other species by raiding the nest, on honeydew and seeds, but its importance as a pest is on account of its taking a wide variety of household foods, preferring those of high protein content. It can enter food containers through extremely small apertures, but it will not take sugar.

TAPINOMA   HOUSE ANT

A number of species of *Tapinoma* are pests in buildings in various parts of the world. *Tapinoma sessile*, known as the Odorous house ant in the United States on account of its odour of rancid butter when crushed, is a common inhabitant of dwellings. The worker size is 3 mm and it is brown to black in colour. The single node of the pedicel is almost hidden by gaster, and there is no circlet of fine hairs at the apex of the gaster.

The nest is often located in a wall or beneath the floors of a house. Out-of-doors it feeds on honeydew, but indoors will take almost any household food, although preferring sweet food. The colonies are large and contain many queens. *Tapinoma melanocephalium*, the Ghost ant, is a well-known nuisance in tropical Australia.

ACANTHOMYOPS INTERJECTUS   LARGE YELLOW ANT

Formerly placed in the genus *Lasius*, the worker size of this yellow ant is 5 mm. Abundant throughout the New England states and the mid-west of the United States, it is mainly a honeydew feeder and does little harm indoors apart from its presence, as it often nests in the walls and beneath dwellings and swarms appear indoors, having made their way through crevices in the floors.

PRENOLEPSIS IMPARIS   HONEY ANT

The workers forage for sweet materials from decaying fruit, nectar, honeydew to anything in the kitchen cupboard. The colony size is small, very seldom located in buildings but the nest entrance is identified by being surrounded by coarse, earthen pellets. The gasters of successful foraging workers are often distended and become amber-coloured. This is a reddish-brown to black, shiny ant of worker size 1.5 to 3 mm in length. It is identified by the antennae in which the scape is much longer than the head; the thorax is constricted around the centre; a single node constitutes the pedicel, and there is a circle of minute hairs around the tip of the gaster.

LASIUS   GARDEN ANTS, ETC.

A number of species of *Lasius* invade buildings when foraging and, when a suitable food source is located, become a great nuisance. Ants of this genus are distinguished by the scape of the antenna being not much longer than the head, the dorsal surface of the thorax bearing deep transverse grooves, the pedicel consisting of a single, vertical node and the presence of a circle of fine hairs around the tip of the gaster.

In central Europe, *Lasius emerginatus*, which is reddish-brown except for the gaster which is nearly black, has a worker size of 3.5 mm and often nests in houses with the cellar being usually the selected location. In Britain, *L. niger*, Black garden ant, is the only ant commonly to invade buildings when foraging for food. The nest is situated out-of-doors but occasionally is in earth closely associated with the foundations. In the latter case, when swarming occurs, a number of 'flying ants' often find their way indoors and fly around the windows.

PARATRICHINA LONGICORNIS CRAZY ANT

Apparently preferring sweet materials in food stores in hotels and apartment houses which it infests, it is practically omnivorous, feeding on insects, meat, seeds, fruit and honeydew. The nest is often sited in walls and foundations of buildings, from which the workers forage over a considerable area. It has occasionally been recorded as present in bakeries and other warm buildings in Britain.

Originally a native of tropical Africa or the Orient, this ant is now established in many parts of the New World. It is well established in Florida and the Gulf States of the United States, the West Indies and, no doubt, in Central America and tropical South America. It is known as the Crazy ant on account of its method of running with sudden, seemingly haphazard, changes of direction.

The size of the workers varies from about 1.5 to 3 mm, and they are dark brown to black in colour and a purplish, metallic sheen may be present. The antennae are very long, the scape being much longer than the head. The legs are also very long and the pedicel consists of a single node. There is a circle of fine hairs at the tip of the gaster.

CAMPONOTUS CARPENTER ANTS

Ants of the genus *Camponotus* are generally large in size, the workers often being up to 12 mm in length. In colour they are usually wholly black or black and red. They are further characterised by the profile of the thorax being evenly convex dorsally, the pedicel consisting of a single node and the presence of a circlet of fine hairs at the tip of the gaster. The eyes are prominent and well developed, as are also the mandibles. The funiculus or terminal whip-like part of the antenna is made up of 12 segments and is not club-shaped. *Camponotus* ants are distributed throughout the world. Their importance in buildings lies in the fact that they excavate galleries and tunnels in decaying or moist wood commonly in buildings. They do not eat the wood but excavate and use it as a shelter and a site for the nest. Subsequent excavations are frequently made into sound and dry timber and *Camponotus* ants are often transported long distances in such wooden structures as packing cases.

Life-cycle

In the United States where six species of *Camponotus* cause a great deal of annoyance (*C. pennsylvanicus*, *C. ferrugineus*, *C. novaboracensis*, *C. herculeanus*, *C. h. modoc* and *C. laevigatus*), mating flights take place in spring and early summer. They occur simultaneously over a wide area, thus facilitating cross-fertilisation between different colonies. After mating the males die and the females select a site for colony formation. This is a crack or other defect in a moist or decaying wooden structure. Sometimes the female breaks off her wings at this point. After a certain amount of excavation the queen seals the entrance with a wood and saliva mixture and then lays 15 to 20 eggs. They are cream in colour and may be up to 3 mm in length.

The resulting larvae are fed entirely by the queen with a secreted fluid derived from her fat reserves and from the wing muscles now no longer required. When fully fed, the larvae spin tough cocoons and pupate within. The period from egg to adult may be as little as 2 months, but under adverse weather conditions it can be as long as 10 months.

The first workers are small, and after removing the entrance plug they commence food collection and enlargement of the nest. The queen lays further batches of eggs which are cared for by the workers, and the resulting larvae are fed by them. The workers also feed the queen. The workers vary greatly in size and are all sterile females.

The food of species of *Camponotus* is extremely wide in variety, but honeydew is probably the chief component. In addition they will feed on living or dead insects and other invertebrates, plant sap and ripe fruit. When they invade a building they readily find areas such as kitchens and pantries where food is stored and they feed on many kinds of sweet substances such as syrup, honey, jellies, sugar and fruit as well as many kinds of meat, grease and fat.

The foraging workers usually digest the food on the site and feed the larvae, queen and general-duties workers by regurgitation. Many species are nocturnal. Wood which has been galleried by *Camponotus* can be identified by smooth and clean surfaces of the tunnels which have a sandpapered effect. All fragments of excavated wood are removed and

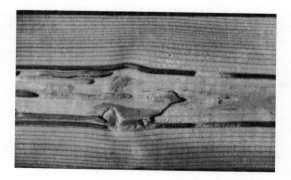

Fig. 15.19 Softwood damaged by *Camponotus herculaneus*, a Carpenter ant.

carried outside, and the small piles can usually be observed. By contrast, the tunnels excavated by subterranean termites are dirty and partly filled with faecal matter and soil particles.

From three to six years after formation of the colony, winged males and females are produced which swarm on their nuptial flight. At this time the number of workers may be from 2,000 to 3,000 and the size of the colony rarely exceeds this number thereafter due to periodic swarming.

In central Europe the species *Camponotus ligniperda* occurs widely, where it often makes its nest in door frames and is an annoying inmate of food cupboards.

Chapter 16

Fleas

SIPHONAPTERA

About 1,000 species of SIPHONAPTERA or fleas are known and they comprise a characteristic and well-defined order. They are all small wingless insects, yellow to almost black in colour, and all are laterally compressed with a hard sclerotised body beset with numerous backwardly directed spines. They are thus well adapted to making their way through hair and feathers. All species are ectoparasitic on mammals and birds in the adult stage and feed on their blood. They leave the host when it dies and move away by walking or leaping; the coxae being large and muscular. They are attracted to warmth and avoid light.

Eyes are present in those species that spend most of their time on the host's body, but in those that usually occur in the nest material or parasitise completely nocturnal animals such as bats or subterranean ones such as moles, they are reduced or absent. The antennae are short and stout and repose in grooves. The mouthparts are modified for piercing and sucking. They consist of a pair of broad, blade-like laciniae which are serrated along the distal two-thirds and form the cutting and piercing organ, and the epipharynx which lies between them. All three form a tube up which the blood of the host is drawn by the cibarial pump. Saliva from the paired salivary glands is injected into the puncture.

In some species a row of almost black, backwardly-curved spines is present on the lower side of the front of the head. This is known as the genal comb and, since it is variable, it is a character which assists in separating one species from another. The segments of the thorax are distinct and are capable of a certain amount of movement, while in a number of species the hind margin of the first segment bears a row of stout spines known as the pronotal comb, the presence of which may be diagnostic.

General life-cycle
The white or cream-coloured eggs of fleas differ from those of most other ectoparasites in that they do not adhere to the hair or feathers of their host. There is one exception. They fall to the ground and are generally found among the litter or hair in the nest itself. Those of *Pulex*

Fig. 16.1 Larva of *Pulex irritans*, Human flea.

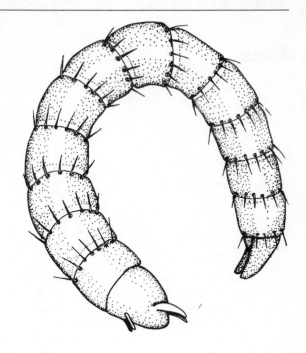

irritans, the flea of man, occur in floor cracks and beneath matting and carpets in unhygienic buildings. The eggs of rat fleas are not infrequently found among litter on the floors of barns and granaries, and another favourable situation for hen flea eggs is among the dried faecal matter and feathers on the floor of fowlpens. The eggs take only from 3 to 10 days to hatch and the larvae are whitish, very active and worm-like, and grow up to about 4 mm.

The larva, although possessing a well-developed head is without eyes and legs. The rather prominent antennae consist of a single segment, the mandibles bear teeth, while the maxillae are brush-like with two segmented palpi. The labial palps are composed of a single segment bearing stout setae. There are 3 thoracic and 10 abdominal segments each being encircled with a band of bristles. There is a pair of anal struts on the tenth segment. The larvae are only exceptionally parasitic on the host, and usually feed on organic debris found in the host's nest or hair, or in its immediate neighbourhood. In some species, however, it has been shown that the larvae require to feed on the blood of the host after it has passed through the body of the adult flea in the form of faecal droppings. When fully fed the larva spins a cocoon to which stick fine dust particles, thus concealing it. The length of the pupal stage is variable in some species depending on vibration or other mechanical stimuli. For example, a person walking across the floor of a deserted house will trigger off the emergence of large numbers of fleas. The newly-emerged flea immediately seeks its host although it can usually remain alive for a considerable time before feeding. Some females require a meal of the blood of their normal host before eggs can be laid.

The length of the complete life-cycle is variable. In Europe *Pulex irritans* takes from four to six weeks. In India *Xenopsylla cheopis* completes its development in about three weeks.

The important diseases in which fleas act as vectors are described in Chapter 2.

Notes of Species of Flea attacking Man

PULEX IRRITANS   HUMAN FLEA

This flea is cosmopolitan and probably parasitises Man throughout the world. Although the primary host of *Pulex irritans* is Man, it occurs also on a number of domestic animals from time to time, such as cats and dogs as well as various farm animals. It is, however, well adapted to feeding on the pig and completing its breeding cycle in pigsties, where it is often found in great abundance. In Britain it is also associated with a number of wild animals such as *Vulpes vulpes*, the Red fox, *Erinaceus europaeus*, the Hedgehog, and *Meles meles*, the Badger.

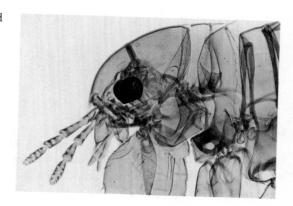

Fig. 16.2 *Pulex irritans*, head of Human flea.

CTENOCEPHALIDES CANIS   DOG FLEA

This cosmopolitan species occurs not only on the domestic dog but also on cats, rabbits and Man. Foxes and other wild mammals fairly closely related to the dog are also parasitised by this species. Flea larvae may swallow the eggs of *Dipyllidium caninum*, the Dog tapeworm, when they are devouring faecal matter. The parasitised adult flea is then swallowed by the dog host when it is nuzzling its fur to allay irritation. The worm is then at the cysticercoid stage (bladderworm), but it then develops to the adult stage in the dog.

CTENOCEPHALIDES FELIS CAT FLEA

Ctenocephalides felis parasitises the domestic cat and is found throughout the world wherever the latter occurs. In addition, it will readily attack Man as well as dogs, and is found also on the wild members of the cat family, FELIDAE. Rats, mice and other small mammals are also attacked, but less frequently. *Ctenocephalides felis* is the species often found in severe flea infestations in dwellings in Europe and in North America.

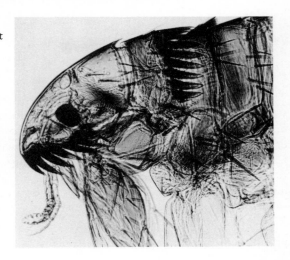

Fig. 16.3 *Ctenocephalides felis*, head of Cat flea.

This flea gives rise to quite serious skin inflammation if the victim becomes sensitised to its saliva. It is also a vector of Plague in Africa.

XENOPSYLLA CHEOPIS TROPICAL OR ORIENTAL RAT FLEA

Throughout most of the tropics this flea attacks *Rattus rattus* and *R. norvegicus*, but readily bites Man and is the most effective vector of Plague. The first record of this flea in Britain came from Guys Hospital – a compliment to their heating system but it is only rarely found in the United Kingdom.

SPILOPSYLLUS CUNICULI RABBIT FLEA

This sedentary European flea, which attaches itself to the ears of wild and domestic rabbits, is often picked up by domestic cats. Unlike other fleas it depends on the hormones of the female breeding rabbit in order to regulate its own breeding.

ECHIDNOPHAGA GALLINACEA STICKTIGHT FLEA

Primarily a parasite of poultry, this flea is found also on cats, dogs, rats, horses and occasionally on Man where it will embed itself in the skin. Those working with poultry are most often attacked. Before the female is fertilised it is much like other adult fleas in habit being very agile and able to move about by leaping. After mating, however, she attaches herself to the host and remains thus for the rest of her life.

Generally, in the case of poultry, the point of attachment is around the eyes and comb, but in dogs and cats the ears are usually selected, especially along the edges. The eggs usually fall to the ground where they hatch into typical flea larvae. These possess mouthparts adapted for chewing, and they feed on faecal matter. In some cases, however, where the female has burrowed into the hosts' tissue, the eggs may be retained, and a group of larvae may be found in the suppurating tissue before finally falling to the ground. When fully-grown, a white cocoon is made. When large numbers of the fleas are present, extensive ulceration may occur which can cause blindness and death. The length of the larval stage is about two weeks, the pupal stage another two weeks. Thus the life-cycle is completed in a little over a month. In the case of dogs and cats, the larvae develop in the bedding material.

CERATOPHYLLUS GALLINAE   HEN FLEA, POULTRY FLEA, COMMON BIRD FLEA

This common parasite of many wild birds of northern Europe and Asia also occurs on domestic poultry, and in their absence it will attack Man and often invades dwellings from nests in the eaves – especially when the fledglings leave the nest.

CERATOPHYLUS COLUMBAE   PIGEON FLEA

Although related to *C. gallinae* it is generally restricted to the domestic pigeon and to *Columba livia*, the Rock dove, from which the domestic pigeon is derived. It will, on occasion, attack Man.

LEPTOPSYLLA SEGNIS

Most usually this cosmopolitan species is found on *Mus musculus*, the House mouse, but it occurs on rodents generally. It rarely attacks Man.

TUNGA, NEOTUNGA   JIGGERS, SAND FLEAS

The mouthparts of the female in these genera are strongly developed. They attack Man and quickly burrow under the skin. In the case of *Tunga penetrans* the site most often selected is under a toenail. When the burrow is complete only the tip of the abdomen protrudes. Several thousand eggs are laid, scattered haphazardly wherever the host walks. The abdomen distends until, at the peak of egg production, it reaches the size of a small pea. This species is widespread throughout tropical Africa as well as Central and South America. At one time it was commonplace

Fig. 16.4 Pictorial key to some common fleas in the United States. (US Department of Health, Education and Welfare)

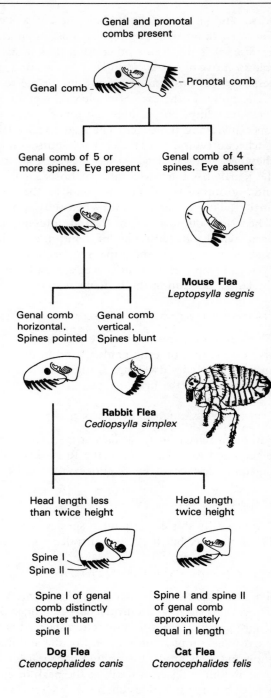

FLEAS

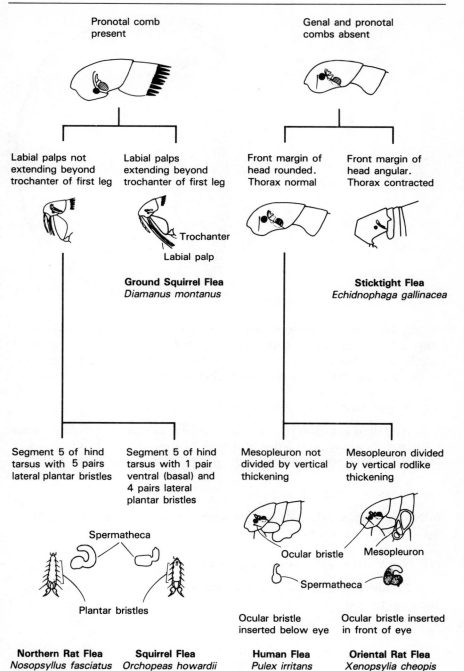

Fig. 16.5 *Tunga penetrans*. Expanded female removed from skin. (After Smit)

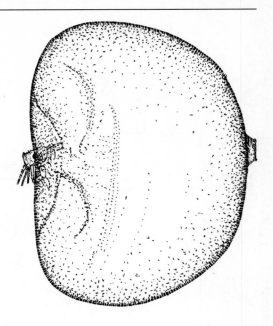

for shoeless, infested persons to work in a dwelling and so cross-infest others. This is much less common today. Inflammation due to *Tunga* infestations is known as *tungosis*, which has been responsible for the loss of toes and sometimes for the whole foot or leg.

NOSOPSYLLUS FASCIATUS   NORTHERN RAT FLEA

This American and European flea is generally associated with rats – *Rattus rattus* and *R. norvegicus* – but is also found on the mouse. It will attack Man. In 1926 it was found to be the most abundant flea in San Francisco.

ORCHOPEAS HOWARDI   SQUIRREL FLEA

In North America, this flea will sometimes attack Man when squirrels make their nests in attics.

CERATOPHYLLUS NIGER   WESTERN CHICKEN FLEA

This North American flea, normally attacking poultry, will transfer its attention to Man.

Chapter 17

Flies

DIPTERA

This important order of insects consists of the two-winged or true flies. It is one of the larger insect orders and the number of species described is probably at least 50,000. They are among the most highly specialised of all insects. Their principal character is the possession of only one pair of wings which are membranous. A pair of special knob-shaped organs known as halteres takes the place of the second pair. The head is a highly mobile and relatively large capsule. The mouthparts are adapted for sucking and sometimes, in addition, piercing also. Mandibles are seldom present but in many groups the labium is expanded to form a pair of fleshy lobes. The mesothorax is very large and to it is fused the small pro- and metathorax. There is a complete metamorphosis. The legless larva often has the head very much reduced and retracted into the thorax. The pupa is sometimes enclosed within the hardened larval cuticle.

The great majority of adult flies are either nectar feeders or are attracted to decaying organic matter whose juices they lap up. Some species are predacious, however, and either catch and eat other insects or parasitise them by laying their eggs on or in them, often the larval stage. In addition, a few species parasitise vertebrates, including Man, the larval stage of the fly taking place within the host. Very many other species suck the blood of living vertebrates, including Man, and it is this habit that has become of the greatest significance to human health. This is because the pathogenic organisms causing such diseases as malaria, sleeping sickness, elephantiasis, and yellow fever, are transmitted to Man through the blood-sucking habit of flies.

Classification

DIPTERA are usually divided into three suborders the first of which, the NEMATOCERA, contains the most primitive groups. The most highly specialised forms are placed in the CYCLORRAPHA while in the second suborder, the BRACHYCERA, specialisation is intermediate.

Nematocera

The most easily recognisable features of this suborder are the long, simple, antennae which are often thread-like and longer than the thorax. The antennal segments are mostly alike and an arista is not present. The maxillary palps are of three to five segments with but few exceptions. The larvae have well-developed heads with biting mandibles which operate horizontally and which are adapted for chewing. The pupa is active and free (e.g. the mosquito) and the adult escapes by means of a dorsal longitudinal slit in the thorax. Usually the adults are small and delicate.

Brachycera

Members of this group show a variety of antennal shape but the antennae are usually short mostly consisting of less than six (generally three) segments. The antennae are usually highly modified and often terminating in a bristle (arista). The palpi are one- or two-segmented. The larvae have an incomplete head with sickle-like mandibles for puncturing skin of prey – not adapted for chewing and they operate vertically. The free pupa is capable of some movement.

Cyclorrapha

In this group the antenna of the adult is three-segmented only, with an arista which is borne dorsally. The terminal segment is the larger. The palpi are only one-segmented. The pupa is not free but is formed without casting the last larval cuticle which hardens and envelops it. This is known as a puparium. The adult emerges through a lid-like aperture produced by a circular slit at the anterior end of the puparium. The head of the larva is vestigial consisting only of a pair of sclerotised mouth hooks (one of these may be atrophied). These articulate with an internal cephalo-pharyngeal skeleton. The mouth hooks scrape at the food vertically.

A highly specialised group of ectoparasitic DIPTERA is known as the PUPIPARA. They appear to fall within the CYCLORRAPHA, but the larvae are nourished singly in the female abdomen and are not extruded until fully grown. They then immediately pupate.

Various authorities have subdivided the DIPTERA in a number of different ways from that given above.

Nematocera

In this suborder a number of families are of the greatest significance to human health.

TIPULIDAE CRANE FLIES

Although a few cases of intestinal myiasis are known involving the larvae (leather-jackets) of this family, they are of little significance in buildings. Sometimes, especially after heavy rain, leather-jackets emerge from the soil in large numbers (see also *Lucilia*) and can become a nuisance underfoot on concrete surfaces, etc., and may invade buildings. Adults of some of the smaller species may be mistaken for mosquitoes.

PSYCHODIDAE

Only rarely exceeding 5 mm in length these flies, variously called moth flies, filter flies, owl midges or sand flies, are very hairy resembling small moths. They possess characteristic venation in which the veins appear almost parallel and the long fringes to the wings serve to identify them.

There are two subfamilies. In the PSYCHODIDAE, the moth flies, the proboscis is short and the female has a horny ovipositor. Species of *Pericoma* are said to bite Man. The larvae of *Psychoda* are greyish-white with hairy bodies, the segments of the body secondarily annulated and mostly furnished on the dorsal surface with sclerotised plates. The larvae feed in decaying organic matter and often serve a useful function when breeding in sewage beds but have been implicated on a number of occasions in intestinal naso-pharyngeal and urino-genital myiasis. In the other subfamily, the PHLEBOTOMINAE, the proboscis is long and the ovipositor is not horny. They are all smaller than 3–5 mm in length and the hairiness is not so apparent as in the PSYCHODINAE. They are usually referred to as sand flies and there is only one genus, *Phlebotomus*, in the subfamily. The legless and caterpillar-like larvae feed among the organic waste to be found in crevices in rocks and stone walls. They are greyish-white in colour except the head which is dark; there are a few short hairs on the body but the most characteristic feature is the two or four long bristles at the hinder end. The latter are visible also on the pupa as the larval skin is attached to the tip of the abdomen.

Sand flies

Adult *Phlebotomus* suck blood and are active at dusk and night. They are known vectors of the following diseases, kala-azar, oriental sore, papataci fever, phlebotomus fever, three-day fever, verruga peruviana and oroya fever (carrion's disease). The various species of *Phlebotomus* are separated only by minute differences and identification is a task for the specialist. Species of *Phlebotomus* are widely distributed throughout the tropical and warm temperate regions of the world.

Some workers give the PHLEBOTOMINAE family status.

Kala-azar

This chronic generalized disease which may affect any or all parts of the body is caused by the protozoan *Leishmania donovani*. It exists in two forms. In Man and some other vertebrates the parasite is oval, without flagella, and only about 3 μm in length. The other form (leptomonad) occurs in insects, chiefly sand flies of the genus *Phlebotomus* where it is spindle-shaped, 14–20 μm in length and with prominent flagella. *Leishmania donovani* dies quickly in soil or water and does not appear to occur outside its mammal and insect hosts.

Although it may occur at all ages, kala-azar is generally a disease of the young. It is widely distributed in many parts of the world although localised. It is very important in certain regions in China north of the Yangtse river and, particularly in Bihar, Bengal and Assam, in India. In southern Asiatic Russia it occurs around Tashkent. It is endemic in the Arabian peninsula as well as Turkey. The disease is found in various areas around the Mediterranean and in a band across Africa stretching from Eritrea, Kenya and Abyssynia, through Equatorial Africa to Nigeria on the west coast. It does not appear to be important in South America although cases have occurred in Argentina, Paraguay, Bolivia, Brazil, Venezuela and Colombia. In the United States the disease has been identified only in soldiers returning from endemic areas.

It is usually assumed that kala-azar is transmitted directly from man to man through the agency of *Phlebotomus*, but nevertheless, in China, dogs are known to be a reservoir of *Leishmania*. These parasites have been found in a dog imported into the United States from Greece.

The symptoms of the disease appear from a few weeks to as long as three years after exposure to the infection, but the actual incubation period is unknown. There is a sudden onset of fever, chills, dizziness and headache, although the disease is sometimes found accidentally in patients showing no symptoms. Sometimes the temperature remains high but in many cases there are daily swings. A number of other symptoms may be present but the enlargement of the spleen and liver is outstandingly important, and is consistent with the visceral and general endothelial location of the disease. In the absence of treatment, the course may last as long as two or three years, and there are often sequels, such as pigmented spots and persistent nodules.

Bartonellosis

This specific disease, caused by a minute rickettsia-like micro-organism named *Bartonella bacilliformis*, was originally thought to be separate infections known as carrion's disease, oroya fever and verruga peruviana. It is characterised initially by an acute, febrile stage in which there is anaemia, the parasite occupying erythrocytes. This is followed a few weeks later (although rarely, as long as a year later), by a nodular, cutaneous eruption. Either stage may be inconspicuous.

The most important regions where this disease occurs consist of deep valleys in the Andes in South America, at altitudes of from 700 to 3,000 m, mostly in Peru but also occasionally in Ecuador, Bolivia and Colombia. The probable vectors are *Phlebotomus noguchi* and *P. verrucorum*. These nodules or verrugas are usually up to 10 mm in diameter, but range up to 40 mm. They occur either singly or may be abundant and commonly erupt on head, hands, feet and lower arms and legs. Sometimes they are found in the mouth or throat. The nodules, which lie half-submerged in the subcutaneous tissue, are enveloped in a bluish epithelium which sometimes breaks down, resulting in ulceration, but healing usually occurs within a few weeks.

CULICIDAE MOSQUITOES

Without doubt the CULICIDAE or mosquitoes have been, and indeed continue to be, one of the greatest scourges of humanity. The females of a relatively large number of species bite Man viciously, suck his blood and cause much pain. Some species transmit the organisms causing malaria, a widespread and debilitating disease occurring in a number of forms. Other diseases of Man, generally tropical in distribution, are transmitted by a number of species, including yellow fever and related diseases, dengue and filariasis. More and more species are being found to harbour a number of viruses. Although in the period immediately following the Second World War control measures carried out on an enormous scale brought about a significant abatement of malaria in some countries, due to the ease with which mosquitoes can be carried unwittingly by aircraft, vigilance must be constantly maintained to prevent this. Many mosquito species enter buildings, some sheltering in dark, undisturbed areas, and others entering at night in order to suck their host's blood. A number of species are urban, breeding in watertanks, tin cans and other disposable containers.

Mosquitoes are small and long-bodied with a pair of long, narrow wings which are folded flat over the abdomen when at rest. The head is globular and the thorax is laterally compressed. The antennae of the female are hairy but those of the male are bushy. The mouthparts are long and thin and project forwards as a proboscis. There are six longitudinal veins in the wings the second, fourth and fifth being forked, and they bear small scales which are present also along the hind margins of the wings forming a fringe. The immature stages of mosquitoes are aquatic, the larvae and pupae breathing atmospheric air through spiracles, those of the larvae rising from the hinder end of the body and those of the pupae from the thorax. The head of the larva is well developed but the enlarged thorax is unsegmented and some of the abdominal segments bear tufts of bristles.

Principal mosquito groups

There are two principal groups of mosquitoes: the culicines and the anophelines. By far the larger number of species belong to the former group which bite readily and painfully. Although none transmits malaria, they can carry the causative factors of yellow fever, filariasis and dengue. They may be easily separated from the other principal group, the anophelines, at the different life stages, as follows:

	Culicines	*Anophelines*
Eggs	1. Air floats absent	1. Provided with lateral air floats
	2. Sometimes massed together in rafts	2. Laid individually
Larvae	1. Hind spiracles at end of siphon, borne on eighth abdominal segment	1. Hind spiracles on surface of body
	2. Hang downwards at an angle with only end of siphon in surface film	2. Rest horizontally beneath water film supported by float hairs
Pupa	1. Respiratory trumpets cylindrical	1. Respiratory trumpets conical
Adult	1. Females with short palps	1. Females with palps as long as proboscis
	2. Abdomen covered with scales	2. No scales on abdomen
	3. Rests with proboscis and abdomen forming obtuse angle with abdomen approximately parallel with surface	3. Rests with proboscis and abdomen parallel but at an angle with the surface

CULICINI

This tribe consists of a large number of genera and species, but only three species (one of which exists as two subspecies) are of importance as disease vectors, although many are painful biters of Man. Brief notes on those species of medical importance are given below.

CULEX PIPIENS FATIGANS

In appearance it is very similar to *Culex pipiens*, the abundant and widely-spread mosquito of the North Temperate region and parts of the African continent. It may be distinguished, however, by the wing venation and by the colour of the thoracic scales which are not as reddish as those of *C. pipiens*. In addition, the male palpi of *C. pipiens* are much longer than the proboscis but in *C. pipiens fatigans* they only exceed the length of the proboscis by the length of the terminal segment. In the larvae the siphon of *C. pipiens* is evenly tapered but that of *C. pipiens fatigans* is convex, being thickest at about one-third of its length from the base. The larvae of *C. pipiens fatigans* are found in all types of containers which can hold water, even temporarily found around human habitation from watertanks to tin cans. It breeds successfully in heavily polluted water such as drainage from manure heaps and cesspits.

Occurring throughout the tropical and subtropical regions this mosquito is the principal vector of bancroftian filariases or elephantiasis, a disease of Man brought about by nematode worms of the species *Wuchereria bancrofti*. *Culex pipiens fatigans* is not involved in the transmission of either yellow fever or dengue.

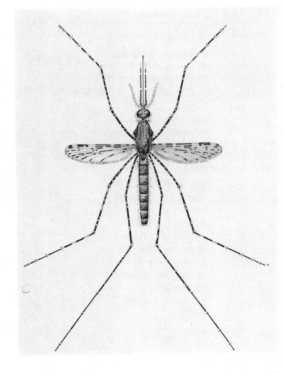

Fig. 17.1 *Anopheles gambiae*. Feeds mostly indoors at night and is very important transmitter of malaria and important filarial worm carrier. Known to transmit a virus disease. Widely distributed. (J. Smith in Gillett 1971)

Fig. 17.2 *Aëdes africanus.* Responsible for maintenance of yellow fever virus in populations of forest monkeys, and occasionally transmits it from monkey to Man as well as other viruses. Occasionally breeds in domestic containers. West and Central Africa. (J. Smith in Gillett 1971)

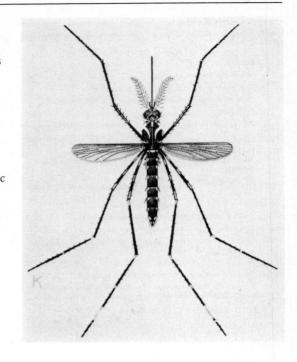

Fig. 17.3 *Culex pipiens fatigans.* The commonest nuisance mosquito in urban Africa from Khartoum to Capetown, feeding on Man inside and outside houses. Principal transmitter of the filarial worm *Wuchereria bancrofti*. Breeds in drains, old motor tyres and domestic water containers. Africa apart, occurs throughout much of the tropics and subtropics. (J. Smith in Gillett 1971)

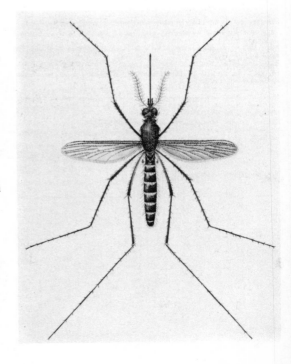

AËDES

Aëdes aegypti can usually be identified by the characteristic 'lyre-shaped' marking on the top of the thorax. It is somewhat variable in colour. The single eggs are laid on wet surfaces but they can withstand extraordinary desiccation, even for many months. Breeding normally takes place in clean rainwater and they are found in typical urban situations. There are a number of varieties described including *v. luciensis* from W. Africa and *v. queenslandsis* from Australia.

The *A. scutellaris* complex occurs in the eastern Malaysian archipelago, as well as in the islands of the Pacific. It has been found elsewhere in the same general area but is not known from India or Australia. It is a transmitter of *Filaria bancrofti*. It has a similar pattern on the thorax as shown by the following species (*A. albopictus*) but there are white longitudinal stripes on the sides of the thorax instead of white patches. This species breeds in small containers as found in urban rubbish such as bottles and tins but also in cavities in palms and vegetable debris such as coconuts. There are a number of described varieties from various of the Pacific islands. *Aëdes aegypti* which is synonymous with *A. calopus, A. argentus* and *Stegomyia fasciatus*, is distributed throughout the tropical and subtropical areas of the world. It is absent from Japan, New Zealand and some Pacific islands, but is found around the Mediterranean. It is an important transmitter of yellow fever in West Africa, Central and South America and the Caribbean, and is responsible also for the transmission of the virus of dengue fever throughout the tropical areas.

Aëdes simpsoni is the main man-to-man vector of yellow fever.

AEDES ALBOPICTUS

This mosquito occurs in the oriental and Australian regions, as well as in Madagascar and nearby islands. It breeds in holes in trees and rocks, and from time to time in water-butts, old tin cans and similar cavities near human dwellings. In South-East Asia, *A. albopictus* is the principal agent for dengue transmission, but does not take part in the transmission of filarial worms. However, this species, as well as a number of other culicines, has been shown experimentally as being capable of 'carrying' yellow fever virus.

Filariasis

This is a disease brought about through an infection by one or more species of thread-like nematode worms. The adult stage of the worm is passed in the human body where the female produces very small, elongate embryos, known as microfilariae. The latter periodically migrate or constantly move through the peripheral blood vessels, from which situation they are sucked up by a certain species of blood-sucking fly.

Inside the insect the microfilariae develop into motile larvae which move into the proboscis sheath of the fly, and become infective. When the fly takes its next blood meal the worm larvae are introduced into the puncture or are deposited close to it. These then puncture the skin and migrate into the localities of the body via the lymphatics and blood vessels, where they will develop. About a year is taken for the worm to mature and produce microfilariae; this is known as the prepatent period.

Bancroftian filariasis

This type of filariasis is brought about by the nematode *Wuchereria bancrofti*. It inhabits the lymph vessels associated with the legs, arms, genitalia and breasts. Blockage of the lymphatic drainage by the parasites leads to the condition known as elephantiasis, shown by a gross enlargement of the limb or organ concerned. The mosquito *Culex fatigans* is probably the chief insect vector, but species of *Anopheles*, *Aëdes* and *Mansonia* are also involved in its transmission. The adult worm lives for about five years in the human body. The microfilariae show a nocturnal periodicity in their migrations to the peripheral blood system where they are found in the greatest numbers between 10 o'clock in the evening and 2 o'clock in the morning. During the daytime the microfilariae are mostly found in the viscera, more especially in the lungs.

Filarial infection can be broadly divided into three types. Children in endemic areas are exposed at an early age. Microfilariae are present in their blood at the age of six. Symptoms are not experienced. When the adult worms die and the microfilariae are no longer present, the only indication is the moderate enlargement of the inquinal lymph nodes. The patient may never be aware of his condition. One-fifth of a group of men from the Caribbean area, passed as fit for military service during the Second World War, were found to be infected with filariae. This type is known as asymptomatic filariasis.

In *inflammatory filariasis* allergic reactions to the presence of living and dead worms take place. This is due to sensitivity to their products, and occasionally streptococcal and fungal infections may be superimposed. This may occur from 1 to 22 months after exposure to infection. There are localised areas of swellings and redness of the arms and legs which may be accompanied by fever, chills, headache, vomiting and malaise, and may continue for several days to several weeks. The affected extremity usually becomes red and hot, as well as very painful.

Elephantiasis, the dramatic but not the inevitable end result of filariasis, is the third or obstructive type. It is slow to develop, often after years of continuous infection, and is of variable incidence. From 1 to 70 per cent of those infected reaching this conclusion to the disease, but this is after years of chronic oedema and repeated inflammatory attacks.

Filariasis malayi

This disease is caused by the parasitic nematode worm *Wuchereria malayi* which resembles *W. bancrofti*. The adult worms live in the lymphatic system of Man, and the microfilariae exhibit nocturnal periodicity in migrating to the peripheral circulatory system, although some remain there during the day. Mosquitoes of genera *Mansonia* and *Anopheles* are the vectors. The symptoms of the disease are similar to those of bancroftian filariasis and elephantiasis is common, but is mostly manifested in the legs, being rare in the genitalia. As the larvae of *Mansonia* derive their oxygen from the tissues of aquatic plants the eliminatian of the latter exercises a considerable degree of control.

Yellow fever

This is an acute disease caused by a virus and which is characterised by its sudden onset, prostration, moderately high fever and a pulse rate which is slow in relation to temperature. Yellow fever occurs in a number of degrees of virulence. Some attacks are so mild as to be experienced almost without incident. On the other hand, when the attack is severe, accompanied by jaundice and 'black vomit', there is then a high risk of death as the end result. Yellow fever is endemic in the areas of tropical rain forest of Africa and south America, but the disease has extended its range by epidemics into temperate areas in summer time.

There are two types of the disease. When it is transmitted from man to man by the bite of the mosquito *Aëdes aegypti* in an urban environment, the disease is known as urban yellow fever. When, however, the disease occurs in a forest environment, transmitted by a forest mosquito (a species other than *A. aegypti*), then it is known as sylvan or jungle yellow fever. In Africa the mosquitoes *Aëdes simpsoni* and *A. africanus* have been found to carry the virus in the wild, the former species being the most important man-to-man vector. In South America, wild-caught mosquitoes of the genus *Haemagogus* and *Aëdes leucocelaenus* have been identified as carrying the virus, the former genus being more important. These are forest canopy-inhabiting mosquitoes and they infect monkeys living in the same environment, although the exact role played by such monkeys in transmission of the disease to Man is not clear, but it would seem probable that they act as an endemic reservoir of the virus. During forestry or road-making operations, canopy-inhabiting and infected mosquitoes bite Man and thus transmit the sylvan type of the disease.

In areas in Africa adjacent to the rain forest where sylvan yellow fever is endemic, epidemics of urban yellow fever are frequent. A severe epidemic occurred in Nigeria in 1946. In Central America an epidemic of sylvan yellow fever has been travelling northwards, originating in Panama in 1948.

The yellow fever virus is classified as a group B arborvirus and is small, being only 17 to 25 mm in diameter. It is able to damage or destroy liver, kidney and heart tissue, as well as the central nervous system. Most yellow fever attacks are mild, exhibiting few of the well-known symptoms. Often there is only a low fever and accompanying headache of short duration. There is first an incubation period lasting from three to six days, then usually a sudden chill. This is followed by fever, severe headache, backache with leg pains, and prostration. Photophobia occurs and the face is flushed. The tongue is bright red both at the tip and the edges. It is at this stage, which lasts about three days, when a bite from *A. aegypti* will infect the mosquito with the virus and it can be transmitted to another human host. In control measures, therefore, it is of the utmost importance to prevent mosquitoes from biting the patient at this early stage of the disease.

Jaundice is absent during this infective period, but then the temperature rises quickly to 40 °C and even higher. The pulse increases to 90 to 100 at first, then slowing in relation to temperature, but it is full and strong. There is nausea and vomiting and epigastric distress and tenderness. There follows a period of intoxication after a remission in the fever which, however, may be indefinite or absent. There is sometimes a deceptive temporary improvement.

Dengue

The causative agent of this disease is a pair of antigenically distinct viruses belonging to the family of group B arthropod-borne or 'arbor' viruses. Dengue is characterised by a sudden rise in temperature with which chills are associated and these are accompanied by localised and general aches and pains. There is excruciating headache with retro-ocular pain, especially when the eyeballs are moved, and photophobia. This is accompanied by backache with pain in muscles and joints. Although the face is flushed and the skin mottled there is no distinct rash. For five or six days there is a fever with temperatures from 103 to 105 °F. Rarely, the temperature returns to normal then rises again. Generally the patient complains of intense discomfort, weakness and depression.

Dengue virus is transmitted by four species of mosquitoes; *Aëdes aegypti*, *A. albopictus*, *A. scutellaris* and *A. Polynesiensis*. After feeding on infected human blood there is an incubation period of from 8 to 14 days before the mosquito can transmit the disease, but then it remains infected for the rest of its life. One bite only from an infected mosquito is sufficient to cause the onset of the disease. The disease is endemic in the southwest Pacific islands, Queensland and New Guinea, Indonesia, the Philippines, parts of India, Malaya, Burma, Indo-China and probably parts of Africa. Newcomers to these countries are at high risk of contracting the disease unless exceptionally strictly-enforced precautions are taken to prevent being bitten by mosquitoes. A certain degree of immunity to reinfection occurs.

During this century a number of epidemics of the disease have occurred, notably on the Gulf coast of the United States, Greece, Australia and Japan. As high a proportion as 40 to 80 per cent of the total population of cities has been affected. Such epidemics are obviously due to the virus being introduced into a non-immune population at a time when the appropriate insect vector is abundant. Dengue is not a lethal disease.

Anopheline mosquitoes

ANOPHELINI

The adults of anopheline mosquitoes can be separated from culicines by the absence of scales on the abdomen, at least on the sternites, by the palps of the male being clubbed at the tip and the palps of the female usually being about as long as the proboscis. The differences between the two in the immature stages and the resting position of the adults are given. Identification of mosquitoes is usually a task for an expert, and keys for this purpose are provided by P.F. Mattingly (1969).

A number of anopheline species are the carriers of malaria parasites and are thus of supreme importance to human health over a large part of the world. The species implicated in the Old and New World are given later.

Mosquitoes of North America

The important common mosquitoes of the North American continent include the following.

CULEX PIPIENS

This introduced cosmopolitan species occurs in the northern states of the United States as well as Canada. It is referred to as the Northern or Common house mosquito and breeds in tin cans, rainwater barrels, flatroofs, sewers, ditches and cesspools. The fertilised females often hibernate in large numbers in street drainpipes, cellars and similar urban locations. It bites at dusk and after dark.

CULEX QUINQUEFASCIATUS

Very similar in appearance and habits to *C. pipiens*, this mosquito is abundant in the urban areas of the southern states where it is a persistent night-biter. The ranges of the two species overlap in Virginia, northeastern Tennessee, North Carolina and intermediate states.

AËDES AEGYPTI

Known in the United States as the yellow fever mosquito, it occurs south of a line from southern Virginia through northern Oklahoma and then southward to El Paso, Texas. It is found also in Arizona and New Mexico but is absent from the Pacific coast. It breeds in relatively clear water in a large number of urban situations, such as cisterns, jars, bottles, old motor tyres, flower vases and urns in cemeteries, unused toilet bowls, water pans put out for poultry and even in dishes of water in which the legs of refrigerators are placed. Its biting activities usually take place in the early morning or late afternoon.

ANOPHELES QUADRIMACULATUS   COMMON MALARIA MOSQUITO

This is the most important malaria carrier in southern, central and eastern United States. It breeds in fresh water containing vegetation. Although some blood meals are taken in daylight, most are taken at night. There are said to be as many as 10 generations annually in the southern states and the fertilised females hibernate during winter in such locations as stables, cellars and similar situations. During warm weather however, they will emerge to take a blood meal.

AËDES SOLLICITANS   EASTERN SALT-MARSH MOSQUITO

This important species occurs along the Atlantic and Gulf coasts where salt-marshes occur.

AËDES TAENIORHYNHCUS   BLACK SALT-MARSH MOSQUITO

This species is widely distributed throughout the United States but is more abundant in the south.

ANOPHELES FREEBORNI

This is the principal vector of malarial parasites on the Pacific coast.

CULEX TARSALIS

This species is the carrier of the viruses causing the diseases St Louis encephalitis and western encephalitis.

CULISETA MELANURA

This is probably the most important carrier of eastern encephalitis, but it is also carried by *Culex pipiens* and *C. quinquefasciatus*.

Malaria

This infectious and febrile disease is one of the chief scourges of humanity and a substantial proportion of the earth's inhabitants have suffered in the past from its debilitating influence. Even today, when the disease has been eliminated from many regions, and the World Health Organisation has undertaken to control this pernicious health problem, when so much of the biology of the causative parasites is known, when so much success has attended the programmes of drug research, there is still much human suffering on its account and large areas of fertile earth remain uncultivated. Malaria often travels hand in hand with malnutrition, each making the other of more consequence. Breaking this vicious bondage will release an incalcuble amount of human happiness.

Causative organisms
The disease of malaria is brought about by four closely-related species of protozoan parasites of human red blood corpuscles. These are species of a single genus *Plasmodium*, namely *P. vivax, P. malariae, P. falciparum* and *P. ovale*. Protozoa are animals consisting of a single cell and the genus *Plasmodium* is classified in the order HAEMOSPORIDIA and the class SPORIDIA. The morphology of the four species is distinct as is their pathogenicity. So far *Plasmodium* spp. have not been artificially cultivated outside their living host.

Transmission
In nature, transmission from man to man is solely by way of the bite of anopheline mosquitoes which play an essential part in the life-cycle of the *Plasmodium*. It is by no means to be considered as a mere mechanical transporter of the parasite from man to man.

In 1959 there were 318 described species in the subfamily ANOPHELINAE and a further 93 subspecies and varieties of these were recognised. A revised *World Catalogue*, now in preparation, will certainly increase this number, possibly by some 20 per cent. For many, their susceptibility to infection by plasmodia has been elucidated, as well as their general importance in malaria transmission and their effect on the incidence of disease in the particular area.

Malaria has also been artificially transmitted through the use of a hypodermic needle, by a drug addict transferring infected blood to another person. Many deaths have occurred by this means. The disease has also been contracted by infected blood being donated in hospital.

Parasite development in the mosquito

For the female mosquito to become infected it must suck blood from an infected human host, and must suck up male and female malarial parasites. The asexual parasites do not survive in the stomach of the mosquito but the sexual organisms known as the gametocytes commence a development cycle there. Fertilisation first takes place and is carried out by the male gametocyte protruding several flagella which break away and come into contact with the female gametocyte. One of the male gametes succeeds in penetrating the female, after which the others retreat. The fertilised cells then make their way through the stomach cells to form cysts or oocysts on the outer wall. They gradually become larger until they mature from the eighth to the tenth day when they are about 75 μm in diameter.

At this stage the oocyst is seen to be full of spindle-shaped cells which are known as sporozoites, and they are released into the body-cavity of the mosquito by the rupture of the oocyst. The sporozoites then make their way to the mosquito's salivary glands where they remain until the mosquito bites its human host when they are injected into the blood system of Man.

Parasite development in Man

Little is known of the early life of the sporozoite after it enters the blood system of Man, and the way in which it attacks the red blood corpuscles (erythrocytes), but in the case of *Plasmodium vivax*, which causes tertian malaria, about 10 days after the introduction of the infection, the 'ring form' of the parasite may be observed. It appears inside an erythrocyte as a ring of cytoplasm with abundant red chromatin. It then develops into the amoeboid stage in which the cytoplasm is arranged irregularly, pigment granules appear and the chromatin is enlarged. The parasitised erythrocyte enlarges, becomes paler, and numerous reddish granules appear (but only in the case of *P. vivax*). The parasite continues to increase in size, together with the quantity of pigment and then the nucleus divides. After 48 hours the interior of the erythrocyte has been consumed and replaced by a rosette of 16 daughter parasites known as merozoites. On rupture of the erythrocyte membrane they escape, and after attaching themselves to another erythrocyte they parasitise it and the cycle is repeated every 48 hours. About 90 per cent of the merozoites, however, are killed by the macrophages in the spleen, liver and bone marrow. The asexual reproductive cycle continues for a number of days, then sexual cells (gametocytes) appear. The male and female gametocytes may be differentiated by staining techniques. The gametocytes appear only in small numbers and persist in the blood for two or three weeks. They are concerned only with the transfer of the parasite to the mosquito.

Plasmodium malaria

This species is responsible for quartan malaria which is similar in a number of respects to *P. vivax* but each asexual cycle takes 72 hours and a 'band' form of the parasite is often observed. Other differences are that the parasitised erythrocyte becomes smaller and the haemoglobin more concentrated. Generally there are only eight merozoites produced at a time, and after being set free they appear to parasitise more mature erythrocytes than do those of *P. vivax*.

Plasmodium ovale

Described as recently as 1923 as a separate species, its chief distinction lies in the fact that the parasitised erythrocyte is oval in shape. Otherwise it is similar to *P. vivax*.

Plasmodium falciparum

The asexual cycle shows first as a small 'signet ring' in the erythrocyte only about half the size of that of *P. vivax*. The more mature stages of the sporozoite are observed only in the case of severe disease. This is because when the infection is acute the ring forms aggregate in the blood capillaries in various parts of the body where they complete their cycle. The time for a complete cycle is probably between 24 and 48 hours. The crescent-shaped gametocytes, however, confirm this species beyond doubt. Young and old erythrocytes are attacked by this parasite. The greater number of deaths due to malaria are due to infection by this species which generally involves the brain.

Symptoms of malaria

Each of the four species of parasite produces different symptoms of the disease apart from the varying periods of sporulation. Tertian (*vivax*) malaria has been the most carefully studied due to its induced infections used in the treatment of neuro-syphilis.

The 'sweating' stage commences abruptly with profuse perspiration and the temperature drops to normal. The headache goes and although drowsy and weak the patient feels well enough to resume work.

Quartan malaria is characterised initially by a lengthy incubation period which may be from 18 to 40 days. The disease begins in a similar manner to *vivax* malaria. The paroxysms occur every 72 hours, this is the period required for the completion of the asexual cycle of the parasite in the blood. The name 'quartan' for this malarial type is due to the counting from the beginning of the day of one chill to the end of the day of the next paroxysm being taken as four days. The quartan paroxysm lasts longer and is slightly more severe than that of *vivax*.

As already mentioned, *falciparum* is the most severe of the malaria types. The incubation period is 12 days and although the paroxysms are similar to those of other malaria types they are irregular and of longer duration. The localisation of the parasites causes a number of different severe symptoms. That of the brain has been referred to but other organs may be involved.

Six days or so following the mosquito bite, mild backache, muscle soreness and a low-grade fever without chillness, occur. It is not until about the fourteenth day that febrile paroxysms occur and they then occur on alternate days coinciding with the parasite segmentation. However, they may occur daily (quotidian), showing that two distinct broods are present, one segmenting on even days and the other on odd days. In *vivax* malaria the paroxysm shows 'cold', 'hot' and 'sweating' stages. The 'cold' stage commences with the patient feeling chilled over the whole body. This increases in intensity until the teeth chatter and a shaking of the whole body becomes uncontrollable. The skin is blue and cold and even though hot blankets are applied the patient cannot get warm. The pulse becomes rapid and weak. This stage continues for about an hour when the body temperature rises.

The 'hot' stage commences with a flushed face then a headache becomes severe and the temperature may rise to as much as 42 °C. The patient may be somewhat delirious and experience a sensation of intense heat, and the skin is hot and dry. This stage lasts for about two hours.

Blackwater fever

Blackwater fever occurs as a frequent complication of *falciparum* malaria after the patient has suffered a number of attacks of it. The consequences are very serious, the mortality rate being about 50 per cent. The urine becomes very dark in colour and there is a heavy amorphous deposit. The fever is high and the temperature reaches 40.5 °C or even higher, and the pulse is rapid. There is vomiting and jaundice.

CERATOPOGONIDAE BITING MIDGES, SAND FLIES

The genus *Culicoides* which is almost worldwide in distribution contains a number of species whose bites are painful and irritating. These midges are very small, being from 1 to 3 mm in length and the females have elongated, biting mouthparts with which many species can cause annoyance as to render some areas virtually uninhabitable. The larvae are found in humus-rich mud and vegetable debris, and are most abundant around tidal marshes. In Britain the following species are serious pests, the first-named being by far the most abundant: *C. impunctatus, C. obsoletus, C. pallidiconnis, C. pulicaris,* and *C. heliophilus.* The well-named *C. vaxans* is an annoying pest in southern England.

In the United States a number of ceratopogonids are classed as severe pests. In South Carolina, *C. canithorax, C. melleus* and *C. dovei* are mentioned. In Arizona and New Mexico, *C. stelifer* is an important pest and in the western United States *Leptoconops torrens,* known as the valley black gnat and *Holoconops kerteszi,* the Bodega black gnat, are of great concern. The biology of *L. torrens* is remarkable. In California the larvae occur in clay adobe soils at a depth of 0.5 to 1.0 m, entrance and egress being dependent on the drying and cracking of the soil. The duration of the larval period is at least two years and the summer is spent in immobile aestivation. If the soil does not crack the larvae enter diapause and this may be as long as three years.

In Australia a number of species of *Culicoides, Leptoconops* and *Styloconops* cause severe annoyance.

The bites of ceratopogonids cause irritation which may last in extreme cases for several weeks and commonly for a number of days. The severe aggravation leads to scratching with consequent bacterial infection. Some are known to transmit viruses to animals, viz. blue-tongue in sheep, and may prove to transmit disease to Man.

CHIRONOMIDAE MIDGES

These delicate and fragile flies resemble mosquitoes but they may be identified by the short proboscis, and the bare or hairy, never-scaly wings. They are non-biters. With few exceptions the larvae are aquatic and a small number are marine. Some are terrestial. The adults are usually observed in large numbers around water and form mating swarms. They are sometimes present in such extraordinary numbers in places such as public parks as to cause considerable annoyance, and may be a nuisance when buildings are being painted.

In such numbers they have caused fire alarms to be raised because of their resemblance to smoke, and they may also cause allergic reactions in the form of asthma, though this is usually due to the wind-borne remains of their exuviae. Fishermen in the African lakes have been suffocated in these huge swarms.

The blood-red, worm-like and wriggling larvae are sometimes found in drinking water, and cause some alarm. This is when the larvae accidentally pass through filters which have been damaged. They are harmless.

CHAOBORIDAE PHANTOM MIDGES

They are attracted to light in such abundance that they sometimes get into the eyes, nose, mouth and ears. The tourist trade can be seriously affected in some holiday resorts, but in such areas funds are usually made available for their control.

Chaoborus astictopus, Clear Lake gnat, of northern California is an example of this, but other species are involved in various parts of the world.

These midges resemble mosquitoes but the proboscis is short, and although the wings bear scales they are almost confined to the hind margins of the wings. Very few or none are borne on the veins. Their importance is derived from the very large numbers which occur locally.

The life-cycle of ceratopogonids varies widely, but generally the cigar-shaped eggs are ridged and are from 0.3 to 0.5 mm in length. The larvae are eel-like, lacking appendages of any sort and while the head is oval and light brown, the body is of a dull, white translucence. Spiracles are absent so that respiration occurs only through the cuticle. The feeding habits are not known precisely, but ALGAE and PROTOZOA are thought to constitute their main food. When fully fed, the larvae seek drier conditions for pupation. The pupa bears a pair of respiratory trumpets similar to those of mosquitoes.

SIMULIIDAE BLACK FLIES, BUFFALO GNATS

Simuliids are not directly associated with buildings but their importance to mankind makes some notes about them essential.

Simuliids carry the parasitic worm *Onchocerca volvulus* which gives rise to the disease onchocerciasis. Simuliid flies are characterised by being somewhat hump-backed in appearance and having broad, rounded wings with pronounced anal lobe and devoid of marks, hairs or scales. They are generally small in size, varying from about 2 to 6 mm in length. The stout body and relatively shorter legs also serve to separate them both from mosquitoes and midges. This family of biting flies is found practically throughout the world, and is of immense importance not only from the aspect of the extremely large numbers which abound but for the role they play in transmitting parasitic worms in a number of tropical areas.

Onchocerciasis   River blindness

The causal agent of this disease is a filarial worm called *Onchocerca volvulus*. The thread-like adults live coiled up in fibrous, subcutaneous tissue, forming nodules which may be as small as a peas to the size of a hen's egg. Some adult worms do not encapsulate to form a nodule but wander freely in the subcutaneous tissue. In the Old World the nodules are commonly encountered on the trunk, thighs and arms, but in the

New World they most usually occur on the head and shoulders. The parasites are widely distributed in Africa and have been found in south Arabia but in America they are confined to the coffee-growing uplands of the Pacific slope of Guatemala and Mexico and also an area in eastern Venezuela.

The males are about 45 mm in length but the females may reach 700 mm and the latter produce large numbers of microfilariae which are found in the nodules and more generally in the skin. Although the nodules cause pain and there are associated itching rashes and dermatitis, the importance of the parasitisation lies in the serious involvement with the eyes. Impairment of sight affects a variable proportion of human hosts according to locality, but may be as much as 85 per cent.

Infection of the eye becomes apparent with photophobia, irritation and conjunctivitis. A number of complications follow with the thickening of the iris and its adherence to the anterior surface of the lens. The eyeball atrophies and when the microfilariae enter the optic nerve sheath, complete blindness is the result.

Apparently simuliid flies are the only transmitters of the microfilariae of *O. volvulus* to Man and it is estimated that nearly 20 million human beings are involved in this disease.

Life-cycle
All species breed in running water and the immature stages are specially adapted for this type of habitat. In the case of *Simulium venustum*, the female crawls underwater to lay her mass of yellow eggs on a subaquatic substrate. The head of the larva is well developed with pigmented eye-spots and small antennae. The mouthparts are specially elaborated as two bunches of bristles which can be opened and closed like a fan.

The abdominal segments are larger than the thoracic, giving the larva a skittle-like shape. There is a proleg on the thorax and a sucker at the end of the abdomen, and there is a circlet of very small hooks around each. Locomotion is effected by looping along like a caterpillar, using the hooks alternately for fixation. Sometimes a silken thread is used as a guideline. Respiration is through a bunch of anal gills. Some species show an unusual example of phoresy where the larvae attach to the cuticle of freshwater arthropods, such as river crabs, and move about by this means.

The larva spins a pointed cocoon and pupates within. The pupa is remarkable on account of the pair of tufts of respiratory filaments as large as itself. When the fly emerges it floats to the surface in an air-bubble.

In Queensland, Australia, *Austrosimulium pestilens* is a vicious biter of Man, as is *S. indicum* (the Potu fly) in the Himalayan region. *Simulium columbaczense* often occurs in vast numbers along some parts of the Danube, but they are frequently scattered long distances by the wind.

ANISOPODIDAE WINDOW GNATS

The adults are gnat-like but possess no transverse suture on the dorsum (the upper surface of the thorax). They are prone to swarming in large numbers. The larvae of this small family feed on decaying organic matter and few species are of importance to Man in his buildings. In Britain, however, *Anisopus fenestralis*, the Window gnat, often causes concern due to the exceedingly large numbers produced from sewage works during the summer season. Normally this species, together with a number of psychodids, is beneficial in that they help in the 'digestion' of sewage. From time to time, however, extraordinary numbers of the flies invade dwellings in adjacent areas and often get into eyes, nose and mouth, causing great annoyance although there is no evidence that disease pathogens are thereby transmitted. The larvae have been rarely involved in cases of intestinal and urino-genital myiasis.

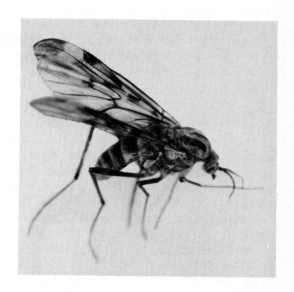

Fig. 17.4 *Silvicola (Anisopus) fenstralis*, Window gnat.

Suborder brachycera

Only the family TABANIDAE is of any importance to health, and no species has association with buildings.

TABANIDAE HORSE FLIES CLEGS

This large family of some 3,000 species of robust, moderate to large-sized, flattened flies is widely distributed thoughout the world.

The bite of female tabanids is exceptionally severe but those of temperate regions are not implicated in the transmission of human disease pathogens. A few tropical species, however, are important carriers of parasitic, filarial worms, including *Loa loa*. They only rarely occur indoors, but may occur in some numbers about buildings close to water, e.g. at ferries, etc. In Europe, *Haematopota pluvialis*, the Cleg, is generally abundant and gives a sharp, painful bite after an unnoticed approach, but never indoors.

Tabanids usually are sombre in colour but the large, laterally extended eyes are often iridescent with bright metallic colours. The mouthparts of the females are adapted for piercing skin and for sucking blood and other liquid tissues, while those of the male only possess the function of lapping up nectar or plant sap. The length of the 'proboscis' varies widely within the family. The wings likewise vary, those of *Chrysops* being banded brown or black, those of *Haematopota* being mottled with grey while those of *Tabanus* are clear.

Life history
The spindle-shaped eggs, about 1.5 mm in length, are laid in vertical batches, sometimes stacked in two layers and are usually placed very near to water or on marshy ground. The larvae are aquatic or live in wet earth. They are long, cylindrical with a small retracted head and consist of 11 segments. The first seven of the abdominal segments bear a ring of fleshy pseudopods near the anterior margin which are usually eight in number. The cuticle is marked with longitudinal striations and sometimes shows dark-coloured transverse bands. A siphon bearing a pair of spiracles is sometimes present at the tip of the abdomen. With some exceptions (some species of Chrysops), the larvae are predacious on insect larvae, worms and small crustacea. The pupal abdominal segments bear bristles and there are six large spines at the posterior end.

Loaiasis   African eye worm

This disease is caused by the presence of a worm *Loa loa*, 30 to 50 mm in length, which travels widely throughout the human body. Loaiasis develops very slowly, the first symptoms generally appearing as long as three years after infection. The worm is transmitted through the bite of *Chrysops dimidiata*, known as the Mango fly. Severe irritation is caused and swellings appear at various parts of the body and these are about hen's egg size (calabar swellings). They are painful but after several days they disappear only to appear in another place. Skin-reddening and dermatitis are usually also present, probably due to reaction to the worm's products of excretion. Deeper pains occur in various parts of the body from time to time, possibly indicating the parasite's migrations, but one of the most painful areas is the face, especially beneath the conjunctiva. Surgical removal of the worm from the area is commonly and effectively employed.

Microfilariae are produced in numbers and around midday they migrate to the peripheral blood vessels. It is at this time that the bites of *Chrysops dimidiata* are most numerous. About 13 million people are thought to be infected. The worm is known to live in its human host as long as 17 years.

SCENOPINIDAE   WINDOW FLIES

Like *Anisopus*, these are commonly found indoors on windows, and are also known as window flies but may be distinguished at once by their black colour, robust appearance and pendulous antennae. The larvae are predacious on the larvae of clothes moths and dermestid beetles. They have been involved rarely in naso-pharyngeal and urino-genital myiasis but can generally be considered as beneficial.

STRATIOMYIDAE   SOLDIER FLIES

HERMETIA ILLUCENS

This fly breeds in a variety of substances, from over-ripe fruit to faeces and cadavers. Its occurrence in privies renders the faecal matter too liquid for the optimal development of house flies. It is occasionally involved in cases of intestinal myiasis and the adults may sometimes appear in small numbers indoors. This is an American species which has become widely distributed via Man's commercial transport, and is now found in Australia, Europe, Africa and Asia. For specific identification stratiomyids require expert examination.

Cyclorrapha

PHORIDAE

The position of this family in the classification of the DIPTERA has been in doubt but they are now generally placed in the ASCHIZA section of the CYCLORRAPHA. The adults are small and sometimes minute, with the anterior veins highly developed and pronounced. They run with agility and exhibit a peculiar hump-backed appearance. The larvae generally feed on decaying organic matter and faecal deposits, though some are parasitic in various invertebrates. They are greyish-white in colour and may be up to 5 mm in length. The integument is covered with bands of small papillae and near the hind end there is a pair of sclerotised processes and the posterior spiracles are situated at their tips. The larva pupates inside the larval skin and the puparium is brown. The hind end is transverse and there is a pair of horn-like respiratory processes on the second abdominal segment.

Puparia of some species, particularly *Paraspiniphora bergenstammi*, are sometimes found in inadequately cleaned milk bottles, and larvae as well as puparia and even adults, have at times passed through the human alimentary canal. (See also *Drosophila*.)

SYRPHIDAE HOVER FLIES, DRONE FLIES

This family contains a large number of species. The adults are usually moderately large and brightly coloured, often mimicking wasps with their pattern of transverse bands of black and yellow. Some species are covered with long hair and are often mistaken for bumble bees when they accidentally enter a building and fly up and down the window panes. An anatomical character peculiar to this family is the thickening of the wing membrane obliquely across the disc of the wing, giving the appearance of a vein known as a 'vena spuria'. It terminates without conjunction with other veins. The name 'hover fly' is taken from their habit of hovering with extremely rapid wing beats while seemingly hanging motionless in the air. The form and habit of the larvae vary widely but only those of the cosmopolitan *Eristalis tenax* and related species concern us here. These occur in stagnant water containing decaying vegetable and animal matter, thus they are often found around buildings in drains and gutters and sewage disposal pits and tanks. They are known as rat-tailed maggots on account of the posterior extension of the abdomen into a long tail, and they are able to breathe air directly through spiracles situated at the tip of it while remaining submerged in the heavily polluted substrate.

The larvae have occasionally been suspected of occurring in the human alimentary canal, although there is always the suspicion that the observer has recovered them from a dirty, contaminated container in which they were breeding naturally. The larvae of other genera have also allegedly been implicated in myiasis. The aphidophagous species (usually subfamily SYRPHIDAE), often slug-like and brightly coloured, may be found on cut flowers or in salads, from which they may be accidentally ingested, indoors.

DROSOPHILIDAE SMALL FRUIT FLIES, VINEGAR FLY

Fermentation processes attract the adults and they appear to arrive suddenly from nowhere. It is thought that the yeasts present may be the principal food but larvae have been found also in faecal matter. They are often known as fruit flies but as this common name is more often applied to members of the family TRYPETIDAE it has been suggested that they are referred to as small or lesser fruit flies.

A common temperate species is *Drosophila funebris* which is attracted to mammalian faeces including that of Man, and as it feeds also on rotting fruit and uncooked food it is possible that it is a mechanical transmitter of disease organisms. The larvae are sometimes concerned in intestinal myiasis. This species also often breeds in dirty milk bottles containing sour milk, and the puparia, cemented to the glass, often pass through the bottle-cleaning plant and the milk-filling process without being dislodged. (See also PHORIDAE.) *Drosophila repleta* is more tropical in distribution although it has become widespread in recent years. It was not recorded in Britain until 1942 but is now common in London. In addition to the usual places where *Drosophila* spp. occur, *D. repleta* is also found in hospital kitchens, and as it is attracted to rotting vegetables as well as to white tablecloths and plates, it could possibly be a carrier of disease pathogens. Several species of *Drosophila*, on account of their small size, are able to make their way through fly-screening of various sorts.

These usually small flies, about 3 mm in length, possess a feather-like arista; red eyes and generally have a distended appearance. They are well known for their habit of hovering, with rapid wing beats, around decaying fruit, vinegar factories, pickle-bottling and cider-making establishments and are common in restaurants, fruit markets, canneries and almost anywhere where food is prepared, including kitchens of dwellings.

As the food material, such as fermenting fruit, is semi-liquid, the advantage of the latter is obvious. The fully-fed larvae leave the fruit or other food material and pupate in a drier environment. The puparium is ovoid with a flat plate at the anterior end from which arise a pair of finger-like or feathery spiracular processes.

Life-cycle
In the case of the well-known species *Drosophila melanogaster* which has been used widely for genetic experimentation, the spindle-shaped eggs bear a pair of filaments which may have a respiratory function. Up to 900 eggs may be laid and the larvae are ll-segmented bearing minute spines. On the anal segment there are three pairs of conical tubercles, and there is an elongate process bearing the posterior spiracles, which is retracted. The life-cycle is completed in about 30 days at 15 °C, 14 days at 20 °C, 10 days at 25 °C and $7\frac{1}{2}$ days at 30 °C. While the length of adult life is only 13 days at 30 °C, it is as long as 120 days at 10 °C.

CHLOROPIDAE   EYE FLIES

Members of this family are small and have a shiny appearance. The wings are without marks, the arista is bare and there is a characteristic venation. A number of species are important pests of cereals but *Siphunculina funicola*, known as the Eye fly, is especially attracted by secretions of the eye, and as these flies are often extremely abundant they cause great annoyance. *Siphunculina funicola* occurs widely throughout the oriental region and is dark brown in colour with the dorsal surfaces darker with a bluish sheen. Several species of the genus *Hippelates* have similar habits and are found in other parts of the world.

Siphunculina funicola and *Hippelates* spp. are thought to transmit mechanically the pathogens causing conjunctivitis. Pseudo-tracheae bear spines which scratch the conjunctiva, thus giving entry to the pathogens carried on the body and brought from persons already suffering from the infection. In Jamaica there are indications that the organisms causing the disease yaws are carried by *Hippelates palipes*.

SEPSIDAE

Sepsid flies, cosmopolitan in distribution, are fairly easy to recognise. In general form they have a delicate appearance and show a 'waist' between thorax and abdomen and the head is globular being either blackish or reddish in colour. Most of them show an apical spot on their wings and, more easily observed, they possess the extraordinary habit of twisting their wings slowly about while they stand still. The function of this is unknown. The elongate larvae in some species possess posterior spiracles situated on tubercles.

They breed in faecal matter and in decaying animal and vegetable material and are concerned in intestinal and urino-genital orifices in unhygienic persons, and the hatched larvae may be swallowed in decaying food, but it is likely that in some cases larvae have been found in the receptacle used for defecation and urination and mistakenly believed to have been voided.

The adults sometimes occur in vast swarms outdoors, but may be found indoors also in some numbers.

COELOPIDAE   SEAWEED FLIES

These flies breed in rotting seaweed and adult *Coelopa* sometimes occur in vast numbers on beaches rendering shoreline buildings, restaurants, refreshment kiosks, etc., uninhabitable. They are particularly attracted to the smell of certain chemicals such as trichloroethylene, and have thus invaded hospitals and dry-cleaning establishments even inland. While they are not known to be transmitters of any disease, their presence in such numbers can be very disturbing. The adults are hairy or bristly and rather flattened.

PIOPHILIDAE   CHEESE SKIPPER

The only species of this family of importance is *Piophila casei*, the Cheese skipper, so-called because of the habit of the larva of 'skipping' along by alternately looping the fore-end of the body towards the hinder-end and catching the latter with the mouthparts, then, the body coming into a state of tension, suddenly letting go the hold. This causes the larva to 'jump' a considerable distance. This may be up to 250 mm in length and 150 mm in height. Eggs are laid on a variety of substrates of high protein content. Although it has been associated with cheese, it is generally a more serious pest of meat. In addition, the larvae have been found in ham, bacon, human faeces, human cadavers, dried bones and moist hog hair, and 52,627 adults have been reared from one dry-cured ham. Over-ripe and mouldy cheese and slightly putrid beefsteak are, perhaps, the preferred foodstuffs. The adult is small with bronze tints on the thorax which is black and shiny, and the head is round with reddish-brown eyes and the antennal arista bare. The wings show slight iridescence and when the insect is at rest they lie flat over the body.

Life-cycle
Piophila casei has been recorded from Europe, throughout the United States, India, the West Indies, Australia, Greenland and Alaska, and may well be cosmopolitan.

Up to about 500 eggs are laid over a period of three to four days and they hatch in from one to four days. The slender larvae crawl by peristaltic action into the interior of the infested foodstuff. They tend to aggregate together but when ready for pupation they wander away, having spent usually about five days as a larva although under adverse conditions this may be up to eight months. Pupation takes place within the larval skin and this stage usually lasts about five days. Two generations per month have been recorded in Washington DC. More in the recent past perhaps than at the present time, larvae of *P. casei* were considered a delicacy by gourmands when they were eating infested cheese, but there are numerous instances recorded of intestinal illness due to the presence of the larvae which had been swallowed with the cheese. They have been recorded also in other parts of the human body and this species is the most commonly reported cause of myiasis.

SPHAEROCERIDAE LESSER DUNG FLIES

These flies breed in dung and various rotting animals and vegetable matter. A frequent indoor pest in Europe is *Liptocera caenosa* which is often associated with leaks in sewage systems, septic tanks, etc.

MUSCIDAE HOUSE FLIES

This very large family includes a number of species of economic and medical importance. Although a number of species of dipterous flies exhibit more dramatic consequences of their attentions to Man, perhaps none bears such a load of responsibility for their annoyance to mankind as a whole, in buildings (especially dwellings) as one species, *Musca domestica*.

MUSCIDAE are small to large in size, the majority of which bear a similarity to *M. domestica*, the House fly. A few species are blood suckers of Man. *Glossina* spp., the tsetse flies of large regions of Africa were at one time included in the MUSCIDAE, but are now usually treated as a separate family, GLOSSINIDAE. These occur in special areas of forest or shrub known as 'fly-belts' and carry the pathogens of the human disease 'sleeping sickness', as well as diseases of cattle. In this way extensive areas of land are rendered uninhabitable by Man. *Glossina* spp. only occur in buildings accidentally.

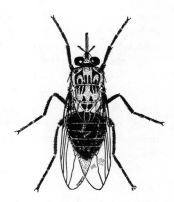

Fig. 17.5 Tsetse fly.

Fig. 17.6 Tsetse fly, diagrammatic section.

MUSCA DOMESTICA   COMMON HOUSE FLY

This cosmopolitan fly, associated with dwellings, is a major scourge of humanity. Only in recent years and around large centres of population has there been any diminution in its abundance. This has been due to mechanical forms of transport taking over from the horse. The House fly's importance in public health lies in the attraction which it has both to human food when prepared and ready for consumption, and to various forms of filth, including human faeces. This habit causes them to be transmitters of a number of diseases such as cholera, dysentery, typhoid, infantile diarrhoea and other forms of food poisoning. The transmission of the pathogens is mechanical only. Additionally, although open to doubt in many instances, a number of cases of the larvae of *M. domestica* are recorded as being concerned in myiasis.

Description

The general shape of *Musca domestica* is almost universally known. The width across the outstretched wings varies from about 13 to 15 mm, while the length from front of head to tip of abdomen is about 6 to 7 mm. The dorsal surface of the grey thorax bears four longitudinal, dark marks, rather blurred at their hinder ends. An important character for identification concerns the wing venation. The fourth vein bends abruptly forwards to almost touch vein three at the margin. The sides of the abdomen near its junction with the thorax are yellowish-buff and are sometimes transparent, more particularly in the male. The rest of the abdomen is dark, the pattern appearing as a central longitudinal mark which broadens to cover the posterior segments.

In various parts of the world there are a number of varieties, subspecies and a few closely related species. *Musca domestica domestica* is characteristic of cool, temperate regions and the coloration is generally dark. *Musca domestica vicina* is intermediate in colour intensity and occurs in warm areas, while *M. domestica nebulo* is pale and found in the tropics. These subspecies also differ in the distance between the eyes in the male. All these subspecies are, however, interfertile.

Musca sorbens, sometimes referred to as *M. humilis*, is a tropical and subtropical species with a distinctive thoracic marking. Unlike *M. domestica*, the larvae often occur in cowdung and human faeces.

The adult

The eyes of *M. domestica* are large globular and each consists of about 4,000 units. This gives to the fly exceptionally good sight with which movement is coordinated. It thus usually eludes capture. The antennae are small, consisting of three segments only, one large and two small, the former bearing a special bristle. Olfactory organs are located in the antennae and these enable the female to detect suitable breeding sites and both sexes to find food sources.

Life-cycle

The glistening white eggs are long and pointed-oval, about 1 mm in length and sometimes slightly curved (banana-shaped). They are laid in batches and up to 150 eggs may be laid in one day. The maximum number of eggs laid by one adult during her life is about 750. They are deposited in crevices in the chosen food material just below the surface. They hatch in from eight hours to two days, according to temperature, the latter being dependent on the heat generated by the bacterial action in the food. On hatching, the young larva immediately burrows into its putrefying food substances, choosing a position at or as near as possible to the preferred temperature, which is very high. On many occasions larvae have been recovered from manure or garbage at a temperature of 45 to 50 °C.

Growth is extremely rapid and the larva moults three times. The larva or 'maggot' is white, legless, about 12 mm in length and in shape is a slender cone consisting of 13 segments. The mouth is situated at the pointed end and consists only of an aperture and a pair of black hooks arranged vertically and articulating with an internal system of black, rod-like parts visible through the translucent skin. Conventional insect mouthparts are absent. A pair of easily visible spiracles are borne on the truncated, hinder end of the body, and form the gaseous interchange component of the respiratory system. Many species of fly larvae can be identified by the size and shape of the spiracles.

When the larva becomes fully-fed, it changes in colour from greyish to creamy-white due to the accumulation of fat reserves and to the internal reorganisation. The larva then leaves its hot, damp environment and wriggles to a drier and cooler situation. Usually it burrows rather shallowly into loose earth but it may tunnel as much as 50 m in order to achieve this. It then assumes a prepupal phase by becoming ovoid in shape, but the last larval skin is not shed but remains as a tough, dry, dark reddish-chestnut coloured covering. The pupa remains entirely within this last larval skin or 'puparium'.

When the adult emerges the puparial skin splits in a regular fashion with a circular rupture around the sixth segment with lateral splits running forwards. The adult fly possesses a special organ, known as the ptilinium, to help clear away obstacles during emergence. This is bladder-like, situated between the eyes, and is alternately distended and deflated.

Myiasis

Myiasis is the name given to infestations of dipterous fly larvae inside the body of mammals, birds, reptiles, and more particularly in the human body. In the case of a number of species in the following genera: *Oestrus, Hypoderma, Cordylobia, Chrysomyia, Gasterophilus* and *Wohlfahrtia*, such myiasis is specific, that is, the larvae are obligate feeders on living flesh, and can only complete their development by doing this. In the quoted genera, species are known to invade human tissues, some commonly, some rarely. Further information is given in the text.

Some dipterous larvae feed generally on flesh which is decaying, but on occasion, only invade suppurating and malodorous wounds and sores, especially when involving body cavities. Such myiasis is referred to as semi-specific and examples are species of *Chrysomyia, Cochliomyia, Calliphora, Lucilia, Phormia, Sarcophaga, Wohlfahrtia, Megaselia* and *Musca*.

In addition, a large number of cases of accidental myiasis are known involving species of about 20 genera. In intestinal myiasis, larvae or eggs subsequently hatching into larvae, are swallowed with food or drink. After a period in the alimentary canal they are either vomited or pass out with faeces. In urinogenital myiasis, eggs are laid in the fouled urinogenital opening or on the pubic area, and larvae are subsequently found in the urine. Very rarely, complete development takes place within the alimentary canal or the bladder. It should be emphasised, however, that accidental myiasis is often reported in circumstances of considerable doubt.

Trachoma

Trachoma is a chronic and contagious disease affecting the eyes as a form of conjunctivitis. The causative agent is a virus of the psittacosis group named *Chlamydozoon trachomatis*, and the principal vector is the Common House fly. Any type of contact with infested eyes, however, can spread the disease.

The disease is characterised by inflammatory granulations present on the conjunctival surfaces of the eyes. Partial or complete blindness is often the result of repeated infestations or acute forms of the disease. In 1970 it was estimated that over 500 million people were affected by trachoma in Asia, South America, North Africa and Central Africa. It is, however, rare in Europe. In India the national prevalence figure for trachoma is 33.6 per cent; trachoma is the cause of from 60 to 80 per cent of the 4.5 million blind in India.

High incidence of trachoma usually occurs in overcrowded areas where there is poverty in a semi-desert environment. Young children become infected early in life at a time when they do not possess an immediate reaction to brush away flies from running noses and eyes. Women are more liable to suffer from it than men as, in tending young children, there is a greater chance of the disease being transmitted to them from infected infants.

MUSCA AUTUMNALIS   FACE FLY

Musca autumnalis is an important pest of cattle in Europe, Asia and the eastern, mid-western and southern states of the United States, where it was introduced in 1950. The adults feed on the exudations from the eyes, nostrils and mouths of cattle, as well as sweat and the blood from wounds and scratches.

It is a pest in buildings on account of its habit of hibernating in company with other species in considerable numbers in such situations as attics, roof-voids and walls. It can be distinguished from *M. domestica* in the male by the eyes almost meeting, and in the female by a silver stripe around the eyes, whereas in *M. domestica* it is golden and the thorax of both sexes is slate-grey in colour.

Fig. 17.7 Flies contaminated with the virus causing trachoma are especially attracted to running noses and eyes. (World Health Organisation. E. Schwab)

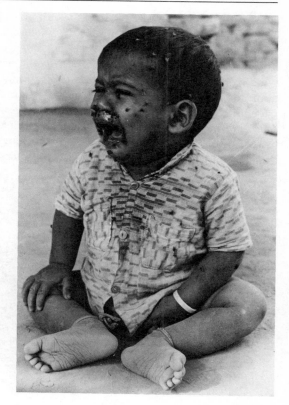

The eggs, which are stalked, are laid on fresh animal dung and the larval stage is completed in from 3 to 10 days. The puparium is whitish, and the total length of the life-cycle is 17 to 18 days. Large clusters of this species emit a sickly, sweet smell, and when a fly is crushed, it is said to smell like 'buckwheat honey'. *Musca autumnalis* is often a component of a hibernating cluster in buildings in Europe, as also are *Muscina stabulans*, *Dasyphora cyanella* and the small yellow, black-marked *Thaumatomyia notata* in the family CHLOROPIDAE. Several additional species are similar in appearance to *Musca domestica* and are important pests of cattle in various parts of the world. Some will bite Man in milking sheds but are not generally found in dwellings. *Haematobia irritans* is known as the Horn fly.

POLLENIA RUDIS   CLUSTER FLY

A number of fly species, but chiefly muscids, have the habit of forming dense clusters in buildings during the hibernation period. They occur mainly in rural or semirural areas and there is usually a history of selection of certain buildings going back over a number of years. During the warm days of autumn they sun themselves on a wall but retire at night into the building, entering through cracks under eaves or tiles. They cause a great deal of annoyance when the temperature rises either artificially or naturally as they fly about in an uncertain manner in a state of semi-torpidity. They may drop on to the hair of persons and hit windows, lights, etc., causing a certain amount of noise. They fall into unprotected water cisterns, often in large numbers.

Pollenia rudis is an abundant clusterer. It resembles a large *Musca domestica*, but it holds its wings flat over the back when at rest. It is brown in colour with golden hairs on the thorax and a dark line down the centre of the abdomen which also possesses a reflective surface.

Fig. 17.8 *Pollenia rudis*, Cluster fly.

Pollenia rudis occurs throughout Europe and is common in Canada and the United States except for the states bordering the Gulf of Mexico. The larvae are parasites of earthworms.

FANNIA CANICULARIS LESSER HOUSE FLY

Fannia canicularis is second only to *Musca domestica* as a major annoyance pest in buildings. In some areas in Britain it is even thought to outnumber the latter.

It is not thought that *F. canicularis* is important as a carrier of disease pathogens, but it is frequently involved in cases of intestinal myiasis. *Fannia canicularis* is often mistaken for *M. domestica* but it is easily identified by its slightly smaller size and by the venation. The fourth vein curves gently away from the third instead of bending abruptly towards it. In addition, when the insect is resting, instead of diverging, as in *M.*

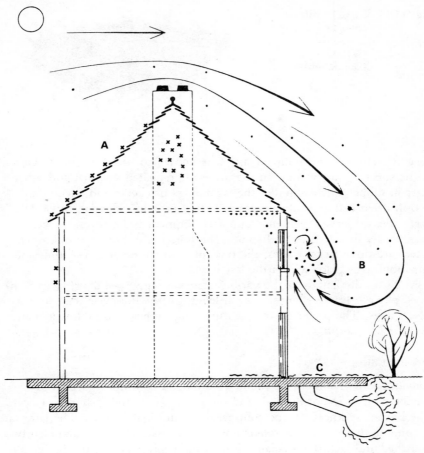

Fig. 17.9 Three ways in which flies 'swarm' into houses: A. Cluster flies and one or two other species settle on the roof on the warm side of the house, crawl inside and cluster on the rafters and brickwork; B. *Thaumatomyia notata* carried by the wind and trapped in eddies under the eaves on the lee side, collect on the ceilings or upper rooms; C. Larvae of *Lucilia* come out of the soil near a drain when a heavy storm causes temporary flooding. (Oldroyd)

domestica, the wings are held parallel and partly folded over one another. *Fannia canicularis* is easy to identify in flight as the males fly backwards and forwards in short, jerky flights in groups of a dozen or so around central pendant electric light fittings and apparently spend hours doing this.

Life-cycle
The 1 mm long, banana-shaped eggs are furnished with flotation ridges and are laid in various sorts of dung although moist poultry manure is

Fig. 17.10 Larva of *Fannia canicularis*, Lesser House fly. (Wellcome Foundation)

very attractive to the adult female. It is known to breed in human faeces, in the soil of poultry-runs, in the urine-impregnated sawdust and other litter in rabbit hutches, under the wrappings of cheese and bacon, and also in dried fish and other types of stored protein-rich foodstuffs. The barrel-shaped larva is flattened and each segment bears several large backwardly directed, curved spines, the function of which is unknown. There are three larval stages, the final of which measures 6–7 mm, and pupation takes place within the last larval skin.

At 27 °C the length of the various stages is: egg 1½–2 days; larva 8–10 days; pupa 9–10 days, and the total length of the life-cycle (egg to egg) is 22–29 days. The preferred temperature range appears to be lower than that for *M. domestica*.

FANNIA SCALARIS   LATRINE FLY

Fannia scalaris is sometimes found indoors and commonly occurs in privies which are not flushed, and the larvae are found in human faeces. The larvae are known to be concerned in intestinal myiasis. The latter are not unlike those of *F. canicularis* but the processes are more abundantly branched and lack the median folds on the ventral side of the body segments. The pupa of *F. scalaris* has an oval anus without the raised V-shaped fold of *F. canicularis*. The adults bear a tubercle on the tibiae of the middle pair of legs which is absent in *F. canicularis*.

STOMOXYS CALCITRANS   STABLE FLY, BITING HOUSE FLY

Very like *M. domestica*, this cosmopolitan species can be identified by its long, pointed proboscis which projects forwards from the head, and the fourth wing vein which curves only gradually. When resting the wings are held at a less acute angle than in the case of *M. domestica*. The arista bears hairs only on the dorsal surface and the small, thread-like palpi can only be seen with a lens.

The larvae live in moist, decaying vegetable matter such as hay, lawn clippings, chicken manure and decaying seaweed on beaches. They are often found in urine-soaked heaps of straw and manure as can be found in a farmyard. They are similar to the larvae of *M. domestica* except that the posterior spiracles exhibit a different pattern.

Stomoxys calcitrans occasionally invades buildings and both sexes bite. In a warm room they will attack in the region of the ankle and easily penetrate socks and stockings. 'The bite is like a sharp stab with a darning needle.' The larva has been recorded in cases of myiasis. *Stomoxys calcitrans* is almost cosmopolitan in distribution and there are several Ethiopian and Asiatic species of somewhat similar habits.

MUSCINA STABULANS   NON-BITING STABLE FLY, FALSE STABLE FLY

Similar to *M. domestica* in general appearance, but larger with the third wing vein lighter in colour and the fourth only gently curving upwards. The larvae occur in horse, cow and pig dung, human faeces and decaying vegetable and other organic matter, and in these situations sometimes prey upon other insects present.

The females often enter buildings and will lay their eggs on a variety of human foods both raw and cooked, and will feed on milk. Up to 200 eggs are laid. The length of the larval stage is up to 25 days and several generations are produced annually. In the United States, overwintering takes place in the pupal stage although sometimes larvae remain dormant during this period. In Britain, and probably in Europe generally, this species often associates with other species for hibernation, in the adult stage, in attics and similar locations in buildings. A number of cases of intestinal myiasis have been recorded, some of which have been protracted. It is almost cosmopolitan in distribution. *Muscina assimilis*, of similar habits, is darker in colour and is less abundant.

CALLIPHORIDAE   BLOWFLIES, GREENBOTTLES, BLUEBOTTLES

The adults are moderate in size, stoutly-built, and commonly the bodies are green or blue with a metallic sheen. The arista is generally plumose. They fly by day and are strongly attracted to, and feed on, sweet liquids such as nectar and honeydew and liquid products of decomposing organic matter. The latter is essential for maturation of the eggs. In some cases reproduction is ovoviviparous.

CALLIPHORA

The several species in this genus are large flies, from 8 to 14 mm in length and usually have blue bodies dusted with grey with a rather bristly appearance when viewed with a ×10 lens. They are popularly known as bluebottles or blowflies and occur throughout the world.

Egg-laying occurs in a variety of circumstances, but animal carcasses are the preferred sites for oviposition, and sometimes newly-hatched larvae are deposited. Only rarely is decaying vegetable matter selected by blowflies. Suppurating wounds on living animals are sometimes infested by calliphorids. Two common species found in many parts of the world are *C. vomitoria* and *C. vicina* (*erythrocephala*). Both species are attracted inside buildings by the smell of exposed meat, but after oviposition they soon leave. Their loud buzzing noise as they try to find an exit is particularly annoying. The larvae feed on tissues which have been liquefied by the action of certain bacteria and partly by protein-digesting enzymes which the larvae themselves have secreted. Shortly before pupation and usually at night the larvae wander away from the foodstuff and bury themselves in loose soil, travelling as far as 30 m in order to find it. If soil is not available, pupation takes place in crevices or under sacking or other material. Meat infested with calliphorid eggs or larvae is known as 'flyblown'. The total length of the life-cycle in slaughterhouse refuse is 19 to 22 days.

LUCILIA   GREEN BOTTLES

Common species which are more or less cosmopolitan in distribution are *L. sericata* and *L. caesar*. They are usually shining, metallic green in colour and are known as greenbottles, although there is some variation in colour from bluish green to copper. They are generally smaller than *Calliphora* spp. being about 10 mm in length. In Europe *L. sericata* is the species generally concerned in sheep 'strike'. This is when the skin of the sheep has become 'scalded' due to being constantly wet with urine, sweat or rain, and some bacterial action has been initiated. The young larvae feed on the serous exudates, but older larvae attack the living flesh. In Australia, the important species is *L. cuprina*.

Larvae of *L. sericata* have been removed from suppurating wounds in Man as well as from body cavities such as the ear when in poor hygienic state. Such infestations are mostly found in shepherding tribes. *Phormia* spp. are rather smaller than *Calliphora* and have similar habits. Larvae of *Phormia regina* and *Lucilia* have been reared under sterile conditions and used for clearing away suppurating tissue and inhibiting bacterial growth in cases of chronic osteomyelitis. Adults of *Calliphora*, *Lucilla* and *Phormia* are very often contaminated with micro-organisms including *Salmonella* and *Clostridium*, but contamination of human food by this means appears rare. *Lucilia* spp. may carry the virus of poliomyelitis and it is possible that this presents an opportunity for transmission of the disease if small wounds are visited by the infected flies.

Larvae in the wandering stage are sometimes found in buildings due to their having dropped from the carcase of an animal, such as a bird or mouse in a chimney or roof void, and which they had been infesting. Sometimes following heavy rain *Lucilia* larvae emerge from the waterlogged soil in large numbers (as do TIPULIDAE) and frequently enter buildings.

CHRYSOMYIA BEZZIANA

This metallic-green blowfly is widely distributed throughout the tropical and subtropical regions of the Old World including Equatorial Africa, India and the Philippines. It is important as laying its eggs in the wounds, abcesses, cuts and sores of Man and animals. It will also infest the nose, eye, ear, mouth and urino-genital passages. This, and related species, is somewhat larger and rather more bluish in colour than *Lucilia sericata* which it appears to replace in tropical regions. *Chrysomyia macellaria* occurs in the southern states of the United States, Central America and tropical South America, and its larva is known as the screw-worm. The larva forms a burrow into the flesh of rats but also sometimes of cattle and Man. This causes a large boil with a central hole through which the larva breathes by means of its posterior spiracles.

Adult *C. bezziana* can be distinguished from related species by the wings which do not have a dark anterior border, and the squamae are waxy-white and bear small, dark hairs, and by the eyes of the male which are not distinctly zoned into two parts. In addition, the lower part of the face is bright orange. The absence of large bristles on the dorsal surface of the thorax also helps to identify them.

Eggs are laid in batches so that the larvae are usually found in numbers. The cream-coloured larva is covered with minute, black spines, mostly occurring in bands around the anterior margin of the segments, and it reaches a length of 14 mm. In the species *C. albiceps* there are processes on the dorsal surface. The young larvae burrow under the skin adjacent to the wound in a mammal and cause a tumour to form which may be of large size according to the number of larvae inhabiting it. When fully fed the larvae drop out of the wound and this is usually from 7 to 14 days after oviposition.

COCHLIOMYIA

This genus is closely related to *Chrysomyia* and a number of species occur in tropical America where they infest wounds, sores, etc., and body orifices in a similar manner to *C. bezziana*.

CORDYLOBIA ANTHROPOPHAGA TUMBU FLY

This human parasite occurs in many areas of tropical Africa. Eggs are laid in the dust of hut floors which are devoid of prepared surfaces, and where human beings usually lie at night. The maggot-like larva is very agile and bites a way into the skin of its host. Where a single larva occurs, an erruption similar to a boil appears which is encircled by an inflamed area. However, it does not give rise to a throbbing pain, and a small hole is present in the centre of the swelling. It is through this that the larva breathes and excretes. When a number of larvae are present together, gangrenous conditions may result.

The larva averages about 14 mm in length and in colour is dirty white. The head end is rather bluntly pointed while the hind end is truncate, but the larva is much plumper than the maggot of *Musca*. The cuticle of the mature larva is covered with spinules. The length of larval life is only from 8 to 10 days, when they emerge through the rupture in the 'boil'. The adult is yellowish-brown in colour and almost reaches 10 mm in length. Dark, longitudinal stripes occur on the thorax, while the posterior segments of the abdomen are almost black and the wings are brownish. Rodents, dogs and monkeys are perhaps more usually parasitised.

AUCHMEROMYIA LUTEOLA   CONGO FLOOR MAGGOT

The adult fly is similar in appearance to *Cordylobia anthropophaga* but is slightly smaller and the second abdominal segment is enlarged when seen from above. The latter is about half the length of the abdomen in the female. The larva is similar in shape to that of *Cordylobia anthropophaga* but it lacks spinules and the body is much folded.

Fig. 17.11 *Auchmeromyia luteola*, Congo floor maggot. (From Castellani and Chalmers, British Museum)

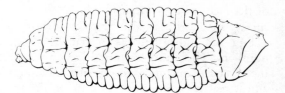

During the daytime the larvae hide in the dust and debris on the floor of native huts but at night crawl about in search of someone sleeping on the floor or on a low bed. They then pierce the skin and feed on the semi-liquid tissue. They are able to live for considerable periods without taking a meal.

SARCOPHAGA   FLESH FLY

The various species of *Sarcophaga* all possess three longitudinal, dark bands on the thorax and a chequered abdomen. *Sarcophaga carnaria*, the common European species, is grey in colour, but other species are brown, brownish-yellow or golden. All sarcophagids, however, have a large-footed appearance due to the distinct claws and large pulvilli. The latter shrink when the specimen is dead. In length it is 13 mm and has a wing span of about 22 mm. It thus has a longer body but a smaller wing span than *Calliphora*. It normally deposits living larvae. The complete life-cycle at 21 °C is 25 days, at 25 °C is 17 days and at 29 °C is 13 days.

Sarcophaga is one of the main genera responsible for wound myiasis in Europe, It has also occurred in cases of intestinal myiasis. The larva is distinctive in having the posterior spiracles sunk in a deep cavity. It does not, however, commonly frequent the inside of buildings but is usually to be found around farm buildings and rubbish tips where there is animal refuse.

WOHLFAHRTIA MAGNIFICA

The adult flies are similar to those of the genus *Sarcophaga*, except that the abdomen is light grey with a pattern of black spots. The larvae are usually found in wounds in a wide range of animal species, including Man. In addition, the larvae infect the natural cavities of the body, such as the eyes, nose, ears and urino-genital passages if these become diseased or, for some reason, cannot be maintained in a clean condition. The larvae may be up to 18 mm in length, and the number found is usually small in any infestation. The adult female deposits living young in the sites.

This is an Old World species, occurring in south-east Europe, the Middle East, Egypt and south Russia, but does not extend as far as India and the oriental region generally. A New World species, however, *W. vigil*, about which little is known, will attack the human body without any previous wounds.

OESTRIDAE   WARBLE AND NOSTRIL FLIES

Mention is made here of a few species of OESTRIDAE, not because they may be found in buildings but because the larvae, when occasionally found in Man, may be confused with species already mentioned. *Dermatobia hominis*, however, is frequently found in Man as a human parasite, the larvae boring beneath the skin as the scientific name implies. It occurs in Central and tropical South America.

Fig. 17.12 *Oestrus ovis*, Sheep nostril fly. Adult, × 4. (From Castellani and Chalmers)

The larvae of several species of the genus *Oestrus* are occasionally found infesting Man. The species concerned is usually *O. ovis*, the Sheep nostril fly. Normally the larvae inhabit the nasal and cranial sinuses of sheep but from time to time are found in the mouth mucosa, nasal passages or conjuctiva of persons attending flocks of sheep. The majority of records of myiasis caused by *Oestrus* spp. have originated in Russia, Italy and North Africa. The young larvae are spiny and the mouth hooks are large. They are creamy-white in colour but when mature become grey.

HYPODERMA   OX WARBLES

The larvae of *Hypoderma* inhabit the bodies of cattle, goats, deer and game animals. When mature, they are usually found underneath the hide on the back, where they may be detected by the presence of a swelling, in the centre of which is a hole through which the larva breathes. The larva is rather rotund with the head rounded, not pointed, and the whole body is covered with rows of minute spines which can be felt as a roughness.

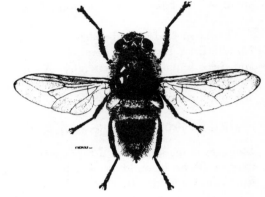

Fig. 17.13   *Hypoderma bovis.* The adult stage of Ox warble, × 2½. (From Castellani and Chalmers)

Human infestations sometimes occur, the larva appearing in the subcutaneous tissues as a 'creeping' eruption appearing in various parts of the body. They occur throughout the world whenever suitable hosts are to be found.

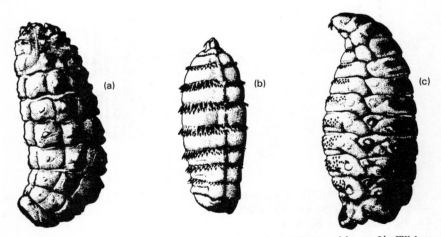

Fig. 17.14   The mature larvae of: (a) *Hypoderma bovis*, Ox warble, × 2½. With the dorsal surface to the left; (b) *Gasterophilus intestinalis*, Horse bot, × 3. With the dorsal surface to the right; (c) *Oestrus ovis*, Sheep nostril fly. With the dorsal surface to the right (the specimen is in an artificially distended condition). (From Castellani and Chalmers, British Museum)

Chapter 18

Butterflies and moths

LEPIDOPTERA

Butterflies and moths are probably the most recognisable of all insects. They are distinguished by the overlapping, powder-like scales which are modified hairs and which cover the body and both surfaces of the wings. The mouthparts have (with the exception of one family) functionless mandibles, and the maxillae are greatly modified to form a long, coiled tube or proboscis.

Both pairs of wings are well developed, and often, in some families, large in relation to the narrow body. The larvae of the LEPIDOPTERA are 'caterpillars' and are remarkably uniform in appearance throughout the order. The presence of prolegs (usually five pairs) on the abdomimal segments, in addition to the three pairs of thoracic legs, is general in the order.

The great majority of LEPIDOPTERA feed in the larval stage on flowering plants, but some live in fungi, while others feed on a variety of broken down organic material. Some feed on animal matter such as fur, hair, wool and silk. There are some wood-eating species.

Many larvae spin silken cocoons in which to pupate and these reach their highest development in the families SATURNIIDAE and BOMBYCIDAE to which the silk moths belong. Larvae of the family PHYCITIDAE, well represented in the stored products fauna, spin prolific threads of silk during the wandering stage. Adult butterflies and moths imbibe only liquid food, water or nectar.

The order LEPIDOPTERA comprises more than 100,000 species, and although the preponderance of species occurs in tropical and subtropical areas, nevertheless, many are found in temperate regions or even in regions less than temperate. At one time, LEPIDOPTERA were divided into two suborders; RHOPALOCERA, butterflies and HETEROCERA, moths. One of the distinguishing characters was the form of the antennae. In the butterflies, the antenna terminates in a club, but in the moths the shape takes a number of different forms.

In the present classification the LEPIDOPTERA is divided into four suborders, the most important of which is DITRYSIA, containing the majority of moth families as well as the seven families of butterflies.

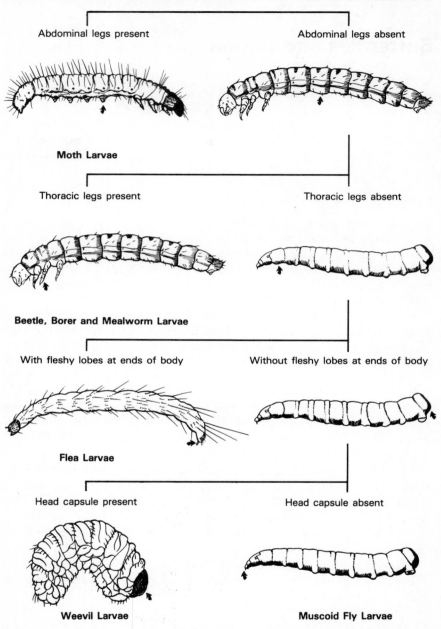

Fig. 18.1 Pictorial key to common groups of household and stored food pests. Larval stages. (US Department of Health, Education and Welfare)

Although the larvae of some butterfly species are agricultural and horticultural pests, none can be considered harmful in buildings. The larvae of *Pieris rapae* and *P. brassicae*, the Cabbage whites, however, often pupate inside buildings, having entered through open windows. Some butterfly species hibernate during the winter months inside buildings – for example the Small tortoiseshell, *Aglais urticae*, but such attractive insects cannot be termed 'pests'.

In size, LEPIDOPTERA vary from as little as 3 mm across the outstretched wings, to as much as 26 cm in the case of some saturniid moths. A number of moth species are associated with stored products, grain, meal, dried fruit, nuts, cocoa, wool, fur, silk, feathers and a number of materials of animal origin when stored in buildings. Note: many moth species are difficult to identify so that the following should be treated with reserve and only as a rough guide.

Stored products pests

Identification
The most common moth species injurious to stored products belong to five families: GELECHIIDAE, GALLERIIDAE, PHYCITIDAE, TINAEIDAE and CECOPHORIDAE. GELECHIIDAE larvae of this family can be distinguished by two hooks or crochets on the prolegs of the abdomen, but recognition of the larvae of other families requires special study. Setal maps showing the distribution of the hairs or setae on the various segments of the body are used.

Although the number of genera and species which damage stored products is small, adults are often difficult to separate and identify. One reason for this is that although a recently emerged moth may show distinctive patterns on the fore- and hindwings, the different coloured scales are readily rubbed off and the pattern becomes obliterated. When that happens it may be possible to place the moth in the family to which it belongs by examining the labial palps and the wings. In many moths the labial palps are much more prominent than the maxillary palps. They may be curved upwards, as in the genus *Ephestia* or straight, as in the genera *Plodia* and *Corcyra*; in the genera *Sitotroga* and *Endrosis* they are long and sickle-shaped.

The wings differ in shape, in the venation, i.e. the arrangement of the veins and in the width of their fringes. The fringe on the forewing can sometimes be discerned when the moth is at rest, as in *Endrosis*, These characters, which can only be clearly seen with the use of a hand-lens of × 15 magnification in a good light, help to separate some genera and species, but in other groups even the genera cannot be separated by them and then it becomes necessary to examine the genitalia.

In the LEPIDOPTERA the important and readily recognisable parts of the male genitalia are the uncus and claspers, and in the female, the ovipositor lobes and bursa copulatrix. These can only be satisfactorily seen by cutting off the abdomen, boiling it in caustic potash until clear, and examining it under a microscope. Once so prepared, accurate identification is usually quite easy.

GELECHIIDAE

The only species of major importance in this family is *Sitotroga cerealella*, the Angoumois grain moth. *Sitotroga cerealella* gets its common name from Angoumois in central France, where it first attracted attention in 1736 as an insect damaging grain.

Sitotroga cerealella does not persist out-of-doors in the UK, and as it attacks the crop when the grain is in the 'milk' stage, it is of minor importance in Great Britain. In southern Europe, North Africa, Central, South and West Africa and in South America it attacks the grain crop before harvesting, and the damage it causes can be serious. Since the larval life is spent inside the grain, detection is difficult at an early stage. *Sitotroga cerealella* is now cosmopolitan.

Sitotroga cerealella is a small moth from 15 to 18 mm in wing span, with pale, golden-yellow colouring. The wings are narrow with wide fringes; the forewings four-sided in shape and the hindwings narrow sharply at the apex. The labial palps are sickle-shaped, curving upwards with the last segment needle-like and commonly black in colour.

The larvae are short and dumpy with the faintly developed prolegs on the abdomen bearing only two crotchets, which distinguish them from all other kinds of larvae in grain.

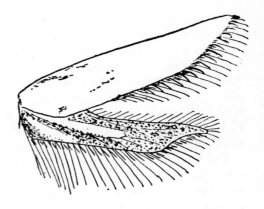

Fig. 18.2 Wings of *Sitotroga cerealella*. (After Hinton and Corbet)

There is need to emphasise that any 'cosmopolitan' insect is only so in areas where the climate allows it to survive and multiply, and that distinction is modified by local establishment in heated premises. Injury to corn (maize) by it is not severe in the north central states of the USA except where maize is stored for more than one season. In the southern portion of the 'commercial corn area' it is likely to be troublesome in some years. Loss can be reduced by prompt harvesting and threshing of the crop. It can be a serious pest of sorghum in Africa. Whether damage continues after the first generation depends on the ability of the female adult to reach the grain. Thus, sorghum stored on the ear is accessible but maize stored in bulk on floors is only so on the surface. Adult *S. cerealella* can find their way to the surface from deep in bulks, arriving there devoid of scales, etc., but they do not go down again but continue to lay eggs on exposed grains at the surface.

OECOPHORIDAE

In this family the hindwings are not so acutely narrow at the tips as in the preceding family, and the species are larger, 12–22 mm in wing span.

ENDROSIS SARCITRELLA   WHITE-SHOULDERED HOUSE MOTH

Recently emerged specimens can be readily recognised by their white shoulders and prothorax, contrasting with the greyish-white forewings marked by dark, almost black patches. The hindwings are narrowed towards the tip but not nearly so acutely as in *Sitotroga*. The labial palps are prominent and curve upwards, 12–22 mm wing span. *Endrosis sarcitrella* is a general feeder and scavanger. It is widespread in northern Europe in warehouses, granaries, farm-buildings and dwelling-houses. Wherever whole or broken grain, flour or other vegetable debris is allowed to accumulate undisturbed over a period, this species is almost certain to be found. It produces several generations annually.

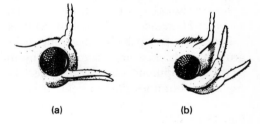

Fig. 18.3 Head and labial palps of two moths: (a) *Corcyra cephalonica*; (b) *Endrosis sarcitrella*. (After Hinton and Corbet)

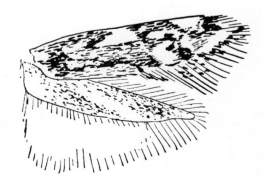

Fig. 18.4 Wings of *Endrosis sarcitrella*. (After Hinton and Corbet)

HOFMANNOPHILA PSEUDOSPRETELLA BROWN HOUSE MOTH

This moth, 15–22 mm in wing span, is bronze-brown in colour with dark brown to black flecks on the forewings. The hindwings are markedly broader than those of E. *sarcitrella*. It is common in warehouses and granaries, preferring damp situations. Except for this, its feeding habits resemble those of the latter insect. In contrast to *E. sarcitrella*, it has only one generation per year. The life-cycle may be prolonged beyond one year since there is a marked larval diapause which may last many months.

The part *H. pseudospretella* plays as a warehouse moth is usually that of an omnivorous scavenger in spillage and dust, but it occasionally becomes a nuisance attacking bagged flour stored for long periods, peas and beans and other stored commodities. Out-of-doors it occurs in birds' nests and these may serve as a reservoir or focus of infestation. The reasons for its persistance and occurrence in very large numbers are its wide food range, its high reproductive capacity and the resistance at all stages, except the larval feeding stages, to adverse conditions. This although the growing larvae thrive only at high humidities, the eggs and the resting (diapause) larvae are indifferent to desiccation. Under most conditions larvae enter diapause, which may be prolonged.

The only biological checks on the increase of *H. pseudospretella* are the Predacious mite, *Cheyletus eruditus*, the Debris bug, *Lyctocoris campestris*, and larvae of the Window fly *Scenopinus fenestralis*.

In addition to its role as a warehouse moth, this species probably occurs in small numbers in almost every private house in Britain. Here it is important as damaging clothes, furnishings and cork floor inlays. Birds' nests are thought to provide an important source of infestation for the entry of *H. pseudospretella* into dwellings.

Larvae are incapable of maturing at humidities below 80 per cent and growth is checked until favourable humidity returns, the duration of larval development varies widely because of the intervention of diapause. At high humidities (90 per cent) development may extend to 250 days

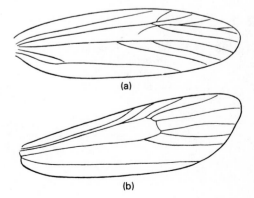

Fig. 18.5 Venation of forewing of two moths: (a) *Corcyra cephalonica*; (b) *Ephestia elutella*. (After Corbet and Tams)

(25 °C) or even to 440 days (20 °C). Under these conditions larvae may, therefore, be found throughout the year. In Northern Ireland, it is stated that adults may emerge and be found at any time, but that two peak emergences occur, one in June and the other in October. Larvae are present throughout the year but are most obvious in February–April, probably reflecting higher indoor temperatures and possibly 'spring cleaning'.

TINEIDAE   CLOTHES MOTHS

Members of this family containing the so-called 'clothes moths' are important to the woollen and clothing industries and much less so to food establishments. Some species are of importance as infesting wool, silk, fur and feathers in dwellings, and are cosmopolitan.

These tineid moths are characterised by prominent hairs or scales on their heads which give them a crested appearance, by their long-fringed wings and by their activity. They have relatively long legs and run rapidly. While in fresh condition the species mentioned can generally be recognised by the wing patterns, but for certain identification the genitalia must be examined.

One member of this family, *Tinea (Nemapogon) granella*, the Corn moth, damages grain. Four other species which occasionally occur in warehouses but more commonly in dwellings are *Tineola bisselliella*, the Common clothes moth, *Tinea pellionella*, the Case-bearing clothes moth, *T. columbariella* and *T. pallescentella*.

TINEOLA BISSELLIELLA   COMMON CLOTHES MOTH

Known as the Webbing clothes moth in the USA (where it is by far the commonest species of clothes moth). The minimum length of its life-cycle is about six weeks. The average length of the adult stage is 28 days for the male and 16 days for the female. The egg is large, about 1 mm long and ivory-white. It adheres to the surface on which it is laid. Larval faecal pellets are often mistaken for eggs by the householder.

The moth is uniformly golden-buff or yellowish-grey; wing span 12–17 mm. Prominent hair scales on head. Very agile but female flies with reluctance. Larva nearly always spins tube of silk. Pest of wool in warehouses and dwellings.

TINEA PELLIONELLA   CASE-BEARING CLOTHES MOTH

Moth dusky-brown with three dark spots on forewing, sometimes feint, 12–17 mm wing span. The habit of the caterpillar of *T. pellionella* of living in a case of silk which includes shreds of cloth fibre and which it drags about as it moves along, is probably the best means of recognising this species. The case is from 6.25 to 9.4 mm long and the larva dies if separated from its case. It rarely spins a web of silk. It is less common than *Tineola bisselliella* and feeds on wool and hair, but it also recorded as feeding on a very wide variety of materials from feather pillows to a number of different drugs of vegetable origin.

NEMAPOGON (TINEA) GRANELLA CORN MOTH

Moth creamy-white in colour mottled with brown patches on forewings; hindwings grey; wing span 10–14 mm. This species is now cosmopolitan in its distribution but seems to vary in abundance in different countries. In the UK and in the USA it is of minor importance. It appears to do more damage in the Baltic region than elsewhere in Europe where it appears to be mainly injurious to rye. This may well be associated with the rather high moisture contents at which grain is harvested and stored in eastern Europe. It is of less importance as damaging wheat and of no importance to oats and barley. At 22 °C its life-cycle is completed in 10 weeks.

PYRALIDAE SUBFAMILY GALLERIINAE

In the moths of this subfamily, sometimes considered as the family GALLERIIDAE, the hindwings have a narrow fringe and are broader than the forewings. The venation of the forewing distinguishes these from moths of the subfamily PHYCITINAE, described later.

PARALIPSA GULARIS STORED NUT MOTH

Until recently this was referred to as *Aphomia gularis*. Although *P. gularis* may be carried into warehouses storing a wide range of commodities, it has become established only in those stored nuts such as walnuts, almonds, hazelnuts and groundnuts; without them it dies. It can survive in northern climates, and its preference for nuts in Britain is striking, for in other countries it attacks a wide range of food commodities. This species has greyish-buff forewings marked in the female with a distinct black patch or spot. The male moth has a smaller spot and a zig-zag, reddish-yellow streak across the hinder part of the forewing. The wing span is 15–30 mm. *Paralipsa gularis* infests almonds and other nuts and occurs in South Europe, South-East and East Asia and the USA, only rarely reaching Britain.

CORCYRA CEPHALONICA RICE MOTH

The Rice moth was first described from the Grecian archipelago, but it is now cosmopolitan in distribution. The Rice moth is buff brown with a tuft or crest of scales on the head. The veins of the wings are slightly darkened. The labial palps (inconspicuous in the male), are prominent in the female and project straight forward. The wing span is 14–24 mm. In the tropics it replaces *Ephestia kuehniella* as the major pest in flour mills. It is particularly damaging to cocoa and confectionery in the USA, while in many countries it attacks rice and other cereals, oilcakes, oilseeds and cocoa beans.

SUBFAMILY PHYCITINAE (SOMETIMES TREATED AS A SEPARATE FAMILY PHYCITIDAE)

In this subfamily of moths the fringes of the wings are also narrow, but adults may be distinguished from members of the GALERIINAE by the venation of the forewings. The PHYCITINAE contains a small number of species which injure grain, cocoa, dried fruits, nuts and tobacco. The most important are *Plodia interpunctella* and several species in the genus *Ephestia*.

In 1956 Heinrich separated *kuehniella* from the other species of *Ephestia* and placed it in the genus *Anagasta*. In 1960 Whalley re-examined the species and proposed that only *elutella* should remain in the genus

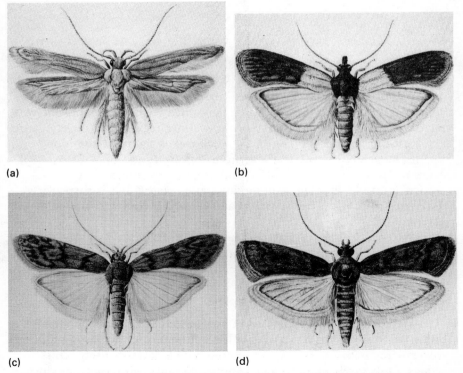

Fig. 18.6 (a) *Sitotroga cerealella*, Angoumois grain moth. (Degesch); (b) *Ephestia elutella*, Warehouse moth. (Degesch); (c) *Ephestia Kuehniella*, Mill moth. (Degesch); (d) *Plodia interpunctella*, Indian meal moth. (Degesch)

Ephestia. He placed the other species in the genus *Cadra*. Today, all species are again assigned to the genus *Ephestia*.

EPHESTIA ELUTELLA   WAREHOUSE, COCOA OR TOBACCO MOTH

This species has been known as injurious to cacao beans since 1884, but its life history has only been known in detail since 1933. There is only one generation in a year, but a small second generation sometimes occurs. The existence of a larval diapause is important since this is how the species passes the winter. The Warehouse moth is a general feeder breeding in grain (tending to concentrate on the embryos), pulse, cocoa, dried fruits, nuts, and tobacco. It can build up huge populations in these commodities and is, potentially, one of the major causes of concern regarding damage to stored products in Great Britain. The wings of the moth are grey with a broad, lighter band at the hind margin of forewings which also possess transverse bands and a dark patch. It flies actively. Fortunately with the development of residual insecticides, *Ephestia elutella* is vulnerable in the wandering, larval stage and is perhaps of less importance now than it once was. Improved standards of hygiene in warehouses and much shorter periods of storage have played their part also.

EPHESTIA CAUTELLA   TROPICAL WAREHOUSE MOTH OR DRIED CURRANT MOTH

This species closely resembles *Ephestia elutella*, and can be distinguished from it by examining the genitalia. Wings of the moth are grey, banded transversely with lighter and darker colours. It flies actively. Larvae are dirty-white with dark brown heads; they often cover their foodstuff with sheets of webbing. It is a native of the warmer regions and ranks next to the Rust-red flour beetle (*Tribolium castaneum*) as the most common species introduced into Great Britain in ships' cargoes. In West Africa it takes the place of *Ephestia elutella* in damaging cocoa beans. A general feeder, it attacks the same commodities as *E. elutella* but is particularly associated with dried fruit currants and cereals. It is now cosmopolitan in distribution.

Three other species of *Ephestia*, *E. figulillea*, *E. calidella* and *E. woodiella* are occasionally imported into Great Britain in dried fruits and carobs from the Mediterranean.

EPHESTIA KUEHNIELLA   MEDITERRANEAN FLOUR MOTH OR MILL MOTH

This species has long been known to be injurious in graneries and flour mills, especially in the latter – hence its common name. Wings of moth are silvery-grey with zig-zag pattern of grey and black transversely across wings. Only rarely does it attack commodities other than cereal products, but in flour mills it spoils far more than it consumes and the webbing which the larvae spin 'clumps' grain and flour which blocks conveyor spouts and

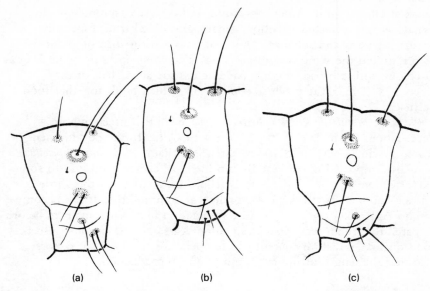

Fig. 18.7 Setal maps of eight abdominal segment for three phycitid moths: (a) *Ephestia cautella*; (b) *Ephestia elutella*; (c) *Ephestia kuehniella*. (After Aitken)

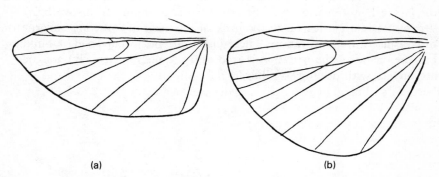

Fig. 18.8 Venation of hindwing of two moths: (a) *Ephestia elutella*; (b) *Pyralis farinalis*. (After Hinton and Corbet)

other parts of the milling machinery. *Ephestia kuehniella* may have five generations a year in warm mills. Larvae pupate in crevices often at some distance. It can develop in flour of very low moisture content, and also in whole grain. Heavy infestations of *E. kuehniella* cause souring of grain and flour.

PLODIA INTERPUNCTELLA INDIAN MEAL MOTH

Plodia interpunctella gets its common name from the United States where it damages meal made from 'Indian corn' or maize and is, perhaps, like maize itself, of South American origin, but is now cosmopolitan. In Britain it first attracted attention as damaging dried fruit, especially raisins, currants and sultanas. This moth has a wing span of about 15.5 mm and the wing fringes are long. The wings are somewhat bronzy in colour with a broad, greyish band across the apical half of the forewings. The larvae produce copious silk which binds together food particles.

Plodia interpunctella with *Ephestia elutella* were the insects whose depredations in the late 1920s and early 1930s led to the development of stored products entomological research in Britain. *Plodia interpunctella* is regularly imported into Great Britain in ships' cargoes and was to some extent endemic in warehouses in Britain, but, apart from this, little is known about it. It attacks a wide range of commodities. It has become rare in Australian dried fruits since fumigation in cargo containers before dispatch to Britain was introduced some years ago. As in most insects, food quality, moisture content and temperature are the main factors affecting the length of life-cycle and rate of reproduction.

Irritating hairs

In several families, but particularly those in the superfamily BOMBYCOIDEA, the larvae live gregariously. The larval hairs, dropping from overhead vegetation in a garden on to the back of the neck of someone underneath, often cause intense irritation. These insects do not occur inside a building, but the origin of the irritation is frequently (quite erroneously) assigned to some insect within a dwelling.

Chapter 19

Thrips

THYSANOPTERA

THYSANOPTERA, or thrips, are small or extremely small, winged insects, usually yellowish-brown to black in colour. The antennae are 4–9-segmented but usually 7–8-segmented and the wings are narrow, strap-shaped with reduced venation and with long fringed cilia. These fringed wings usually make identification relatively simple, although a few species do not possess wings. They have asymmetrical, piercing and sucking mouthparts, and the majority feed either on vascular plants or fungi, although a few are predatory on other small arthropods. Some species may be so abundant that they cause serious harm to cultivated crops. Most winged species flex the abdomen upwards in preparation for flight, although many may be reluctant to fly. Some, notably *Limothrips cerealium*, the Corn thrips, in Europe, migrate in very large numbers in late summer to take up winter quarters, usually during sultry weather - conditions. During these periods they may enter dwellings and subsequently secrete themselves in minute crevices. Grain thrips have the habit of crawling into any opening which is small enough to bring their upper and lower surfaces into contact with their surroundings. Individuals find their way into such places as cracks, behind wallpaper, picture frames and even into jewellery.

In the United States the Onion thrips, *Thrips tabaci*, is apt to bite people, causing a pinching sensation followed by a slight itching, and the Grass thrips, *Anaphothrips obscurus* migrates into houses. Sometimes thrips collect on blankets hung outside and these bite when taken inside.

Chapter 20

Bugs

HEMIPTERA

The HEMIPTERA, known as 'bugs', are characterised by the form of their piercing and suctorial mouthparts. Two pairs of wings are normally present, the anterior pair being tougher than the posterior. In the suborder HOMOPTERA the forewings are uniform while in the suborder HETEROPTERA the apical part of the forewing is more membranous than the remainder. There is a gradual metamorphosis which is only rarely complete. A dorsally-grooved sheath, which is the labium, receives two pairs of stylets in the form of bristles which are the modified mandibles and maxillae. This order, the dominant group of the exopterygote insects, is of the greatest importance to man on account of the vast amount of damage caused to vegetation, either directly or indirectly. This is brought about through feeding by the sucking of sap and in some cases by the transference of disease viruses by doing so. In certain groups a remarkably high reproduction rate contributes to this.

In the HOMOPTERA, in addition to the character given above, the wings generally cover the abdomen and are usually sloping tent-wise although wings are often absent in some groups. The pronotum is small and the number of tarsal segments is between one and three. In spite of the great significance of the HOMOPTERA to man's economy in relation to his crops it is of little importance in buildings.

Heteroptera

In the suborder HETEROPTERA a number of families feed on animal matter being predacious on a variety of other animals. The families REDUVIIDAE and CIMICIDAE are of the greatest importance as far as man is concerned in buildings.

REDUVIIDAE

There are over 3,000 species in the REDUVIIDAE and they show a remarkable variation of form. Most of them are predacious on other insects, catching them then sucking their liquid tissues. They are sometimes called assassin bugs. Some species attack man and transmit a trypanosome, causing serious disease. REDUVIIDAE are characterised by the presence of a prosternal stridulatory groove, a three-segmented rostrum and filiform antennae, *Rhodnius prolixus* and species of *Triatoma* carry *Trypanosoma cruzi* which brings about human trypanosomiasis in South America which is often fatal.

The REDUVIIDAE is a family showing an exceedingly wide variation in form. Members of it are characterised by possession of filiform antennae of four or five segments, a three-segmented rostrum and a stridulatory groove on the prosternum but scent glands on the metathorax are absent. Many are predators, generally sucking the blood of other insects but some prey upon higher animals while a few species attack man. They are known as 'assassin bugs', 'conenoses' and 'kissing bugs'.

Chaga's disease

Several species, but principally *Panstrongylus megistus*, *Triatoma infestans* and *Triatoma gerstaekeri* which suck the blood of Man are carriers of the trypanosome, *Trypanosoma cruzi*, the pathogenic organism responsible for Chaga's disease. It occurs in Central and South America from Mexico to the Argentine and in certain of the West Indies. In these regions it poses an important public health problem.

The bugs generally favour certain hiding places when resting. In Panama the bug *Rhodnius pallescens*, for example, is generally found near beds in cracks in walls; in folds in bedclothes; in crevices in the bedstead; underneath hanging clothes on the wall near the bed and in the roof thatch over the bed.

Several million cases of the disease occurred between 1909 and 1961 with thousands of deaths. Occasionally cases of the disease are recorded from Texas in North America, and the trypanosome has been demonstrated in wild mammals or *Triatoma* species in 8 states, but potentially it could occur in 27 states. In addition to the species given above, other actual or potential carriers of the trypanosome are *T. recurva, T. lectularius, T. rubida, T. sanguisuga, T. protracta* and *T. neotomae*.

The bugs are usually found in unhygienic buildings where at night they bite the faces of human inhabitants. The bugs are infected by sucking blood from mammalian reservoirs such as dogs, cats, pigs and rats, as well as from wild animal hosts such as the armadillo, and also other human beings.

All stages of the bug may transmit the infection but the adult stage is the most important. After two or three weeks the parasites assume the infective metacyclic form and the bug remains infective for the rest of its life.

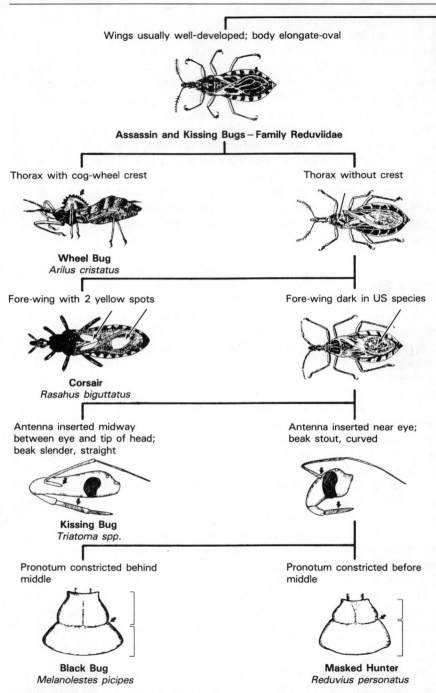

Fig. 20.1 Pictorial key to some adult bugs that may bite Man. (US Department of Health, Education and Welfare)

BUGS 241

Wings reduced; body broadly-oval

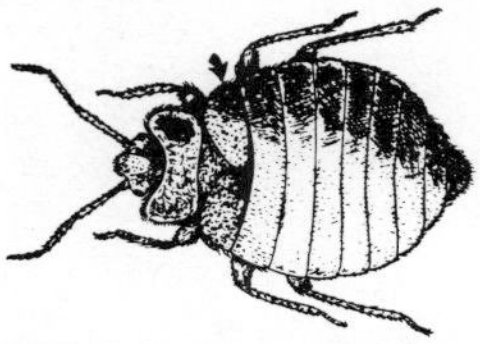

Bed Bugs – Family Cimicidae

Middle coxae nearly touching. Beak reaching 2nd coxa

Poultry Bug
Haematosiphon inodorus

Middle coxae widely separated. Beak not reaching 2nd coxa

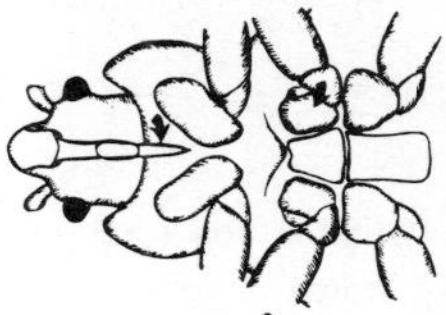

3rd and 4th antennal segments equal

Barn Swallow Bug
Oeciacus vicarius

4th antennal segment shorter than 3rd

Fringe hairs on pronotum longer than, or equal to, width of eye

Bat Bugs
Cimex adjunctus E. N. AM.
Cimex pilosellus W. N. AM.

Fringe hairs on pronotum shorter than width of eye

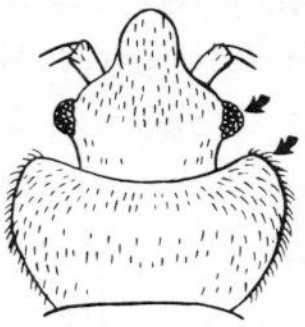

Pronotum with anterior margin moderately excavated

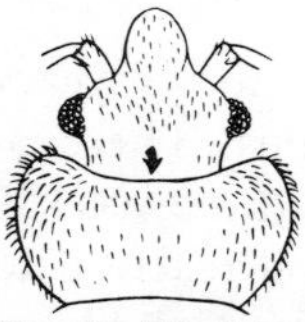

Tropical Bed Bug
Cimex hemipterus
SO. US. AND TROPICS

Pronotum with anterior margin deeply excavated

Bed Bug
Cimex lectularius
TEMPERATE AREAS

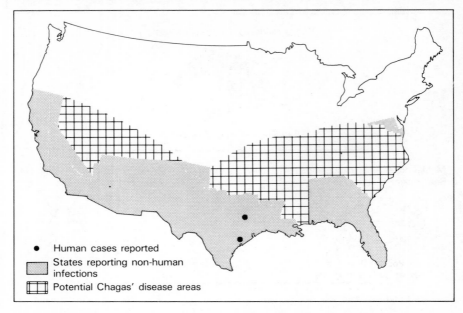

Fig. 20.2 Chaga's Disease distribution in the United States. (US Department of Health, Education and Welfare)

After being bitten the site of the bite is scratched and thereby the organisms present in faecal matter are introduced into the puncture. Generally this is around the eye, nose or mouth, but may be rarely found anywhere over the body where the skin is exposed. Children are most commonly infected and transmission across the placenta is recorded as well as infection from swallowing the bugs. Artificial infection of the bedbug has occurred but this insect is of no importance in natural transmission of the trypanosome to man.

After the bite a small subcutaneous swelling occurs at the puncture. The organisms spread via the lymphatics to the lymph nodes which enlarge. They then reach the bloodstream within a few days and enter tissue cells and these multiply in the leishmanioid forms. This is repeated until large numbers of the organisms cause damage, especially to cardiac muscles. The death rate is heavy in children.

The acute disease is commonest in children from one to six years of age but it may occur in much older non-immune persons. There is an incubating period generally lasting from a few days to three weeks when the onset of the disease is usually sudden, although local reaction to the infective material may occur. There is a small raised plaque which is red and this is surrounded by an area of loose oedema. This is known as the *chagoma* and it is extremely painful. It takes three to four days to reach maximum size and then subsides. Patches of oedema occur elsewhere, particularly on the face and spread over the eyelids and neck. The

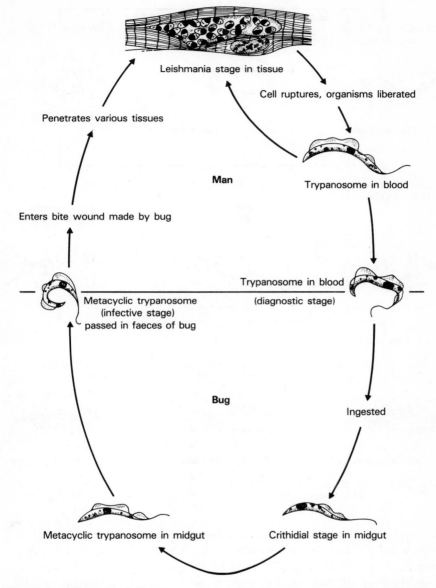

Fig. 20.3 Life-cycle of *Trypanosoma cruzi*. (US Department of Health, Education and Welfare)

lacrymal glands are usually swollen and there is excessive lacrimation. The conjunctiva is intensely congested and the eyelids are often glued together with dried secretion. A moderate remittent fever persists for some weeks and the pulse rate is very fast. When the myocardial damage is severe, fatal cardiac failure occurs within a week of the onset.

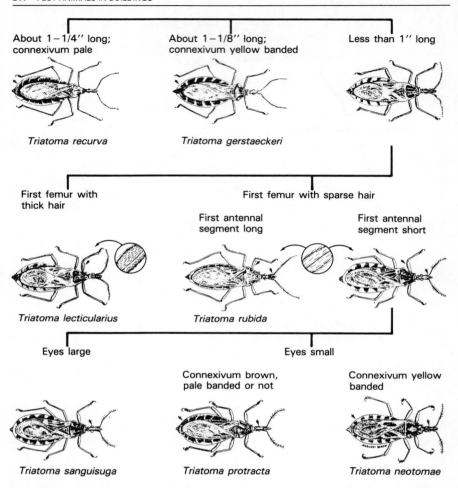

Fig. 20.4 Pictorial key to some common species of *Triatoma*, Kissing bugs, in the United States. (US Department of Health, Education and Welfare)

Generally, however, there are increasing signs of heart involvement which may take place over several weeks. The usual terminal event is dilatation and failure brought about by the cardiac muscular damage. Confirmation depends on the demonstration of the trypanosomes in the blood. Treatment is generally unsatisfactory as drugs which are successful in African trypanosomiasis appear to be ineffectual. Control depends on the elimination of the insect vectors.

CIMEX LECTULARIUS   BED BUG

Cimex lectularius, the Bed bug, probably originated in the Middle East and it has been known in the Mediterranean region since the earliest recorded times. However, it did not reach Germany until the eleventh century, England until the sixteenth century, whereas it was not found in Sweden until early in the nineteenth century. A second species of *Cimex*, the Tropical bed bug, *C. hemipterus*, occurs in tropical areas only, whereas *C. lectularius* is distributed throughout most regions of the world. Other species of CIMICIDAE parasitise bats and birds of the family HIRUNDINAE (swallows and martins).

The adult stage of *C. lectularius* is about 6 mm in length and leaf-like, the width of the abdomen being twice that of the saddle-shaped pronotum. The colour varies from amber to more reddish or 'mahogany'. The antennae are well developed, although consisting of only four segments. The hemispherical compound eyes consist of about 30 facets each. The stylets (mandibles and maxillae) are carried in a groove on top of the three-segmented labium. This is known as the proboscis and is usually held bent backwards under the head. When preparing to feed, however, it is extended forwards. Wings are absent in both sexes, the vestiges of the forewings appearing as a pair of scales which almost cover the mesonotum. The legs are well developed with very efficient tarsal claws. The top of the abdomen is pointed in the male but more rounded in the female.

Another character by means of which the adult female can be identified from the male is the presence of a small, nick-like cleft at the posterior margin of the fourth apparent abdominal segment. This is the opening of the copulatory pouch.

Habits

During the day bed bugs secrete themselves into cracks and joints in furniture around the position taken up by a sleeping person. At night they wander about seeking their host, but it is not known how this is accomplished except at short range (50–75 mm), the stimulus appearing to be warmth.

When the bug feeds, the skin is pierced by the stylets while the labium maintains contact with the skin surface. The outer stylets (mandibles) are needle-like and pierce the skin and lacerate it, there being about 20 teeth near the tip of each. The inner pair (maxillae) are broader as well as longer, and are grooved on the inner side. However, basally they knit together to form two canals. The larger, dorsal channel is the one through which the blood is drawn up, while the smaller, ventral channel conducts saliva to the wound in order to lubricate the process. A full meal of blood is taken from 5 to 10 minutes and there is sufficient for the requirements of each instar.

Eggs

Eggs are laid in large numbers in cracks in furniture or in crevices near to where people sleep. The total number of eggs laid by one female at 25 °C is about 345 but about 5 per cent will be sterile. With two blood meals per week continuous egg-laying takes place at about three eggs per day.

They are covered with a quick-drying cement which fastens them down permanently so that long after hatching the egg shells remain. The egg is bag-shaped, with the 'open' end covered with a circular cap. They are variable in size from 0.8 to 1.3 mm in length and from 0.4 to 0.6 mm in width. Some time before hatching the eyes of the embryo bug can be discerned as two pink spots, then the egg-cap is forced off and the young bug struggles out. Whereas the unhatched eggs are opaque, the empty shells are translucent.

The young nymphs resemble closely the adults in shape and share many of their habits. They hide in crevices during the day, and because of their extreme flatness they can secrete themselves into very narrow apertures. At night they emerge, and when, wandering about, they come across a sleeping person or a mammal, they take a meal of blood. There are five nymphal instars, all resembling the adult shape. The cuticle of the abdomen, however, is thinner than that of the adult, and through it the colour of the partly digested blood in the gut can be made out. After each blood meal, the abdomen expands greatly, as the quantity taken

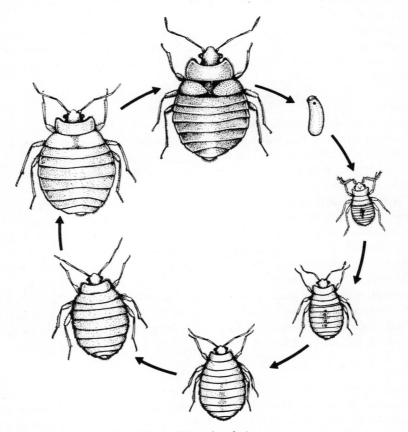

Fig. 20.5 Life-cycle of Bed bug, *Cimex lectularius*.

may be from two to six times the original weight of the bug.

The faecal matter is a viscous liquid varying in colour from pale straw to nearly black and when deposited on walls in the vicinity of their hiding place, gives a characteristic speckled appearance. On the back of the abdomen in the nymphs, but beneath the thorax in the adults, are 'stink-glands' which give off a characteristic unpleasant odour when the bugs are disturbed. This, together with the odours of an unhygienic bedroom can be thoroughly nauseating.

Bite

The human reaction to bed bug bite varies enormously. Whereas some individuals are not affected others are extremely sensitive. As a rule a small white hard swelling is produced accompanied by severe itching or inflammation. Bites on the face and neck sometimes lead to swelling and temporary disfigurement. The irritation is brought about by the secretion of the salivary gland which contains blood anticoagulants. Bed bug bites can be differentiated from those of the flea as the latter are indicated by a central red spot surrounded by a circular reddish halo fading into the normal skin colour. Immediately after feeding the bug withdraws, then defecates the remains of its previous meal in the vicinity of the minute wound.

No pathogens of human disease are known to be habitually carried by bed bugs or by any other species of CIMICIDAE which attack man from time to time. Bed bugs, however, are loathed and abhorred not only for their painful bites but also because of their association with unsanitary and unhygienic living conditions.

Life-cycle

The process of fertilisation in *C. lectularius* is unusual. The male genital organ is sickle-shaped, curving round the tip of the abdomen to the left. It is not introduced into the female's egg pore but into a curious copulatory pore underneath the abdomen to the right. The sperms bore their way through the walls of the bag-like end of the pore and make their way to the oviducts.

The length of the adult life is temperature dependent. At normal room temperatures (18–20 °C) this is from 9 to 18 months but at 27 °C this is reduced to about 15 weeks and at 34 °C to only about 10 weeks. Bed bugs show fair resistance to short periods of low temperature. Only about 25 per cent are killed when exposed to one hour at −17 °C, but two hours kills all of them. Considerable mortality of eggs and immature stages occurs at prolonged low temperatures prevailing in unheated rooms in winter in Britain (0–9 °C). In these conditions the first two instars take from 100 to 200 days and the eggs fail to hatch after 30 to 60 days. At high temperatures eggs are killed on being exposed for one hour at 45 °C.

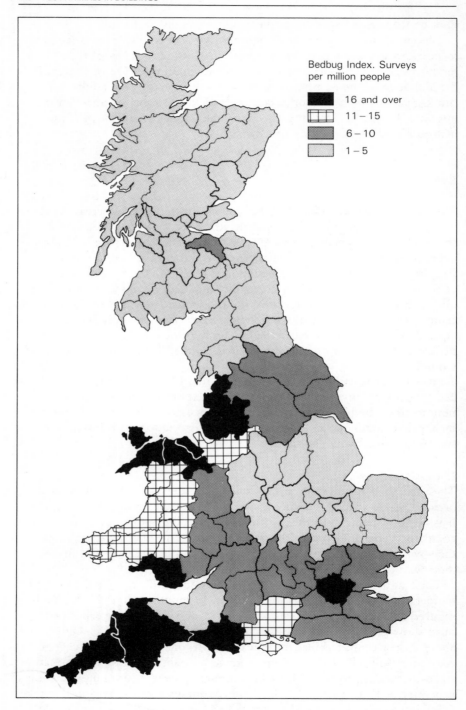

Fig. 20.6 Bed bug index. Surveys per million people in Britain.

Predators

A number of insects and other invertebrates are recorded as predators of bed bugs. House-invading FORMICIDAE have been seen to fall on them and, after dismembering them, carry them away. Chief among these are *Monomorium pharaonis*, Pharaoh's ant, and *Iridomyrmex humilis*, the Argentine ant. Another hemipteran, *Reduvius personatus*, the Cone-nose, has been recorded as sucking nymphs and adults. It is thus feeding on human blood at secondhand. *Thanatus flavidus*, a web-spinning spider will take 30–40 bed bugs in a day, and *Chelifer cancroides*, a pseudo-scorpion will also prey on them.

Distribution in Britain

Surveys by the largest pest control firm in Britain show that bed bugs are now of relatively minor importance ranking only thirteenth compared with other undesirable species in buildings. Between 1967 and 1972 the numbers remained reasonably constant. The seasonal variation in numbers showed an increase during the summer months which is thought to show holiday movements of people. Over 60 per cent of the infestations were in domestic properties and 25 per cent in holiday situations. The remaining 15 per cent were spread equally among hospitals, public meeting places, shops and manufacturing buildings. Bed bugs occurred most frequently in seven different areas, Cornwall, Devon, Dorset, Lancashire, London, Glamorgan and North Wales. There were few infestations of this insect reported from the major cities of Scotland and the Midlands. Infestations arise from the movement of people and it has been suggested that the infestations in South Wales and South-West England are the long-standing result of the evacuation of children from the major cities during the Second World War.

CIMEX HEMIPTERUS

Cimex hemipterus, the Tropical bedbug is separated from *C. lectularius* on account of its lack of wing-like extensions to the pronotum and the hairs fringing the pronotum curve backwards instead of being relatively straight. It occurs widely in the tropical areas of Africa, Asia and America.

A number of other species of CIMICIDAE are known sometimes to attack man in buildings although their usual hosts are bats and birds. *Cimex pilosellus*, the Bat bug often preys upon bat infestations in roof voids and attics in the western United States, while in the east *C. adjunctus* is the species concerned. *Cimex pipistrelli* will also on occasion turn its attention from bats to man. *Cimex columbarius* is commonly associated with feral pigeons, starlings and other birds, both in Europe and North America, but can also feed on man. A small bug, only 3 mm in length, parasitises the Chimney swift and occurs in the eastern United States as well as the middle-west, and isolated instances are recorded of its attacks on man.

Oeciacus vicanus, the Swallow bug of North America, breeds in the nests throughout the summer, but in the autumn, sometimes even before the birds migrate, the bugs scatter and often invade adjacent living areas. For a few days they bite humans viciously, sometimes being the cause of red areas, 50–75 mm across, which itch incessantly. Occasionally those bitten exhibit swellings on the mouth, arms and face.

Oeciacus hirundinis, the European barn swallow bug, is recorded as biting human beings.

Haemotosiphon inodorus, the Poultry bug, has a resemblance to the bed bug but the legs are longer and it usually shows greater activity. It also lacks the bed bug odour. This bug occurs in the south-western states of the United States and Mexico where it is an important pest of poultry. Hiding during the day, at night it emerges from the crevices in which it has hidden and sucks the blood of the roosting fowl. Where fowlpens are close to dwellings it can be a serious human pest.

LYCTOCORIS CAMPESTRIS

Three species in the LYCTOCORINAE are found in warehouses where food is stored. They feed on larvae of moth and beetle pest species. *Lyctocoris campestris*, the Debris bug, is European and often bites warehouse workers. *Xylocoris galactinus* and *X. flavipes* are tropical and widely distributed, often occurring in large numbers in stored grain and peanuts, where they feed on the immature stages of intesting insects.

Chapter 21

Sucking lice

SIPHUNCULATA (= ANOPLURA)

Phthiraptera

Members of the order PHTHIRAPTERA are known as lice and there are two important suborders which concern us here. They are sometimes considered as separate orders. These are the MALLOPHAGA, the chewing or biting lice, and the SIPHUNCULATA (ANOPLURA), sucking lice.

All species in the order are wingless, dorsoventrally flattened and with either mandibulate mouthparts or piercing and sucking. All are ectoparasitic on birds or mammals and the entire life is spent on the host. About 3,000 species are known and the range of size is from less than 0.5 mm to 10 mm. In addition to the characters given above, which are modifications for their life among feathers and fur, the tarsi are modified for clinging to the host. Whereas some species are covered with long setae, others are almost bare and generally they are well sclerotised. Eyes are either much reduced or absent and the antennae are from three- to five-segmented.

Siphunculata

There are about 250 species of this suborder of sucking lice distributed throughout the world. They are ectoparasitic on many mammal species and are even found on seals which spend the greater part of their time in the water. They have almost transparent flattened bodies but the legs are relatively large and robust with strong exaggerated tarsal claws adapted for gripping hair; they are prehensile. The tracheal respiratory system opens on to the upper body surface. The antennae are short and the mouthparts, which are in the same line as the remainder of the body, are adapted for piercing skin and sucking blood. There are no traces of wings but eyes are present in the family PEDICULIDAE although absent in other families of the suborder.

There are two species of sucking lice parasitic on man but one occurs as two subspecies. These are *Pediculus humanus capitis*, the Head louse and *P. humanus corporis*, the Body louse. The second species is *Phthirus pubis*, the Crab louse or Pubic louse.

The two subspecies of *Pediculus humanus* differ little from each other except in average size and proportion. The extremes of measurements overlap. Body lice are usually from 10 to 20 per cent larger than head lice and the latter have relatively stouter antennae, the indentations between the abdominal segments are deeper and the abdominal musculature is not so well developed. The physiology of the two subspecies is similar although the body louse lives longer, has greater resistance as far as starvation is concerned and lays a greater number of eggs. *Pediculus humanus corporis* lives among the underwear whereas *P. humanus capitis* inhabits the hair of the head.

PEDICULUS HUMANUS CAPITIS   HEAD LOUSE

The piercing mouthparts consist of three components. The upper and lower stylets form the channel up which blood is drawn while the intermediate stylet passes saliva into the puncture from the salivary glands. The resulting irritation is due to the action of the louse's saliva. The louse is adapted to walking along a hair almost like a tightrope. The long, sharp terminal claws hook around the hair and press against a peg-like projection on the apex of the adjacent segment, and thus it is able to gasp the hair. The sexes can be differentiated by the naked eye, as the apex of the abdomen of the female consists of two triangular projections whereas in the male the abdomen terminates in a single point. The claws of the front legs of the male are enlarged and with these he grasps the hindlegs of the female at the commencement of copulation. The male is then beneath, and the elaborate penis is inserted into her genital aperture by curving the abdomen. Copulating pairs walk about and also feed.

Life-cycle and habits
Eggs are laid on the hairs of the head close to the scalp and are cemented at the base of the individual hair. The adhesive is remarkably persistent and the empty egg shell remains in position long after hatching and only leaves the vicinity of the scalp with the growth of the hair.

The egg of the Head louse is known as the 'nit'. It is oval, 0.8×0.3 mm, comparatively large considering the size of the louse. Before hatching they are yellowish white and pearly, but afterwards they become opalescent and translucent. Whether an egg is hatched or unhatched can, with practice, be determined by the naked eye or with a $\times 10$ lens. On hatching a circular cap falls off, allowing the young louse to emerge. The first stage nymph is light yellow in colour and feeds shortly after emergence when the blood can be seen through the transparent cuticle. In older lice the gut is purplish-black, and feeding occurs at least twice daily. After the third moult the louse is adult – shown by the presence of external sexual organs.

There is considerable variation in colour, ranging from dirty white to greyish black, according to the colour of the hair in which the louse has spent the immature stages. Light-coloured lice occur in people with light-coloured hair and dark on dark. When moving, the antennae are moved from side to side, and although the eyes are poorly developed, the louse shuns the light. The mouthparts are not usually visible when the insect is not feeding as they are withdrawn into the head. When feeding takes place, however, the mouth pouch is pressed against the skin when recurved teeth hold it in position and allow the piercing organs to enter.

PEDICULUS HUMANUS CORPORIS

In most respects the life history of the Body louse is similar to that of the Head louse. The eggs, however, are not usually cemented to hairs but to fibres of clothing. Body lice congregate in large numbers along the seams of clothing next to the skin, and it is in this sort of location that eggs are laid. Body lice are attracted to each other by their smell and the smell of their excrement. From this situation the lice move towards the skin to feed. This is usually when the host is sitting or lying down, and they become very active when the host is hot. On heavily infested persons a few lice may be seen on the outer clothing, although most infestations consist of only about a dozen lice. Infestations consisting of several hundreds of lice are comparatively rare, whereas those involving one or two thousand lice are very rare.

PHTHIRUS PUBIS   CRAB LOUSE

The life history of *Phthirus pubis* is similar in many respects to that of *Pediculus* spp. The egg is somewhat smaller, however, and the cap is more convex. Nymphs and adults do not move about to anything like the same extent as the Body and Head louse, generally staying in one position with the legs of both sides grasping hairs and blood is sucked from time to time. The large club-shaped claws are adapted to holding hairs of large diameter and the holding of the body flat against the skin requires the hairs to be more widely spaced than those on the head. The Crab louse occurs, then, most commonly in the pubic and peri-anal areas. More rarely it is found on the thighs, abdomen, axillae, eyebrows, eyelashes and around the scalp.

LINOGNATHUS SETOSUS   DOG-SUCKING LOUSE

The general shape of this louse is similar to that of *Pediculus humanus*, except that the abdomen is broader and eyes are absent. The male is about 1.5 mm in length and the female is about 2.00 mm. This louse infests long-haired dogs in many parts of the world, and often occurs in large numbers, usually on the back of the dog, on the flanks and at the base of the tail. The attentions of the louse cause irritation and loss of condition, shown by a lifeless-looking coat. Little information is known

Fig. 21.1 *Pediculus humanus corporis.*

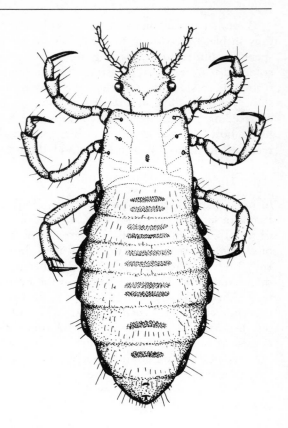

Fig. 21.2 *Phthirus pubis.* (British Museum, Natural History)

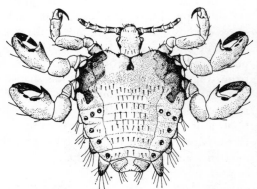

of the biology of this species, but all stages are passed in the fur of the host. The immature stages of the Dog tapeworm, *Dipylidium caninum* are passed in this louse.

Transmission of body lice
It is impossible to be dogmatic as to the manner in which Head and

Fig. 21.3 *Linognathus setosus*, Dog-sucking louse.

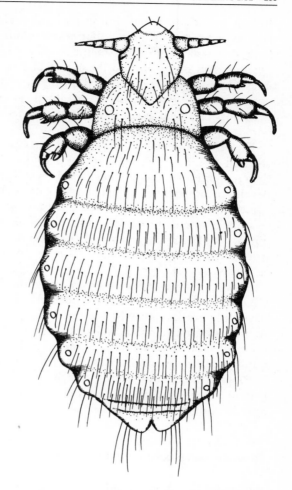

Body lice are passed on from one person to another, as by the nature of the transmission any method would be difficult to substantiate. But as both subspecies are unable to survive for any length of time away from the human body, transmission must usually take place during an association of a moderately intimate nature with an infested person. Bedlinen, blankets and clothing are probably implicated to a much less extent for the reason given above. From work carried out in Britain it has been found that the Head louse is common among girls and young women in some industrial areas, even though of clean habits. It has been found also that women and girls are more likely to be infested due to their longer hair, although present fashions among the male population may cause this to be rewritten to read that long-haired persons are at greater risk in this than those with shorter hair!

Head lice are occasionally acquired by children at school when the infestation will often spread through the family with rapidity. Children in

Britain are subject to medical inspection regularly, including examination for the presence of parasites, but when disinfected often become reinfested again from other members of the family. Persons infested with *Phthirus pubis* occur only to the extent of 1 in 30 or so of those louse-infested. It is usually distributed during sexual intercourse, but occasionally the parasite occurs in situations where this could not have been possible, such as on the scalp of infants and on eyelashes of persons without pubic infestations.

The environmental temperature of the Crab louse is equable as it lives adjacent to the skin in areas of the body usually covered with clothing. In these conditions the egg takes 7–8 days to hatch and nymphal development occupies a further 13–17 days. The length of the adult stage is rather less than a month. There are few records of egg-laying but it is at a lower rate than *Pediculus*. One female laid three eggs each day for nine days. The adults are unable to survive for any length of time away from the human body. Of 200 taken away only one survived for as long as 24 hours.

Bites of *Phthirus pubis* are characterised by blue spots from 0.2 to 0.3 cm in diameter with an irregular margin and they persist for several days. It is not known whether this louse plays any part in disease transmission but judging from its sedentary habits it would appear to be minimal. *Pediculus humanus*, on the other hand, is associated with a number of important diseases of Man. It is the carrier of the organisms causing exanthematic typhus, trench fever and a relapsing fever. *Pediculus humanis corporis* is the subspecies concerned, although *P. h. capitis* has been found capable of transmitting the pathogens in the laboratory.

Epidemic louse-borne typhus

Epidemic louse-borne typhus fever has been one of the most important epidemic diseases inflicting the human race, certainly within the last four centuries. Following in the wake of wars, famines and human misfortunes of all kinds, it is said often to have had a more decisive effect on military campaigns than the actual battles themselves. It has thus played a major role in the history of mankind. Even in this century at least 3 million deaths from this cause took place in eastern Europe and Russia between 1918 and 1922.

Typhus is a generic name for three distinct diseases which clinically and pathologically are almost the same. These are epidemic louse-borne typhus fever, Brill–Zinsser disease, and murine flea-borne typhus fever.

Epidemic louse-borne typhus fever is caused by the organism *Rickettsia prowazeki* which has been demonstrated outside the laboratory only in man and the human louse.

Rickettsia

A number of most important human diseases are caused by a group of micro-organisms in the family RICKETTSIACEAE. They occupy a position somewhere between the bacteria and the viruses and, thus, are of a low order of complexity. Little is known of them with the exception of their association with disease which is not surprising as many research workers and medical personnel have died when investigating them. However, they possess four common features as follows:

1. They exist in a different number of forms, coccobacillary in general shape and are visible under the light microscope.
2. They multiply only within certain types of cell within the infected animals.
3. They occur in certain arthropods such as the human Body louse, the Rat flea, ticks and mites.
4. They are the causative organisms of acute, febrile, self-limited diseases of Man, which are usually accompanied by a skin rash.

Micro-organisms of identical appearance to pathogenic rickettsiae occur as harmless symbionts in a number of different arthropods, but it is possible that a rickettsial disease situation may occur whenever Man, rodents and arthropod ectoparasites common to both, are closely associated.

The rickettsiae are present in the blood of a typhus patient especially during the first few days of the febrile period. When a louse feeds on a person the introduced saliva into the puncture causes irritation which causes scratching. As the louse sucks blood so it defecates and the typhus rickettsiae gain entrance to the bloodstream through the skin abrasions and punctures.

A louse becomes the host for the rickettsiae by feeding on the blood of a typhus patient when they are taken into the gut of the louse. They enter the living cells of the intestinal tract and multiply. The cells enlarge and eventually burst and the rickettsiae pass out with the faeces. After a few days, the louse gut becomes obstructed with rickettsia-filled cells, and it dies. Eggs laid by rickettsia-inhabiting lice are not infected. Lice crawl away from a person with a high fever, and also from one who succumbs, so that this is the reason for the transmission of the rickettsiae from person to person during epidemics. Not proved, but generally accepted as another method of transmission from louse to Man, is by contact with infected louse faeces from garments or bedlinen used by a typhoid patient. If the clothing is shaken while the rickettsiae remain viable, they may enter the respiratory tract or conjunctiva of a person. It is thus possible for a non-louse-infected person to become typhus-infected.

In Man, the rickettsiae invade the cells forming the inner lining of small arteries, capillaries and venules. The affected cells proliferate and the injury results in thrombus formation, with small areas of necrosis, and accumulations of phagocytic cells. These produce the typhus 'nodules' which are distributed throughout the organs and tissues of the body, but which are found especially in the skin, the brain and the heart muscle.

The incubation period is about 10 to 14 days, but the onset may be preceded by lassitude and headache. Generally, however, the onset is abrupt with a severe, unremitting headache and general aches and pains. Within two or three days the temperature rises to 40 °C (104 °F), and remains above this for 10 to 14 days. If recovery takes place the temperature will subside in three to four days. Shaking chills may occur, as also may an unproductive cough. Between the fourth and the seventh days, but usually on the fifth day, a rash appears on the back and chest, spreading over the abdomen and the limbs although not on the face, palms and soles, except in severely ill individuals. The rash consists of spots, 1–4 mm in diameter with irregular outlines. At first they are pinkish to reddish, fading on slight pressure, but later are fixed. The rash remains as long as the fever. In the second week, the spots become dark red or purplish with small haemorrhages. In severe cases destined to end fatally, the rashes are purpuric and confluent.

A number of complications may occur, such as skin necrosis, gangrene of the toes, ears, tip of nose or genitals, in the second week of the disease, and bouts of stupor or delirium are followed by coma. Death takes place in 20 per cent of all cases between the ninth and eighteenth days. In children, however, the fatality rate is less than 10 per cent, but increases with increasing age. For persons over 50 years of age the fatality rate is about 60 per cent. 'Adequate immunisation' is possible today on a wide scale. It probably reduces the actual incidence of typhus fever among exposed persons, but it definitely reduces to almost zero the mortality rate as well as reducing the severity of the illness.

Trench fever

Trench fever is an eruptive infection first recognised during the First World War, when it became one of the most important medical problems. It almost disappeared within a few years after the war, but reappeared on a small scale only during the Second World War. The organism concerned is a *Rickettsia* (*R. wolhynica, P.Pediculi, R. quintana*), found extracellularly in the louse gut, and is present in the louse faeces, where it remains viable for several months. The rickettsiae are introduced into the bloodstream by scratching them into the itching louse punctures and abrasions.

After an incubation period of from 10 to 20 days usually, or from 5 to 38 days in extreme cases, there is a sudden onset of fever with temperature rising to 39 °C (103 °F). There is a severe headache especially behind the eyeballs, and pain on rotating them. There is a sometimes nausea and vomiting, but pain and soreness in the muscles, bones and joints is general.

Conjunctivitis and photophobia are also associated with this disease. The characteristic rash appears during the first 24 hours and comes and goes with bouts of fever. The rash consists of red spots from 2 to 4 mm in diameter first appearing on the chest and abdomen and usually not appearing elsewhere. Only occasionally are the limbs involved but never the face. In about 85 per cent of cases a return to work is possible within two months or so. No single fatal case is known.

Brill–Zinsser disease

Another disease recognised as a separate entity since 1910 is, in effect, a recrudescence of epidemic louse-borne typhus fever. It is sometimes called recrudescent typhus on this account. The causative organism is identical with that of epidemic louse-borne typhus fever, *Rickettsia prowazeki*, and has remained in the body of the patient from a previous attack of this disease. Body lice can become infested by sucking blood from a sufferer from Brill–Zinsser disease. The disease is similar in many respects to epidemic louse-borne typhus fever, but is shorter in duration and somewhat milder.

Chapter 22

Feather, biting, chewing or bird lice

MALLOPHAGA

MALLOPHAGA are small, sometimes very small, dorsoventrally flattened insects adapted for an ectoparasitic life, the whole of which is spent on the bodies of their hosts. There are about 1,700 species known, the great majority of which are found on birds, and a small number infest mammals. The principal interest in this group in the present work is on account of their occurrence on the domestic dog and cat which may spend a substantial part of their time in buildings in close proximity to Man.

The mouthparts of MALLOPHAGA are modified for biting and chewing the epidermal products and feathers on which they subsist. The mandibles are toothed and sharp for cutting off lengths of feather barbule or loose or flaking tissue. The maxillae are single-lobed, the maxillae being represented by minute, forked rods sometimes barely visible, similar to the 'picks' in psocids. In the superfamily AMBLYCERA, which takes blood and serum in addition to feathers, the maxillary palpi are four jointed, while in the superfamily ISCHNOCERA, which consists mainly of feather eaters, they are absent altogether. The antennae, which are waved about as the louse moves, are used as tactile organs, except for some of the AMBLYCERA, where they are small and enclosed by a fold in the head. In the ISCHNOCERA, the male often clasps the female with the antennae, which may be specially modified for the purpose. The head and body are generally well sclerotised.

There is considerable variation in the body shape adapted to the specialised areas of the bird's body which they generally inhabit. The body is usually elongated in those species inhabiting the larger feathers of the body and wings, whereas those found on the head and neck are short and round. The comparatively uniform legs are adapted for clinging to feathers and hair, and the lice can move rapidly backwards and forwards across and around the feathers or through the hair, according to species.

The egg is furnished with a cap or operculum, and hatching takes place by the nymph sucking in air and forcing off the cap by air pressure. There is no metamorphosis. The young nymph resembles the adult except in size, colour and the absence of reproductive organs. There are three moults before maturity is reached.

No diseases are recorded as being transmitted by the MALLOPHAGA of birds, but *Trichodectes canis*, which parasitises the dog, is the intermediate host of at least one species of cestode (tapeworm) occasionally found in man.

It is almost certain that no species of bird is without one or more species of mallophagan, and many of these exhibit a high degree of host specificity. Any birds introduced into a building, either as pets or for any other reason, can be expected to harbour a number of them. Small birds generally carry only a few – a dozen or so – but large birds are often the host to several hundred. When a bird dies or is killed, the lice soon move away from the cooling body, and are seen on the outside of the feathers. Lice will usually die unless they are able to transfer themselves to a similar, living host within a few days. Phoresy or 'hitching a lift' is known among mallophagans. They hold on to another animal, such as a hippoboscid fly, which may take them to another host.

When freshly-killed birds, such as the domestic chicken or duck are plucked, the pluckers often find the lice crawling up their arms, and they have to be brushed off. The Common Chicken louse, *Menopon pallidum*, is well known in this respect, and the duck is usually infested with *Anatoecus dentatus*, as well as a number of other species.

TRICHODECTES CANIS   DOG-BITING LOUSE

This is by far the most common biting louse found indoors, and of the two species of louse occurring on dogs it is the most abundant. It is brownish in colour and up to about 1.5 mm in length, while the head is flat across the top and bears a bulbous lobe on each side. It is easily differentiated from *Linognathus setosus*, the Dog-sucking louse, which is greyish in colour with a smaller head and with a longer and more pointed abdomen.

The egg is cemented to the base of a hair, and the nymphs are very similar to the adult in shape. The entire life-cycle takes place on the dog's body.

The Dog-biting louse is parasitic throughout its life and subsists on bits of dead skin. They are not uniformly distributed over the dog's body, but take up certain specific positions. The most usual situation is from the ears down to the base of the neck, but other sites are the saddle, the flanks and the base of the tail. They are only transferred by bodily contact. Their presence can be diagnosed by the dog's irritable behaviour. It will suddenly try to reach high up on its flanks with its nose, or, standing on three legs, it will endeavour to reach the same part with a hind paw. An infested dog will often sit down quickly and then urgently try to sink its nose into the base of its tail, or try to allay an irritation at the back of the ear with a hind paw.

Fig. 22.1 *Trichodectes canis*, Dog-biting louse.

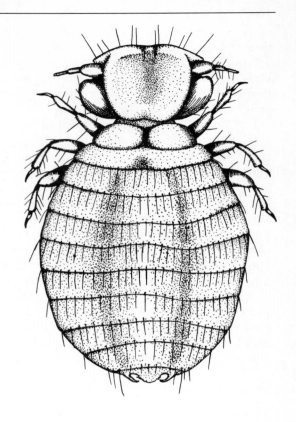

Parasites

The lice and fleas of dogs and cats serve as the intermediate host of *Dipylidium caninum*, a common cestode (tapeworm), parasitic on those animals. Occasionally it occurs in Man, although usually in young children. The worm takes about two months to mature. Transmission of the parasite takes place when an infected louse or flea accidentally enters the mouth of a child, when its face is licked by a dog which has been trying to allay irritation by biting or licking itself.

Two other species of cestode, with an intermediate stage inhabiting insects, parasitise Man, and whereas mallophagan parasites of domestic animals have not been specifically implicated, they cannot be ruled out entirely as transmitting the cestode. These species are as follows:

1. *Hymenolepis nana*, the Dwarf tapeworm is the commonest tapeworm of Man. It is also the smallest, being only a few centimetres in length. Up to several thousand may be present in one person. It is characterised by bearing a crown of hooks and four suckers on the head. The eggs which are liberated in the small intestine are passed

out with the human faeces. If the egg is swallowed by Man a six-hooked embryo is released which develops into a cyst. The larval cestode then emerges and develops to maturity in two weeks. Thus the presence of an intermediate host is not obligatory. At other times, insects such as fleas, and beetles whose larvae subsist on grain, are used as intermediate hosts.

2. *Hymenolepis diminuta*, the Rat tapeworm, often infests Man, several hundred cases having been reported. The human incidence is certainly much greater as its presence may remain undetected. The intermediate stage of the cestode is passed in fleas and lepidopterous and coleopterous larvae feeding in meal grain. Transmission takes place when the latter is eaten uncooked. Again, the possibility of other insects (such as MALLOPHAGA) acting as intermediate hosts cannot be ruled out.

3. *Felicola subrostratus* parasitises the domestic cat throughout the world. As far as is known, it neither acts as an intermediate host for a cestode, nor does it transmit any disease to Man.

Chapter 23

Booklice or psocids

PSOCOPTERA

This order of very small or minute insects consists of about 2,000 described species and is of worldwide distribution. Very few reach a length of 10 mm. They are characterised by the large mobile head with long thread-like antennae of from 12 to 50 segments. The mandibles are asymmetrical, the maxillae possess rod-like lacinia known as 'picks' and the post-clypeus is enlarged or bulbous. The labial palps are reduced, the tarsi are either two- or three-segmented and cerci are absent. Generally the body of psocids is soft and the abdomen has a plump appearance. Often they are whitish or light in colour. Most species are winged but in some species, especially among the LIPOSCELIDAE the wings are either rudimentary or entirely absent. The wings, which are delicate and membranous are, in most species, held tentwise over the body, but are sometimes horizontal and overlapping. In some families the wings are covered with scales. The forewings are larger than the hind and are always coupled. The wings of some psocids are green, some yellow, while some scaly-winged psocids have gold patches or entirely gold wings. The wing pattern of many psocids, e.g. MYOPSOCIDAE, is cryptic. Cerci are absent.

Psocids have been known as present in dwellings for several centuries, and the distribution of some species appears to be almost universal. Often they appear in very large numbers shortly after a house has been erected when plaster is in the course of drying out. In those situations they feed on hyphae of surface-growing fungi, often invisible, or in moulds of damp wallpaper. They are often associated with the so-called 'plaster beetles', *Cryptophagus acutangulas, C. dentatus, Enicmus minutus* and *Coninomus nodifer*

In ill-ventilated situations they feed on farinaceous materials, such as paste as well as glue. In libraries, where such conditions may prevail, they feed on binding adhesives and hide between the leaves or down the spine. This type of infestation has given rise to the common name of 'booklouse' now applied to the whole order. Identification to species is generally difficult and best left to the specialist.

Common species found in dwellings in Britain are *Trogium pulsatorium*,

Fig. 23.1 *Liposcelis* sp., Book louse.

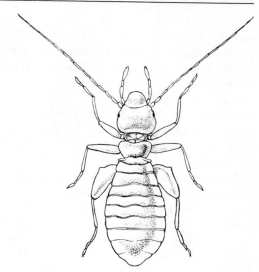

Nymphopsocus destructor, Lepinotus inquilinus, L. patruelis, Cerobasis annulata, Lachesilla pedicularia, and *Lepinotus reticulatus.*

Insect collections in museums are often psocid-infested if no preventative measures have been taken, and considerable damage occurs.

Psocids are sometimes found in buildings, in extraordinary numbers when conditions are particularly favourable to them. In nature, although they occupy a wide variety of habitats, including, for example, caves and birds' nests, generally they can be considered inhabitants of woodland. Here some live in the leaf litter or alternate between this and the leaves of broad-leaved trees. Other species live on the bark of trees or under the bark of dead trees. The food consists of unicellular algae, lichens, fungal hyphae, spores and fragments of plant and insect tissue.

The adults of some species spin webs while the nymphs of some others are made inconspicuous by debris particles adhering to glandular hairs. Distribution is sometimes effected by tree-inhabiting forms being caught in air currents and becoming part of the drifting fauna of the upper air. On the other hand many species have become cosmopolitan by being carried in the holds of ships with infested foodstuffs.

A substantial number of psocid species is known from stored foodstuffs in warehouses, silos, food depots, flour mills and from ships' holds. This information is largely due to an extensive survey carried out during 1945–53 at English ports. In all, 30 species were found, classified in 5 of the 17 families and, in addition, another 7 have been recorded elsewhere. Although widely associated with stored foods only rarely do psocids cause damage directly. Usually they occur when other insects or rodents have been present.

A number of points of exceptional interest emerge from the survey noted above. Firstly, the great majority of the species show wing reduction or their complete absence. The 12 species of *Liposcelis* are entirely without wings as is *Ectopsocus vachoni*. Most of the species listed are not indigenous British insects, one species alone, *Lachesilla pedicularia*, occurs in natural habitats in Britain. The country of origin of some of the cosmopolitan species is unknown.

Although a large number of psocid species has been associated with Man in his buildings, comparatively little has been accomplished in the way of detailed life histories. However, detailed observations on *Liposcelis granicola*, have been made, and the following account deals solely with this species.

Although a somewhat elaborate courtship occurs in many psocids this is not the case with *L. bostrychophilus* since it is parthenogenetic, as are several other psocids living in buildings. The eggs are extremely large, being one-third of the body length of the adult. The egg is smooth and has a bluish-pearly lustre. Its relatively large size is a common feature in psocids, although this is not so pronounced as in the winged species. When first laid, the egg is sticky and adheres to the substrate. Then particles of detritus and faeces fetched by the adult are stuck on to it. Initially about three eggs are laid each day, but this rate gradually diminishes until only one each week is laid. Immediately before hatching the egg turns brown, and the red pigment of the eye is visible. A peculiar hatching organ is present, known as an egg burster.

This is a saw of seven teeth held over the head and attached to a membrane which completely covers the embryo. As the front of the head pulsates so the teeth of the saw cut into the egg shell. Such a mechanism is present in most psocids, but also accurs in a small number of unrelated arthropod groups such as spiders, scorpions, fleas and earwigs.

The young nymph in generally similar to the adult but is white and more delicate in appearance, and the eye consists of two ommatidia only. At 25 °C and 76 per cent RH, the egg hatches in 11 days, and then, after four moults the adult stage is reached in 15 days. The length of adult life is about six months, during which time about 200 eggs are laid. The extremely short developmental period, followed by a long adult life are the reasons why populations build up with phenomenal speed, when the environmental conditions are optimal.

In the United States, psocids have been the cause of heavy losses to property owners, not, of course, from any destruction of buildings or furnishings, but from broken leases and lawsuits entered into to recover damages for fancied loss of health and property. Fear based on the erroneous association of psocids with true lice and disease, and the mental uneasiness of the uninformed were important factors in arousing public interest, and the subsequent initiation of research into the biology of psocids.

At least one species, *Trogium pulsatorium*, found in houses, produces a ticking noise. This is produced by a very rapid elevation and depression of the abdomen, so that it taps on the surface of which it is resting, rather in the manner of some species of termite workers when under some degree of stress. For about 200 years the ticking psocid was confused with the Death Watch beetle, *Xestobium rufovillosum*. Species of *Liposcelis* do not appear to make any sound.

Chapter 24

Termites

ISOPTERA

Introduction

All members of the insect order ISOPTERA are known as termites and about 2,000 species are known. All termites live in social organisations of greater or lesser complexity known as colonies. Social organisations can exist only when there is a behavioural response between individuals, which can be likened to language. This has brought about a coordination of activity and a task specialisation in the termites of a remarkable order, paralleled only in the HYMENOPTERA, and by Man himself. The communication between termites during feeding, defence, nest-construction, foraging, etc., is of an extremely high order, especially when it is considered that these phenomena almost always take place in total darkness, so that visual signals are denied them.

Within the colony the termites occur in a number of different forms or castes. Principal among these are the female and male reproductives known as the queen and king, respectively. When a new generation of reproductives is produced each possesses two pairs of large flat overlapping wings, they are then known as alate reproductives. But when, after a brief flight, they fall to the ground, the wings are quickly shed by breakage at a well-defined suture and they enter a crack or hole in pairs in order to form a new colony. After copulation, eggs are produced, at first in small batches, then at a rapidly increasing rate until, as has been observed in one species, between 2,000 and 3,000 are laid each day.

In the early stages of the new colony only wingless, sterile castes are produced from the eggs. These are the workers and soldiers.

Workers
The workers are of unspecialised structure, of pale colour, and although the mandibles are strong and sclerotised the general body integument is not. They generally occur in large numbers and numerically they are the strongest. They are responsible for most of the general tasks associated with the colony. They forage for food and bring it back to the nest and

feed all the other castes at their different stages of development, and also feed each other. They repair and enlarge the nest and the gallery system as necessary, and generally care for the eggs and immature stages.

The worker caste is smaller than the reproductive to which a general resemblance is borne. They are, however, always wingless and have fewer antennal segments. Almost always they lack eyes and ocelli. Those workers that live entirely within the nest, shelter-tubes or wood, are usually unpigmented, but those workers that forage above ground in

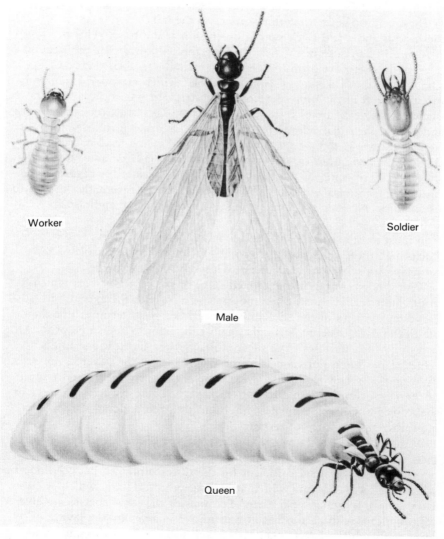

Fig. 24.1 Stages of *Reticulitermes flavipes*, Eastern Subterranean termite. (Shell Chemical)

more or less open situations, are generally pigmented. The true or 'terminal form' worker is not capable of differentiation into soldiers or alates, is incapable of moulting to produce a larger worker, and is sterile.

Soldiers
The soldier caste is present in all termite genera except two. It is characterised by the head being structurally modified for a defensive function. The head is usually much larger and usually much longer than is found in other castes and it is heavily sclerotised. Often it is yellowish or chestnut in colour. In certain species in the KALOTERMITIDAE, however, the head is short and thick with a protrusion at the front. This is an adaptation of the use of the head to plug the hole made for the ejection of frass. In the more primitive termites, including HODOTERMITIDAE, the soldiers mostly possess very large mandibles which are often extensively modified, and in the more developed RHINOTERMITIDAE, head and mandible size may be of two quite distinct types. In the RHINOTERMITIDAE, and also in the highly developed TERMITIDAE, another form of soldier is found, known as the nasute. The frontal gland in the head is much enlarged and the head capsule accommodating it is large and often pear-shaped. The opening of the frontal gland in the nasute is situated at the tip of the front end and is used for ejecting a sticky, aromatic fluid on to its enemies. Nasutes are usually small in size and the mandibles are vestigial.

In most termite species male and female soldiers occur, but in the TERMITIDAE there are some species with male soldiers exclusively, and some species with female soldiers exclusively. All are sterile.

With the exception of the winged stage of the reproductives which fly in the light for, at most, a few hours, and also with the exception of the harvester termites of the subfamily HODOTERMITINAE, termites live their life in the dark, reacting strongly against the light. This is one of the ways in which termites have successfully resisted desiccation, and is achieved by the use of tunnels in the wood and by the covered runways, by means of which they are able to control the temperature and humidity of the microclimate within their living space. These covered runways or shelter-tubes are of the greatest importance to the building surveyor as indicating the presence of termites.

With the exceptions given above, termites are blind, so that in finding their way about they must rely on senses other than sight. Termites possess a sense of smell sufficient for them to distinguish between odours which are acceptable and which attract them, and odours which are not acceptable and which repel them. Termites of the same species are able to communicate with each other in a primitive manner by the laying of scent trails which attract other termites to follow along the same trail. These trail-laying secretions, known as pheromones, play an important part in the activity organisation of the colony. This means of communication, chemical odour reception, taking place entirely in the dark, prevents wandering, dispersal, intrusion of other species, and perhaps predators. It fosters social grouping and keeps them together in well-knit aggregations necessary for grooming and oral food exchange.

Colony foundation
The young, winged adults congregate in batches at different locations in the nest, and it is probable that due to some behavioural change, the workers are now antagonistic towards them. Workers are known to kill young, pre-flight adults, which stray from or return to the nest. It has been conventional to refer to the flight from the nest of the winged reproductives as the 'nuptial flight', but a better term would be the 'dispersal flight'. This occurs at certain seasons, local critical climatic conditions being most important. At the mass flight from a nest, thousands of individuals emerge in a steady stream and take to the air. The slow, blundering flight of termites is characteristic. The slow, directive motion of the body propelled by the fast-beating wings is distinctive. In most cases the termites fly off individually, but in *Pseudocanthotermes* and *Microtermes*, the male fastens himself to the female's abdomen and they fly off together. In all cases, however, when the termites reach the ground the wings are shed. This is known as de-alation, and it is accomplished by twisting the abdomen around when the main shaft of the wing becomes detached from the stub at the humeral rupture line. After a mass termite flight it is quite usual to see the ground littered with large numbers of the shed wings. After the wings have been shed, the female arches up her abdomen and produces an attractant pheromone. The male then takes up a position immediately behind. He then follows her, keeping in contact by tactile responses from the antennae. This tandem position is a feature of the behaviour of newly de-alated reproductives, as is a certain amount of erratic movements and jumping caused by the vibration of the muscles of the now non-existent wings.

The king and queen
The termite pair referred to as the king and queen seek a crack in the ground and enter it, the female taking the major part of the excavation required. Advantage is often taken of exit holes made by other insects. Copulation does not take immediately on pair formation, but, as in *Zootermopsis angusticollis*, in North America, it may be as long as two weeks afterwards that it occurs. In the above species, the pair move slowly past each other in opposite directions with the genitalia distended and curved towards each other. Coupling lasts for a few seconds only. The termite pair do not move far away from each other, and during the nest construction they are gradually built into a chamber. The reproductive organs of the queen increase in size to an extraordinary degree so that the abdomen takes up a large part of the total length of the

Fig. 24.2 The shelter tubes of *Reticulitermes* termites forming a bridge between the earth and groundfloor joists. Miami, Florida. (Truly Nolen)

body (the phenomenon known as physiogastry), and she can never escape from the nuptial chamber. Here she produces vast numbers of eggs, but at the same time acts as a nerve centre for the whole colony through the control she exerts with her pheromones. The king does not increase in size.

Drywood termites
Termites are divided into three main groups according to the location of their colonies in respect of the availability of water. Members of KALOTERMITIDAE are known as drywood termites. This is a primitive group in which true workers are absent although soldiers are present. Other than soldiers, all members of the colony are capable of development into winged reproductives.

They are referred to as drywood termites on account of the fact that the colony is able to exist without any association or communication with moisture. In nature they are usually to be found in dead trees or dead branches. Only abnormally do they construct shelter tubes. Drywood termites are generally more common in coastal areas and

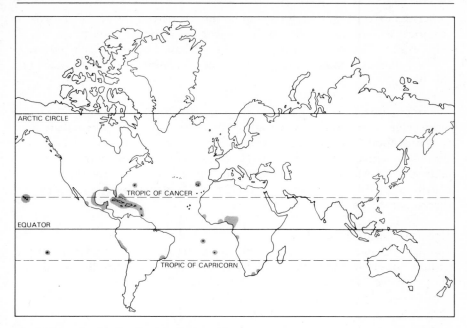

Fig. 24.3 World distribution of *Cryptotermes brevis*.

islands; indeed they are generally abundant in such localities in tropical areas. The paired reproductives first find their way into a building. A crack or crevice is sought into which they soon disappear to found a colony.

Although a number of species are capable of biting a way into a piece of roughened wood, the line of least resistance is usually taken. Nowhere do drywood termites cause so much economic loss as in the state of Florida, especially along the heavily populated coastal belt, and its termite fauna is so exceptional that faunistically it is placed in the Neotropical

Fig. 24.4 Drywood termites working in the timbers of a roof void can be easily identified by the piles of 'poppy-seed' faecal pellets. Here shown marked round with white chalk. Mombasa, Kenya.

rather than the Nearctic region – quite different from the rest of North America. In Florida, housing development has assumed fantastic proportions. Drywood termite control has become an important industry. The substantial list of termites which occur there includes nine species of this group.

Subterranean termites
Subterranean termites possess fixed nests from which workers radiate in search of food. The nest may be in or on the ground, or above the ground in a tree, but all need direct access to ground moisture. Colonies consist of a large number of individuals and are usually tropical in distribution. All are classified in the family TERMITIDAE which includes three-quarters of the world's termites.

Dampwood and moistwood
Dampwood and moistwood termites construct their nests under or on the ground, often in buried timber. The TERMOPSIDAE are generally found in cool climates, while the RHINOTERMITIDAE, usually referred to as moistwood termites, are widely distributed.

Classification
Termites are divided into six families, five of which represent natural groupings, but the TERMITIDAE, consists of an aggregation of genera with little affinity with each other. With one exception, (TERMOPSIDAE), the names of families end with -TERMITIDAE and subfamilies with -TERMITINAE, with two exceptions, although TERMITINAE is the name of a subfamily without any prefix. There are altogether 141 genera of termites, and with only four exceptions to the convention, all their names end with *-termes*. However, one genus is named *Termes*.

It will be seen that in the key to families, which follows, there are three sections – alates, soldiers and workers. Keys to the last named, however, are the least satisfactory, so that it is desirable, if not essential, when collecting termites for subsequent examination and identification, to ensure that alates or late nymphs and soldiers are obtained.

Key to families of alate termites

1. Tarsi distinctly 5-segmented, with pulvillus;   MASTOTERMITIDAE
   antennae with about 30 segments; hindwing
   with anal lobe Tarsi 4-segmented, viewed from   2
   above; antennae rarely with more than 27
   segments; hindwing without anal lobe

2. Anterior wing scales large enough to cover   3
   posterior scales; wings reticulate

Anterior wing scales short, not reaching to base of posterior scales; wings not wholly reticulate	TERMITIDAE
3. Ocelli present	4
Ocelli absent	5
4. Fontanelle present	RHINOTERMITIDAE
Fontanelle absent	KALOTERMITIDAE
5. Pronotum saddle-shaped; tarsi definitely 4-segmented Pronotum flat; tarsi viewed from below seen to possess a rudimentary fifth segment	HODOTERMITIDAE
	TERMOPSIDAE

Key to families of worker termites

1. Right mandible with a subsidiary tooth at base of first marginal tooth	2
Right mandible without subsidiary tooth	3
2. Relatively large insect usually pigmented	TERMOPSIDAE
Relatively small and white	RHINOTERMITIDAE
3. Eyes Present	HODOTERMITIDAE
Eyes absent	TERMITIDAE

Note: A true worker caste is considered absent in the MASTOTERMITIDAE and KALOTERMITIDAE; the older nymphs can be determined from the key to alate characters.

Key to families of soldier termites

1. Tarsi 5-segmented	MASTOTERMITIDAE
Tarsi 4-segmented, rarely with a rudimentary fifth segment	2
2. Pigmented eyes and large abdominal cerci	3
Pigmented eyes and rudimentary abdominal cerci	4
3. Head rounded, generally subconical	HODOTERMITIDAE
Heat flattened, more angular	TERMOPSIDAE
4. Fontanelle present	5
Fontanelle absent	KALOTERMITIDAE

5. Pronotum flat without anterior lobes   RHINOTERMITIDAE
   Pronotum saddle-shaped with anterior lobes   TERMITIDAE

Note: The soldier caste is absent in the genera *Anoplotermes* and *Speculitermes*.

MASTOTERMITIDAE

There is only a single species, *Mastotermes darwiniensis*, in this family, which is confined to northern Australia. Over a million individuals are reported as being present in fully developed colonies, and the nests are in hollowed-out tree trunks, poles and posts near ground level. Relationships with the cockroaches are shown by the presence of anal lobes on the hindwings and the laying of eggs in pods containing about 20. It is important economically although of local distribution.

KALOTERMITIDAE

These are drywood termites. The colony does not usually reach large proportions and is always housed in infested wood. Flagellate PROTOZOA are present in the intestine. The more important genera are *Kalotermes*, distributed throughout the tropics and subtropics: *K. flavicollis* in southern Europe and *Cryptotermes* which includes species spread across the world through commerce and of exceptional importance. *Cryptotermes brevis*, the West Indian drywood termite, and *C. domesticus* attack wood in buildings in the Pacific and Oriental regions, and *C. dudleyi* in Central America, South-east Africa and East Africa.

The faecal pellets of drywood termites are characteristic and because of this are of the utmost importance to the building surveyor. They are dry and seed-like and are often known as 'poppy-seed'. They usually occur in small piles where they have been pushed out of temporary holes constructed for the purpose.

HODOTERMITIDAE

These are the harvester termites. There are only three genera and they tend to colonise arid regions – mostly in Africa, Arabia, Iraq, Persia and north India. It is not usual for them to be found in buildings.

TERMOPSIDAE

These are known as dampwood termites. Like the KALOTERMITIDAE they are found only in small communities and possess intestinal flagellate protozoa. In buildings they are found in situations where there is damp wood not only in contact with the ground, but often at considerable heights above ground in the vicinity of leaking tanks or pipes, or where defective guttering allows rainwater seepage to take place on to woodwork.

RHINOTERMITIDAE

This family contains six subfamilies, all species of which are small and subterranean in habit. A character of the family is the presence of a fontanelle or opening of the frontal gland, which in the soldiers is used as a defensive organ. When alarmed they secrete a sticky fluid from the gland.

TERMITIDAE

This is by far the largest family, the great majority of termite species being included in it. In general they are subterranean in habit or are mound builders, but some, e.g. the *Nasutitermes*, construct nests in trees or on posts.

Feeding
With few exceptions the main food of all termites consists of wood, the constituent chiefly utilised being cellulose. With the exception of a number of species of *Coptotermes*, however, living wood is not eaten. Dead, sound wood, even if still attached to a living tree, is usually selected, especially by the more primitive KALOTERMITIDAE, but also by the more highly developed genera such as *Amitermes*, *Microcerotermes* and *Nasutitermes*. However, many termite species eat, and sometimes are especially attracted to, wood invaded by certain fungal species, either exclusively or in addition to sound wood.

A few species feed on soil and extract the decaying vegetable matter, humus, which it contains.

With the exception of the TERMITIDAE, all termites harbour symbiotic PROTOZOA, and the hind-gut is relatively larger in consequence, in order to accommodate them. The PROTOZOA secrete the cellulose-digesting enzyme cellulase. In the absence of the PROTOZOA, the termites die.

Termites feed in a number of different ways. In the first place, worker termites gnaw wood substance, usually from the walls of their galleries or tunnels within the wood. The worker termite will then feed soldier termites, reproductives in various stages, and young of their caste. This is done in various ways. The passage of food from one individual to another (known as mutualistic feeding or trophallaxis) is widespread, if not universal, among termites. This phenomenon is of the greatest importance in the economy of all social insects, giving rise to efficient social integration and division of labour. Obviously the soldier termite, with grotesquely shaped mandibles, or the physogastric female, would find it impossible to take nutrient material through their own efforts.

Food exchange takes place in two ways. Firstly, from mouth to mouth, or stomodaeal, and secondly, from anus to mouth, or proctodaeal. In stomodaeal feeding the prospective recipient uses its antennae to caress the head of the prospective donor, or it may touch the throat region of the latter with its mandibles.

This action stimulates a food droplet to be disgorged which is then passed over. There are two types of stomodaeal food. Firstly, a clear or nearly clear, viscous fluid which is a product of the salivary glands and secondly, a paste consisting mostly of a suspension of wood fragments.

In proctodaeal feeding, on the other hand, the prospective recipient stimulates the region around the anus of the prospective donor with its antennae, palps or sometimes the front legs. If faeces are present in the rectum, these are then voided, but, if not, a drop of rectal pouch contents is excreted which is immediately taken up by the recipient. The significance of this is that, not only is food material transferred, but, in addition, living PROTOZOA able to digest the cellulose are also passed from one individual to another.

Trophallaxis is of the greatest importance, not only to the individual termites receiving their nourishment in this way (and the means to digest it), but also to the colony as a whole. By this means intercommunication between individuals makes possible the supression of some parts of the colony and the proliferation of others. Much of this is brought about by pheromones – chemical substances passed with the saliva or food droplets, one function of which is responsible for bringing about changes in the physical structure of termites, thus determining what castes they are to be. Trophallaxis brings about the delicate balance between the numbers of the various sorts of individuals in the colony.

Wood-rotting fungi

Fungi compete with termites for the plant tissues which constitute their food and the resulting competition has brought about many associations between the two, some simple, some more complex. Termites have thus adapted their feeding habits and digestive processes to the consumption of partially fungally-decayed food. The two main constituents of wood, cellulose and lignin, are not easily broken down into single nutrients, so that fungi which are able to do this by the secretion of specific enzymes are generally in a favourable relationship with termites. Additionally, certain wood-rotting fungi produce an attractive substance which leads the termites to the food source. On the other hand, some fungi are known to be toxic to termites.

Termites are not specific, as a rule, with regard to the tree species whose wood they attack, although, of course, the geographical distribution of termites and tree species is of importance. Especially important is this in a number of species in the TERMITIDAE which do not possess symbiotic PROTOZOA in the gut.

Cultivation of fungi

The extraordinary phenomenon of fungus cultivation is practised by termites of the subfamily MACROTERMITINAE. This horticultural activity provides an indirect method of cellulose and lignin utilisation. In the nests of the termites concerned, globular or ovoid cavities occur usually from walnut to coconut size and occupying the cavities are nodules of a friable but hard consistency. These are known as fungus gardens or combs. They are perforated in a random way by innumerable galleries of such a size that termites can move around inside them. The combs are constructed of the termites' faecal matter and inoculated with spores of fungi of the genus *Termitomyces*. The latter, of which about 19 species are known, are not known to grow wild, that is, away from termite combs, and it is assumed that a true symbiosis exits. The mushroom-like sporophores of the fungus emerge from the ground above the termite nest and the spores are obviously carried down to the combs when they are being made.

Although it was thought that the termites browsed on the mycelial nodules of the fungus growing from the surface of the comb, it is now believed that the importance of the comb lies in the fact that the termites are constantly nibbling the undersurface of the comb while, at the same time, renewing it. Many combs are concave underneath. The material which the termites consume is wood substance which has passed through the gut, has had a substantial part of the cellulose removed and is thus rich in lignin. The latter has passed through a stage of fungal degradation which has led to the lignin having been rendered into a state more acceptable as a nutrient. It is probably richer in nitrogen than is fresh wood substance, and it probably contains vitamins. In the MACROTERMITINAE, the fungal degradation of the comb of faecal origin probably takes the place of proctodaeal feeding, general among termites in other families.

Cellulose digestion

A high proportion of the wood cellulose is utilised by termites, from 74 to 99 per cent having been determined by experiment. With regard to the importance of the other principal wood constituent, however, there has been some conflicting opinion. At one time it was thought that this other constituent, lignin, was wholly excreted by the termites, but it is now believed that some is utilised. Up to 83 per cent has been found for *Reticulitermes lucifugus santonensis*; up to 26 per cent for *Kalotermes flavicollis*; from 14 to 40 per cent for *Heterotermes indicola*; and up to 52 per cent in the case of *Nasutitermes ephrata*.

Economic loss

Wood utilised by Man in his structures has become one of the most important materials in world commerce. In almost every country of the world the numbers of dwellings, commercial buildings, harbour and sea defence works, communication systems and other fabrications which

Fig. 24.5 Records from French Government archives destroyed by *Reticulitermes lucifugus santonensis*.

incorporate timber, are increasing. However, man-utilised wood is vulnerable to attack by the same organisms which occur in the natural forests where their activities are mostly beneficial.

Over a large area of the world ISOPTERA constitute the most important of the wood-destroying insects and are responsible for heavy economic loss. Other cellulose substances generally manufactured from wood, such as paper, are often destroyed.

When Man uses timber where termites abound they are likely to carry out the same activities in a building as would be their natural role in forest or savannah. The various termite species, however, not only vary in their habits in respect of the particular plant tissue requirements for their nutrition, but also in climatic conditions or microclimate essential for their well-being and development. Not all termite species are undesirable either in buildings, in agriculture or in forestry. Many species occupy particular niches in Nature such that they are never likely to conflict with Man's interests. On the other hand, some species, particularly of the so-called drywood types (but not confined to them), have travelled involuntarily through Man's commerce and are established in many new regions because conditions have proved particularly favourable.

Beneficial role of termites
Other than when wood is used to satisfy human needs, termites, together with other wood-destroying organisms, play a wholly beneficial role. They then function as scavengers, boring into, breaking up and digesting woody tissue. The products of their activity are available either directly

or indirectly through the activities of other organisms as a contribution towards the nutritional requirements of a succeeding generation of trees. If we examine the total biosystem of tropical and subtropical areas, the number of termites present is of the greatest magnitude, and other activity and effect are likewise immense. This great importance of termites in the biodeterioration of woody tissue in the areas of the world where they abound is not always apparent to the untrained observer, due to their light-avoiding behaviour – excepting the brief period of their nuptial flight and dispersal.

It is not unreal to suggest that the continuation of the biosphere as we now comprehend it depends more on maintaining termites alive and in their present numbers in the tropical, tree-growing areas of the world than on destroying termites where Man utilises wood in his own social economy. In the event, the biologist must seek to satisfy both sides of this concept, allowing the maximum natural fertilisation of forest soils through the decay of unwanted woody tissue (presupposing that termites are beneficial in this respect, except for those which cause damage to living trees), but at the same time denying to termites the availability of wood in use by Man.

World cost of termite damage

More is known of the economic aspects of termite damage in the United States than in any other country. Undoubtedly this is because the high standard of living enjoyed by a substantial proportion of the people has led to a high cost of timber renewal (where buildings are mostly of all-timber construction), and also to the ability to pay for the freedom from the anxiety associated with the presence of noxious insects and decay. Fifteen species of subterranean termites are considered to be of great economic importance and of these, *Reticulitermes flavipes, R. hesperus, R. hageni* and *R. virginicus* are responsible for most of the damage. It is reported that termites cost the United States $500 million annually, and that they should be placed among the most important groups of pests from a world standpoint.

It is a matter of the greatest difficulty to compare costs from one country with those of another. This is even more difficult when one country is the United States and comparison has to be made with a developing country in a tropical area where the fruits of a man's labour, using simple hand tools, are to be compared with the products of highly sophisticated automatic machinery! But all countries strive to lessen this discrepancy, and it does seem that the preservation of utilised wood must come fairly high among priorities, otherwise the decay of wood by termites must cause a strain on human and consequently on national economics.

In 1966 it was stated that a world figure for termite damage must lie between £250 million and £500 million annually, but in the United States alone, termites caused damage to the first given sum, so that the world cost in respect of timber damage alone cannot now be less than £500 million. Detailed statistics concerning termite damage cost are hard to come by, but Australia can be cited as an example of a country of relatively small population size where considerable termite damage occurs.

The total cost of damage was estimated in 1971 to be about A$7 million. Australia is a country of 12 million inhabitants and almost all of its 3 million dwellings are timber-framed, so that termites pose an important problem.

Plastic damage
To the widespread and heavy cost of damage which the tropical and subtropical countries must sustain in respect of damage by termites to man-utilised wood, must be added a further indefinable amount attributable to termites on account of damage to non-cellulosic substances.

No satisfactory explanation exists as to why these highly organised insect communities should forsake cellulosic material in order to break up, gnaw, chew or otherwise destroy, chemical substances which we may designate in the broadest sense as 'plastics'. There are inumerable recorded examples of this. The most costly of such cases, out of proportion to the amount of physical deterioration brought about, is when underground, electric, or communication cables are penetrated. This has occurred on a substantial number of occasions even when an appreciable thickness of lead has been used as a covering. Four species of *Coptotermes*, *C. formosanus*, *C. acinaciformis*, *C. havilandi* and *C. niger*, are known to cause this type of damage as well as the Australian *Mastotermes darwiniensis*. Apart from underground damage, attacks on cables also occur when they are located close to heavily attacked woodwork above ground by either subterranean or drywood species.

The first record of plastic attack by termites was as far back as 1896 when rubber was found to be damaged. Then in 1955, natural rubber, polyvinylchloride and neoprene coatings for underground cables were damaged, while cellulose acetate resisted attack. As a plywood adhesive phenolformaldehyde was considerably more resistant than ureaformaldehyde. On the other hand, another investigator found that cellulose acetate was more attractive to the drywood species *Cryptotermes brevis* than the wood on which it had been deposited as a film. Polyvinyl acetate was also found to be more attractive for the termites to chew than wood. From 1953 until 1959 in Nigeria it was found that polyethylene, plasticised polyvinylchloride and a phenolformaldehyde laminate were attacked from a range of polyethylenes, cellulose plastics, phenolics, polymethyl methacrylate, urea and melamine-formaldehyde, polystyrene, polyvinylchloride polyesters and polyamides, all buried in the ground in an area of high termite activity.

In 1958 a number of instances of expanded ebonite being damaged were recorded, and in 1962 under laboratory conditions polyethylene sheeting and polyurethane foam were damaged even when suitable wood was present.

It is certain that plastic and similar materials do not serve in any way as a source of nutrition for termites. One reason for damaging or penetration of these non-cellulosic substances seems simply to be the removal of obstacles from the path believed to be leading to cellulose material. Such substances are not passed into the alimentary canal, but are rejected or taken off and built into the covered runways. Buried nylon monofilaments have been chewed up and the resulting fragments built into nearby underground galleries.

The word 'plastic' is used here in the widest sense, and includes all rigid and semi-rigid compounds other than cellulose used as films, bonding adhesives, containers and coverings. Investigations have been made in order to find which plastic types were susceptible to termite attack, and information is now accumulating. Product thickness, surface hardness and type of finish have been investigated, but some disagreements have shown up in the results, obviously due to varying design of experiments and also to the different termite species used.

In 1962 a comprehensive investigation into the phenomenon was undertaken in Australia. *Coptotermes acinaciformis* was the most important species damaging plastic materials, but, *C. lacteus* and *Nasutitermes exitiosus* were also capable of this, although the latter species to a much lesser extent than the other two. The main results were as follows:

Less liable to termite damage
Polyvinylchloride (unplasticised)
Epoxy resins
Polyester resins
High density polyethylene
Polycarbonate
High impact polystyrene
Phenolformaldehyde polymides

Liable to termite damage
Polyvinylchloride (plasticised)
Low density polyethylene
Polysulphide rubber
Polyvinylidene chloride
Polyester-isocyanate
Cellulose acetate butyrate

In 1963 more information became available, tabulated as follows:

Resistant to attack
Phenolformaldehyde
Melamine-
  formaldehyde
Aniline-
  formaldehyde
Epoxides
Polyvinylchloride
  (unplasticised)
Polymethyl
  methacrylate
Polyvinyl carbazole

Slightly attacked
Polychlorotrifluoroethylene
Polytetrafluoroethylene

Heavily attacked
Polyethylene nylon
Polyvinylchloride

Polyvinyl butyral
Polypropylene

The physical properties of materials exert a strong influence on whether attack takes place or not. The type of surface finish is not of importance, but the thickness of the material is of great importance. Materials of relatively high intrinsic hardness suffer the least amount of damage. Exposed ends and edges are consistently more liable to attack than flat or curved surfaces.

Geographical distribution
Termites are exceedingly abundant and are found throughout the tropical and subtropical areas of the world. In some areas they extend into the temperate regions. Two species, *Reticulitermes lucifugus* and *Kalotermes flavicollis*, occur in Europe but do not thrive further north than Paris. A few colonies of a third species, *Reticulitermes flavipes*, maintained themselves in the basements of warehouses in Hamburg until a few years ago. Another species, *Zootermopsis agnusticollis*, is found extensively on the Pacific coast of Canada, where severe wintry conditions often prevail. However, the nearer to the Equator one travels so the number of species and the total number of termites increase, so that, in very large areas of the world, by far the greater part of the biodegradation of vegetable material is brought about through their agency. Evidence has been put forward that in the United States some termite species have increased their range northwards and now inhabit buildings in areas which would formerly have been climatically hostile to them. This is probably due to higher standards of heating in buildings now prevailing, and further shifts of distribution can be anticipated.

Another aspect of termite distribution which may become of great importance in the future concerns those species carried about in ships. Reference has already been made to a number of drywood species where this occurs but *Coptotermes formosanus* in the RHINOTERMITIDAE has in the last few years given rise to much anxiety concerning its introduction into the southern United States.

Chapter 25

Cockroaches

BLATTODEA

Cockroaches now constitute a separate order of insects, the BLATTODEA, previously, as the suborder BLATTARIA, they were joined with the mantids to form the order DICTYOPTERA. Before that they were associated additionally with grasshoppers, crickets, and stick insects, in the old established order ORTHOPTERA.

This worldwide distributed order of insects contains about 3,500 species but it has been stated that more than as many again are still to be described.

Of all the insects known to Man, cockroaches are considered to be among the most obnoxious and loathsome. This is not because of their association with specific human diseases, nor to ectoparasitic habits, but to their repellent, characteristic odour with which they taint food and indeed everything with which they come into contact, fouling it with their excrement. Many people are frightened or startled by the discovery of cockroaches in a room when lights are switched on. Most of the species are relatively large in size, and the rapidity with which they run, together with the extremely large numbers in which they occur, cause feelings of disgust. Cockroaches are nocturnal in habit, and their flattened, leaf-like shape enables them to hide during daylight hours in cracks and crevices or in dark voids which may only be entered through the latter. Several species are of cosmopolitan distribution and it is most probable that their specific names bear no relation to their country of origin.

The different species of cockroach show a range of variation in the extent of their dependence on buildings. The physical presence of Man, however, is not essential, although cockroaches usually feed on prepared or discarded human food. Although essentially tropical or subtropical insects, the pest species of cockroaches have extended their range far into temperate regions by populating heated buildings such as kitchens, warehouses and industrial premises. Some species are able to live outdoors during the summer months in such places as rubbish tips, but for survival during winter the protective environment of premises which are warm is essential.

Seven cockroach species are closely associated with Man. These are the ones most commonly encountered in buildings and four of them are cosmopolitan. These are *Blattella germanica, Blatta orientalis, Periplaneta americana* and *Periplaneta australasiae*. The other three species are North American in distribution, only being found rarely elsewhere. They are *Periplaneta brunnea, Periplaneta fuliginosa*, and *Supella supellectilium*. In addition, a number of other species are only occasionally associated with Man in his buildings.

Cockroaches (BLATTODEA) are easily identified by their flattish appearance, the horny modification of the forewings to form tegmina, the long thread-like antennae of numerous segments, and by their quick-running habit.

They are generally medium in size but a number can be considered large. Apart from the characteristics mentioned above, well-developed compound eyes are present as well as a pair of poorly developed ocelli which are referred to as the ocelliform spots. The mouthparts are unmodified and the toothed mandibles are very strong. They serve to bite and chew. The large prothorax is shield-shaped and usually overlaps the head. The spiny legs are large and strong with a five-segmented tarsus with a strongly clawed terminal segment, and all are adapted for running. The tegmina and hindwings are richly veined, the former folding over flatwise and protecting the latter. The hindwings exhibit a large anal lobe which is folded in a fan-like manner when at rest. The terminal segments of the abdomen are characterised by the presence of a pair of blade-like cerci arising from the hinder end of the tenth tergum, and a pair of smaller styles borne by the ninth sternum.

A simple courtship takes place during which the female has been seen to lick the secretion from tufted dorsal glands of the male. The eggs are produced in an ootheca, a bag-like case provided with a median partition in which the eggs are arranged in two rows in a series of pockets. The number of eggs in each ootheca varies with the species but is usually of the order of 12 to 40, and the oothecae are variously shaped and ornamented according to species. The incubation of the eggs in the ootheca also takes place in different ways. In some cases the ootheca is carried by the female protruding from the abdomen for a short period. In other cases the ootheca is carried by the female for most of the incubation period, while in the remainder, the ootheca is retained in an enlarged brood sac by the female until the nymphs hatch.

BLATTELLA GERMANICA

Blattella germanica is generally known as the German cockroach or 'Steamfly', and although named thus is thought to have been endemic to the area bounded by the great African lakes, Ethiopia and the Republic of Sudan. It commenced its wider distribution many centuries ago, reaching southern Russia through the agency of trade. It then progressed slowly westwards across Europe, but when the great ports were reached it became distributed by commerce much more rapidly, and has now reached practically all parts of the world. It is believed to have been introduced into England by soldiers returning from the Crimean War, although it was certainly present in Denmark as long ago as 1797. It is found throughout the United States, and is found as far north as Ontario, Manitoba and Alberta in Canada. It is also recorded from Australia in 1893.

The common name of 'Steamfly' refers to the warm, moist conditions which it requires and the reason for its presence in kitchens, restaurants and industrial premises where steam is used. In Britain only rarely does it survive outdoors, but it occasionally does so in the north-central United States. In the more tropical areas such as California, it is found outside buildings more frequently.

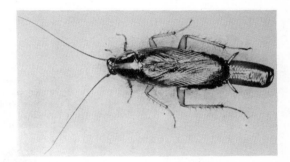

Fig. 25.1 *Blattella germanica*, female.

Blattella germanica is small, the largest specimens reaching a length of 15 mm. It is pale buff to tawny in general colour, and the pronotum bears a pair of dark, parallel bands along its length. It is possible that in the United States the colour is rather darker. In some cases the colour is a dark, golden-brown, devoid of a reddish hue, to nearly black. These bands are broader and continue on to the meso- and metanota in the nymphal stages when they are not covered by the tegmina.

The sexes are distinguished by the male having a slender body with the terminal segments of the abdomen visible. There are conspicuous gland openings on the seventh and eighth abdominal tergites, and the cerci are made up of 11 segments. On the other hand, the body of the female is stouter, and the abdomen is entirely covered by the tegmina. There are no conspicuous glands on the abdominal tergites, and the cerci are made up of 12 segments.

Life-cycle

The ootheca is large, relative to the size of the adult, and is carried until the eggs hatch. The shiny surface of the ootheca is ridged, indicating the positions of the eggs.

Fig. 25.2 Pictorial key to some adult cockroaches in the United States. (US Department of Health, Education and Welfare)

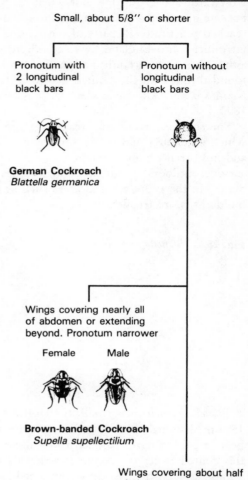

Mating takes place a few days after the final moult, and although repeated copulation takes place one successful mating is sufficient to fertilise the total number of eggs produced by a female. The ootheca is produced two to four days after copulation. When it first protrudes from the abdomen it is white, but in a few hours it changes to pink, then to

COCKROACHES

Medium to large, longer than 5/8"

- **Wings absent, or shorter than abdomen**
  - **Wings absent** (Female) — **Oriental Cockroach** *Blatta orientalis*
  - **Wings shorter than abdomen** (Male) — **Oriental Cockroach** *Blatta orientalis*
- **Wings covering abdomen, often extending beyond**
  - **Pronotum more than 1/4" wide**
    - Front wing without pale streak. Pronotum solid colour, or with pale design only moderately conspicuous
      - Pronotum solid dark colour. General colour dark brown to black — **Smokey Brown Cockroach** *Periplaneta fuliginosa*
      - Pronotum usually with some pale area. General colour seldom darker than reddish chestnut
        - Last segment of cercus not twice as long as wide — **Brown Cockroach** *Periplaneta brunnea*
        - Last segment of cercus twice as long as wide — **American Cockroach** *Periplaneta americana*
    - Front wing with outer pale streak at base. Pronotum strikingly marked — **Australian Cockroach** *Periplaneta australasiae* (Pale streak)
  - **Pronotum about 1/4" wide with pale border** — **Wood Roach** *Parcoblatta* spp.

light brown, and after a day or so it is chestnut-brown. It is almost always turned sideways so that the keel is to the left or right. Three or four days before hatching a green band appears down each flat surface of the ootheca, and increases in intensity until hatching.

The ootheca is carried for from 6 to 16 days, with an average of 10 days and it is deposited shortly before hatching occurs, although occasionally hatching takes place with it still attached to the female. An average of seven oothecae are produced but the number may vary between four and eight, and about 22 days elapse between successful oothecae. The number of eggs in the ootheca varies, but is usually about 37 of which 28–32 hatch.

Under optimal conditions development from hatching to adult takes 6 weeks but may be long as 130 days at a lower temperature. Five to seven moults take place and the total life span of males is about 128 days and of females about 153 days. The rudiments of wings first appear in the penultimate nymphal stage but become conspicuous in the last nymphal stage. Although the adult is capable of gliding flight it very seldom ventures into the air, but it runs swiftly. It may sometimes be found in daylight.

BLATTA ORIENTALIS

Blatta orientalis is known as the Oriental cockroach, although it is most likely to have originated in North Africa. It is established that it first occurred in Holland early in the seventeenth century, and in 1624 was recorded in wine cellars in England.

Description
In all species of *Blatta* the tegmina and wings of the male almost reach the tip of the abdomen while the tegmina of the female are very short and the wings are either absent or are greatly reduced. The pronotum is broad and in both nymphs and adults an arolium or pad between the tarsal claws is absent. This has the effect of preventing them from climbing very smooth surfaces. *Blatta orientalis* is large, being about 20–24 mm in length and dark reddish-brown to almost black in colour. Neither sex is capable of flight.

Life-cycle
The ootheca also is large, being 10 × 5 mm, with egg sacs feebly defined by depressions. It is dark reddish-brown and although soft at first becomes hard and brittle, and black in colour, after deposition.

Five to ten oothecae are produced and when the young nymphs emerge the legs, antennae and mouthparts adhere together. These are known as pronymphs but this stage lasts for only a few minutes when the first moult takes place. The young nymph is white except for the black eyespots and for the teeth of the mandibles which are sclerotised and brown. The life-cycle at 28 °C is completed well within a year. At 25 °C the period is 533 days, and at 30 °C it is 316 days.

In unheated situations the maximum period of development is probably about two years, the oothecae remaining dormant during the winter months. The length of adult life varies from two to nine months according to temperature. Males mature more rapidly than females and when the nymphs are reared in groups in the laboratory they complete their life-cycle more rapidly than when reared in isolation. Unfertilised females produce oothecae and a few females ultimately result.

At 30–36 °C the oothecae incubation period is 44 days. There are up to 18 eggs per ootheca of which 15 usually hatch. The males mature in 164 days taking 7 moults, while the females take 10 moults over a period of 282 days. From Spain it was transported to Argentina and Chile where it became established. It has now invaded almost all the temperate regions of the world where it has become one of the most undesirable of the INSECTA, although it does not appear to have established itself in the humid tropics.

Blatta orientalis is the most abundant cockroach in Britain while in the United States it has a wide distribution except in the most northern parts.

Habitat

Blatta orientalis has a preferred temperature range of 20–29 °C whereas that of *Blattella germanica* is considerably higher at 33 °C. The Oriental cockroach requires a source of drinking water so that it is usually found in kitchens behind appliances and radiators, and in basements and cellars where conditions are somewhat cooler than the areas favoured by *B. germanica*.

It is also commonly found in toilets, behind baths and sinks and, whereas it is most frequently encountered below ground level, in the United States there is an increasing number of records of it occurring in upper storeys.

Occasionally this cockroach survives the winter out-of-doors in Britain and is often found during summer months in refuse dumps. In the north central states of the United States it is commonly found out-of-doors on summer evenings where it may occur in such numbers that distribution to other buildings must occur. Also in the United States the Oriental cockroach is found in refuse dumps, in sewers and in ships' cargoes.

PERIPLANETA AMERICANA

Although known as the American cockroach, the specific name given by Linnaeus in 1978, it seems certain that its origin is in tropical Africa. It is believed that the slave ships provided the means of distribution of this insect from West Africa to South America, the West Indies and the southern United States. It is now one of the most important of the harmful insects in tropical and subtropical areas, and has been distributed well into the temperate regions of most of the world. It is exceptionally common in India where it is almost ubiquitous. In temperate areas it is well etablished in the City of New York, and in Britain it is found around the main ports where it occurs in warehouses and manufacturing premises. From these foci it has been introduced to industrial buildings inland where it flourishes, if somewhat tenuously.

Description

Periplaneta americana is large, from 28 to 44 mm in length and in colour is mainly shining red-brown but with a pale brown to yellowish area around the edge of the pronotum. The wings, when fully developed, extend well beyond the abdomen in the male but in the female they only just overlap it. Flight is possible only over short distances and it is sluggish and rarely undertaken. *Periplaneta americana* is easy to identify except that *P. brunnea* is rather similar. In the former species the cercus is stout basally but tapers markedly, the last segment tending to be parallel-sided, two or three times as long as the basal width. In the case of *P. brunnea*, however, the cercus is stout, spindle-shaped, the last segment triangular, less than twice as long as the basal width. Other characters by means of which the two species may be differentiated are given in the description of *P. brunnea*.

The adult male may be separated from the female by the possession of a pair of ventral styles on the last abdominal sternite but the male does not have a ventral keel with a slit running along it which is borne by the female.

The pale patches on the pronotum do not appear until the sixth nymphal stage; until then the colour is a uniform pale brown but the sexes of all the early instars can be made out by the shape of the posterior margin of the ninth sternite. In the female there is a sharp median notch, but in the male it is only slightly indented.

Life-cycle

After copulation, the production of the first ootheca occurs in from 3 to 7 days, and this is from 7 to 36 days after the final moult to adult stage. The interval between successive oothecae is from 4 to 12 days and the total number of oothecae produced varies from 10 to 84, while the ootheca is carried from 6 hours to 6 days. The number of eggs contained in each ootheca varies from 6 to 28 of which an average of 13.6 hatch. The number of moults ranges from 7 to 13.

The rate of nymphal development again varies widely according to the ambient temperature. At 24 °C, 17 months; 28 °C, $6\frac{1}{2}$ months; 30 °C, 7 months; 30–36 °C 5 months for females and 13 months for males, and another experimenter found that at 25–30 °C the period of nymphal development was generally $4\frac{1}{2}$ to 5 months, but some individuals took as long as 15 months. The length of adult life varies from 102 to 700 days, the males being shorter-lived than females.

At the third or fourth nymphal stage wing pads appear and by the last stage the venation of the tegminal pads is distinct. In all nymphal stages styles are present on the ninth abdominal sternites of males and females except the last nymphal stage of the female. In the last two nymphal stages in females the styles are hidden beneath the seventh sternite.

The ootheca is brown at first but blackens within one or two days. In size it is 8 × 5 mm. It is slightly indented between the egg sacs and there are 16 teeth along the ridge, each of which shows a minute opening at the apex.

In this species a wide variation in biological data occurs according to temperature and accounts for apparent discrepancies in observations by various observers. In addition, *P. americana* develops faster when reared in groups then when in isolation. At 29 °C the total life-cycle from egg to death of adult ranges from 630 days to as long as 1,243 days. Under extremely adverse conditions the period may be as long as four to five years.

Habitat
A warm moist environment is preferred and *P. americana* is similar to *Blattella germanica* in that the upper limit of preferred temperature is 33 °C. It is common outdoors in tropical and subtropical America. The adults and nymphs congregate around a preferred temperature of 28 °C, and in the summer swarm in many localities. The American cockroach has been found in mines in various parts of the world: India, Sumatra, South Africa, and in coal-mines in South Wales. In buildings it commonly occurs in restaurants, bakeries and grocery stores, usually in basements. In the United States it is sometimes found in latrines and sewers. Although it was once common in the galleys and messrooms of ships its place in these situations has been taken by *B. germanica*, although it is a frequent inhabitant of the holds among the cargo.

PERIPLANETA AUSTRALASIAE

The Australian cockroach, like *P. americana*, probably originated in the tropics or subtropics of Africa from whence it was transported to America by the slave trade. Subsequently, it reached most of the warmer inhabited areas of the world through commerce. In addition, it often occurs in temperate regions when it gets a foothold in warm, moist buildings such as greenhouses and manufacturing establishments where steam is used, particularly in the United States.

Habitat
Generally, as an inhabitant of buildings, it is almost as ubiquitous as *P. americana* and is almost as important as a worldwide inhabitant of buildings. It is the most frequently encountered cockroach in Brisbane, but is less common in southern Australia. At the beginning of this century it was the most abundant and troublesome species in Florida, occurring both indoors and out-of-doors. As it is less resistant to cold than is *P. americana* it is not a problem in conventionally heated buildings

in the northern United States, occurring only in places with high humidity and higher than ordinary temperatures.

In India it is found mostly in the south as well as in Ceylon. Elsewhere in buildings it is known from tropical Africa, Ecuador, Puerto Rico, the Philippine Islands, the West Indies and Brazil, and from the last two regions it is frequently carried to Britain on cargoes of bananas.

Description
Although generally somewhat smaller than *P. americana*, it is a large cockroach being from 30 to 35 mm in length. It is reddish-brown in colour and the wings, which are fully developed, overlap the tip of the abdomen in both sexes. The female is distinguished (as in *P. americana* by the possession of a ventral keel at the end of the abdomen. Also similar to *P. americana*, is the pale edging to the pronotum, but this is much more distinct than in the latter species. It also differs from *P. americana* (and also from *P. brunnea*) in having the basal margins of the tegmina pale. There are yellow spots on the thorax and abdomen of the nymphs.

Life-cycle
The ootheca shows slight indentations and is 10 × 5 mm. At 30–36 °C the female produces about 20–30 oothecae and the subsequent nymphal stage of development takes about a year. In the laboratory, when reared in groups they mature more rapidly than when reared singly. At the above temperature it normally takes 24 days after attainment of adult stage before the first ootheca is produced, and about 10 days elapse between successive oothecae. The incubation period is 40 days. In each ootheca there are 24 eggs of which about 16 hatch. Eggs are occasionally produced parthenogenetically, but after hatching do not reach maturity.

Prefering moist, warm conditions, *P. australasiae* is circumtropical but a somewhat higher temperature for development than that for *P. americana* is required. In a number of its tropical localities it occurs out-of-doors. Its distribution in Britain follows closely that for *P. americana*.

PERIPLANETA BRUNNEA

Like most important pest species of cockroach this species is also thought to have had its origin in north or central Africa. However, commerce has spread it around the world, but in a rather narrower band than in the case of *P. americana*, almost throughout the tropics. It is known as the Brown cockroach and was first recorded from Illinois in the United States in 1907. Ten years later it was found in Texas and in 1938 it occurred in Chile and Guyana. Today it is common in the southern United States, in Florida, North and South Carolina, and from Georgia westwards to Texas. It is more abundant in some areas than is *P. americana* and has been found in buildings as far north as Philadelphia.

Habitat

This obnoxious insect is found typically in grocery stores and pantries. and occurs in army camps, privies, sewers and the basements or crawl into spaces under residential or commercial buildings. It is capable of infesting buildings in temperate regions provided the required conditions are present, and is exemplified by infestations at London International Airport in 1965 and 1966.

Description

Rather smaller than *P. americana*, it is 31–37 mm in length but is similar in general appearance being reddish-brown with a yellow margin to the pronotum. The paired blotches in the centre of the pronotum, however, are less conspicuous than in the case of *P. americana*. Compared with this latter species the tegmina and wings are not as long, relative to the abdomen, and do not differ in length between the sexes quite so much. *Periplaneta brunnea* is capable of gliding flight.

The most useful features which distinguish the two species are as follows:

	Periplaneta brunnea	*Periplaneta americana*
	Adult stage	
Cercus:	Stout, spindle-shaped, the last segment triangular, less than twice as long as the basal width	Stout basally, but tapers markedly, the last segment tending to be parallel sided, two or three times as long as the basal width
Ootheca:	Large (12–16 mm long), less rounded laterally, securely glued when deposited, containing an average of 24 eggs	Smaller (8 mm long), not so securely glued when deposited, containing an average of 16 eggs
	First stage nymph	
Antennae:	The first 8 and last 4 antennal segments conspicuously white, the intermediate ones brown	Antennal segments uniformly brown
Mesothorax:	A median translucent area allows light to pass through	This area is absent
Abdomen:	Faint cream-coloured spots on dorsolateral margins of 1st and 2nd segments	These segments entirely brown
	Intermediate stage nymphs	
Abdomen:	Cream-coloured spots on dorsolateral margins extend from 2nd to 6th segments	These segments entirely brown

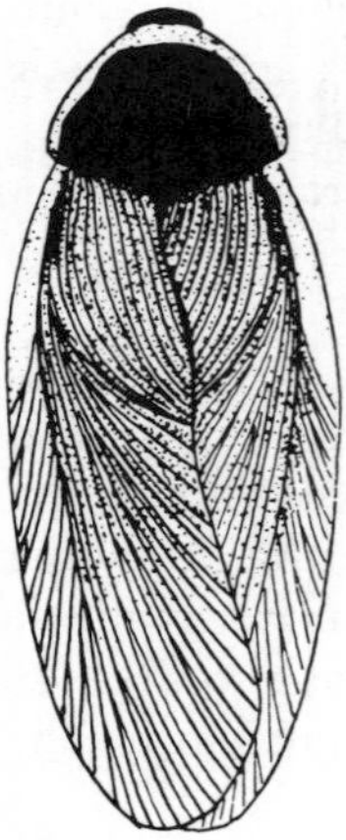

Surinam Cockroach
Pynoscelus surinamensis

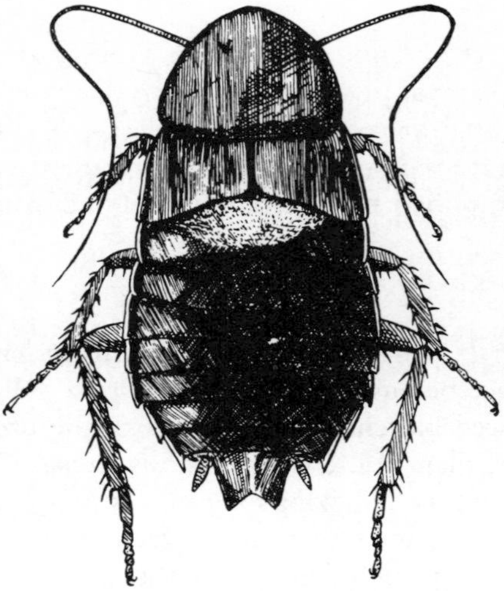

Large Florida Cockroach
Eurycotis floridana

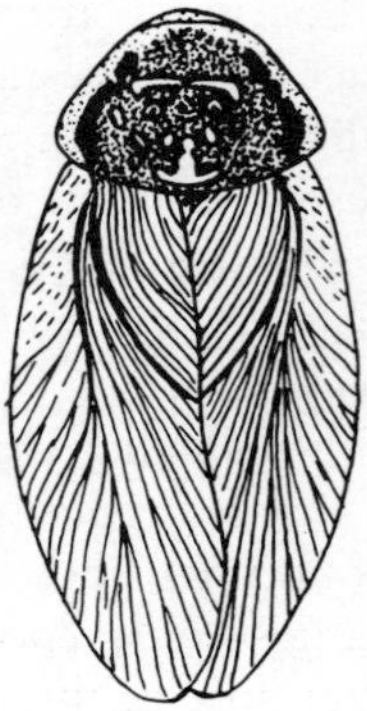

Cinereous Cockroach
Nauphoeta cinerea

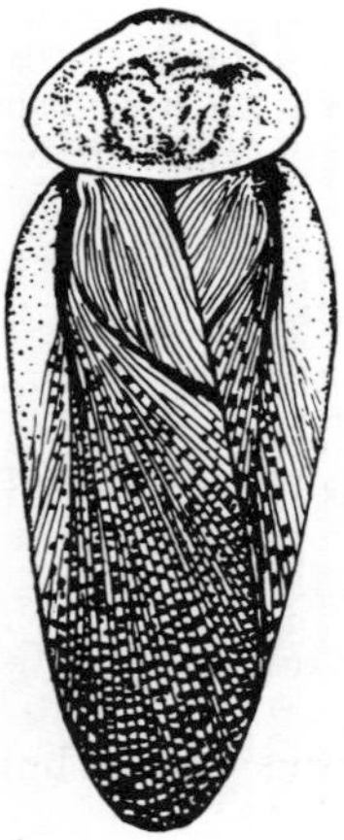

Madeira Cockroach
Leucophaea maderae

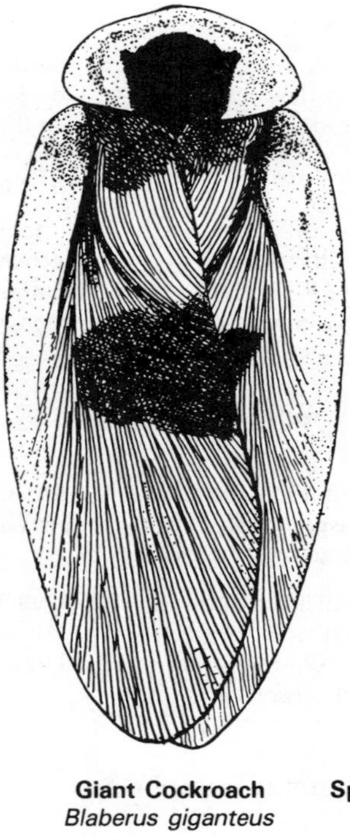

Giant Cockroach
Blaberus giganteus

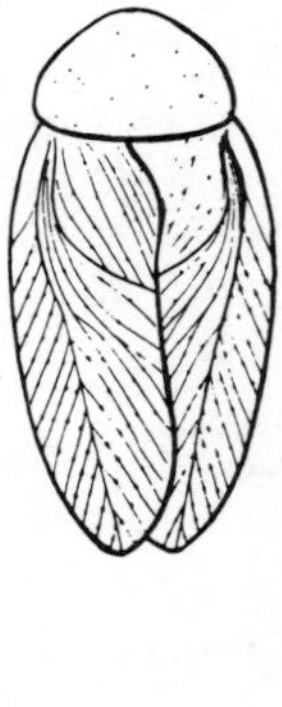

Spotted Mediterranean Cockroach
Ectobius livens

Fig. 25.3 Cockroaches: some species less commonly found in houses in the United States.

PERIPLANETA FULIGINOSA

This subtropical species is known as the Smoky-brown cockroach and was first recorded from North America in 1839. It is most abundant in the southern states from Georgia and northern Florida then westward to Texas. In the coastal areas of the latter state, it lives out-of-doors, often on roofs and in gutters where bird-droppings as well as plant debris accumulate. In the south-east it is common in outbuildings and porches, flying to light at night. In some parts of its distribution it is established indoors. In south-west Georgia it is second only to *Blattella germanica* in abundance in buildings. It is sometimes found further north in heated buildings.

Description
Periplaneta fuliginosa is about the same size as *P. brunnea*, being 31–55 mm in length. It is completely shiny brownish-black but not as dark as *Blatta orientalis*. In both sexes the wings cover the abdomen.

Life-cycle
Temperature response is similar to that of *Periplaneta americana* although temperature plays little part in rate of nymphal development. Hatching from the ootheca takes place in 37 days at 30 °C. At 30–36 °C the production of the first ootheca after the final moult to adult stage is 16 days, and the interval between successive oothecae is 11 days. There are 20 eggs in each ootheca and in the laboratory the percentage of hatched insects that reached maturity was 51 per cent. There are 9–12 moults. The period of nymphal development varied widely according to whether the insects were reared in isolation or in groups. In the case of the former, males undergoing 9 moults to maturity took 474 days, while those taking 11 moults took 586 days. Of those reared in groups, males reached adult stage in only 179 days and females in 191 days.

SUPELLA SUPELLECTILIUM (S. LONGIPALPA)

Known as the Brown-banded cockroach this species was first described from Mauritius. It is assumed to be of African origin. It is now distributed widely over the tropics and subtropics of the Old World, and has become widespread and abundant in many states of the United States. In the New World it appeared in Cuba in 1862, doubtless brought there by the slave trade, but in 1903 it appeared in Florida and in 1930 it was recorded from Georgia, Alabama and Texas. It was not reported in dwellings, however, until 1929 when it was found in such situations in Nebraska. By 1937 it was in Chicago and Indianapolis. In 1940 it had spread to Washington DC, Illinois, Wisconsin and South Dakota, and in 1954 it was present in no less than 20 states. It now occurs in every state in the United States except Vermont. In 1924 it was reported from Queensland, Australia. In Britain a few infestations have occurred and it has now become established.

PYCNOSCELUS SURINAMENSIS

The Surinam cockroach is of oriental origin but is now associated with buildings and other man-made structures, giving it shelter in various islands of the West Indies, the Philippines, Tanzania, Hawaii and the southern United States. In common with a number of related species it is not cold-tolerant. It is recorded that 95 per cent of a laboratory culture died overnight at a temperature of 2–4 °C, but in the hot, humid tropics it is often abundant out-of-doors although it may not often be found in buildings. A number of colonies are known from Britain, Germany and the northern United States from heated greenhouses.

Description
This is a medium-sized cockroach, being from 18 to 24 mm in length. In colour it is a shining dark-brown to black. The wings in both sexes extend well beyond the abdomen. The posterior margin of the pronotum is strongly sinuate and has a pale band along the anterior margin. In general shape it is broad in relation to its length.

Males of this cockroach are known only from the oriental region, females only occur in the New World as also in the small number of colonies known in Britain. This is the only species of cockroach in any way associated with buildings which exhibits parthenogenesis to any degree.

Life-cycle
In addition to the exclusive occurrence of parthenogenesis throughout the New World *P. surinamensis* exhibits false ovoviviparity. This is the production of an ootheca which is extruded then withdrawn into the brood sac and the eggs are developed internally. This feature is also shown by *Leucophaea maderae* and *Nauphoeta cinerea*. Under laboratory conditions at 18–24 °C the parthenogenetic strain gave the following information. Seven days elapse between the attainment of adult stage and the production of the first ootheca and the eggs hatch after a further 35 days. The average number of ootheca produced is three, the broods being produced in 48 and 82 days after the first. Twenty-six eggs are contained in each ootheca.

There are between 8 and 10 moults and at 30 °C those taking 8 moults develop in 127 days, while those taking 10 moults require 184 days. These periods refer to insects reared in isolation whereas those reared in groups take 140 days. The length of adult stage is about 307 days.

LEUCOPHAEA MADERAE

First described from the island of Madeira in the Atlantic, *Leucophaea maderae*, known as the Madeira cockroach is thought to have originated in West Africa. Sometime before 1800 it had been transported to the West

Indies and the Brazilian coast, and from there it has become distributed to many parts of the tropics, where it has become established, as well as to some Mediterranean areas. It is often abundant in dwellings, warehouses and other buildings. It occurs throughout almost the whole of the West Indies, the Bahamas, and in many parts of South America. Here it is found not only in those countries where tropical conditions prevail but as far south as Argentina. It has been known in the United States since the early nineteenth century and in the 1950s became established in the basements of buildings in the Haarlem district of New York, but is no longer found there. In the Mediterranean it occurs in Morocco, Spain and Corsica. In Asia and the Pacific area it is less widespread having been recorded only from Java, the Philippines and Hawaii. It does not occur in India, Australia, southern China or the greater part of Malaysia.
Although imported into Great Britain on a number of occasions, usually with bananas, it has not succeeded in becoming established.

Description
This is a very large cockroach being from 40 to 50 mm in length. It is larger than *Periplaneta americana* but not as large as *Blaberus giganteus*. The tegmina and wings cover, or almost cover, the abdomen, and are pale brown to tawny-olive in colour, with two dark brown longitudinal bars basally, and the distal areas are mottled. In the male, the dorsum of the second abdominal segment bears a specialised organ. Along the dorsal posterior margin of each segment in the nymphs there is a conspicuous row of short microscopic spines.

Although the adult moves slowly it is an active flyer. When disturbed it produces a disagreeable odour and it can produce sound by stridulation, moving the posterior margin of the pronotum over the mesonotum.

Life-cycle
In Brazil mating occurs during the warm rainy season. False ovoviviparity is shown. An ootheca is first produced and extruded but is then drawn back into the abdomen where the eggs are incubated. The number hatching at one time ranges from 25 to 32 but only 18 are recorded from laboratory-reared insects.

The first ootheca is produced 20 days after reaching adult stage and incubation takes 58 days at 30–36 °C (the temperature range in which all the laboratory data quoted here were observed). The number of moults is seven for males and eight for females, the former taking 121 days and the latter 150 days. When reared in isolation these periods are some days longer.

NAUPHOETA CINEREA

In northern Europe it is imported into Britain and Germany with fruit and other foodstuffs, and is sometimes recorded from these countries in heated buildings. It was, however, originally described from Mauritius specimens and its original home is believed to be East Africa. Through commerce, in historic times, it is now spread throughout the tropical areas of the world, occurring in buildings (although it generally prefers outbuildings) and commercial premises. It occurs in Cuba, Hispaniola, Mexico, Brazil and the Galapagos Islands. In the East and the Pacific it is found in the Philippines, Sumatra and Singapore. It was not known to breed in the United States until 1952 when it was found in Florida, and it was recorded from Australia in 1918. In Hawaii it was rare until 1943 when it became extremely abundant.

It does well when subsisting on animal feeds, especially those containing fish oil, and it has predacious habits for it is known to kill and eat the Cyprus cockroach, *Diploptera punctata*.

Description

This cockroach, of from 25 to 29 mm in length, is ashy-grey in colour and has a fanciful lobster-like marking on the pronotum. The wings of the male are slightly longer than those of the female but do not cover the abdomen. The males stridulate by moving the side and hind margins of the pronotum against the costal vein at the base of the tegmina.

Life-cycle

The ootheca contains from 26 to 40 eggs, but false ovoviviparity occurs. The nymphs hatch on extrusion of the ootheca. On shedding the embryonic membrane the nymphs eat it as well as the ootheca itself. They then crawl beneath the female for an hour or so for shelter before scattering. The first ootheca is produced 13 days after the final moult, and hatching takes place 31 days later. The average number of broods is 6 and the incubation period is 36 days. There are 36 eggs in each ootheca, of which 31 usually hatch. Males undergo 7 moults, taking 72 days, while females take 85 days in 8 moults. When reared in isolation the periods are somewhat longer. The adult life span is 364 days in the case of the female, and 365 days in the male. The above data are based on laboratory rearing at 30 °C.

NEOSTYLOPYGA RHOMBIFOLIA   HARLEQUIN ROACH

Originating in Indo-Malayan region, where it is generally abundant. *N. rhombifolia* is especially common in the Philippines. It occurs also in Hawaii, and in the Indian Ocean in Madagascar, Mauritius and the Seychelles. Along the east coast of Africa it has penetrated deep into the old trade routes – into Malawi and the Zambesi valley. In the Atlantic area it is found in Madeira, and in the New World it was first recorded from Mexico, Venezuela and Argentina in 1865. It was found in Brazil in 1895 and a few years later became established in southern California; a move northwards from Mexico where it had long been established.

Fig. 25.4 *Neostylopyga rhombifolia*, Harlequin roach.

Description
It is medium in size – from 20 to 25 mm in length, the males being smaller than the females. In both sexes the tegmina are reduced to small lobes 4 mm in length, and the hindwings are entirely absent. It has a striking and characteristic colour pattern of shining brown-black, marbled with yellow.

Life-cycle
Males are less common than females but parthenogenetically produced eggs do not hatch. When reared at 27 °C and 70 per cent RH, oothecae averaged 22 eggs, and males took 302 days to develop and females 286 days when reared in groups.

BLABERUS

A number of the species of the genus *Blaberus* which are mostly confined to the American tropics (but sometimes extending to temperate South America) occur in buildings. In nature they are found under logs, in caves or under vegetation in tropical rain forests. *Blaberus atropus, B. boliviensis* and *B. discoidalis* are associated with bananas, the latter species occasionally being found in shipments in Britain. *Blaberus craniifer* is established in southern Florida, and is known as an inhabitant of households in Cuba. It occurs also in Mexico and Belize. *Blaberus discoidalis* has been found in restaurants in Ecuador and in dwellings and fruit stores in Hispaniola and Puerto Rico. It is also known from Cuba and Jamaica, and gets an occasional temporary foothold in the United States.

Blaberus spp. are characterised by the pronotum being large and leaf-like, usually elliptical and hiding the head. They are often very large, *B. craniifer* being up to 70 mm in length. The tegmina are broad, covering the abdomen, and in the latter species extend by about one-third of the body length. In *B. discoidalis*, however, the tegmina and wings are only slightly extended beyond the end of the abdomen, and it is much smaller.

Life-cycle

Little information is available on the biology of species of *Blaberus* but information is available on *B. craniifer* reared at 30 °C. The ootheca contained 34 eggs but only 20 hatched, and only half of the nymphs reached maturity. Of those which underwent 10 moults, the males took 257 days and the females 277 days to reach the adult stage. The length of the adult stage averaged 14 months for males and 16 months for females. *Blaberus giganteus* has been reared in the laboratory at 30 °C and 60 per cent RH; 140–200 days were required for development to adult stage, and seven or eight moults were undertaken. The adult stage lasted for as long as 20 months.

Accidental invaders

EURYCOTIS FLORIDANA

The genus *Eurycotis* includes a number of species occurring in tropical America and the West Indies, and *E. floridana* is an example of those species which normally live out-of-doors in tropical situations. However, they may enter homes either by being attracted to light or by being introduced with firewood, etc. In Florida they are called palmetto bugs suggesting a common harbourage in palms and palm-like plants. *Eurycotis floridana* is a large cockroach, the males being up to 35 mm in length and the females sometimes reaching 40 mm. The extremely small tegmina reach only just beyond the posterior margin of the mesonotum, and hindwings are absent. The general coloration is a rich red-brown to black, but the pro-, meso-, and metanotum of the late nymphal stages often bear conspicuous broad bands of pale yellow. When alarmed, the adults of both sexes emit a greasy liquid of a most repellent odour. The first ootheca is produced 55 days after the moult to adult stage, but successive oothecae are produced at about eight-day intervals. They contain about 21 eggs and at 30–36 °C, hatch in 48 days. There are from six to eight moults and development takes 100 days for males and 113 days for females when they moult seven times.

PARCOBLATTA

A number of species in the genus *Parcoblatta* are distributed in North America. The females are characterised by the possession of reduced tegmina and wings. They are important as fortuitous intruders into homes in the United States and southern Canada, and about a dozen species are involved. They are known as wood cockroaches. The best known and the most widely distributed is *P. pensylvanica*. Where there is woodland the males are attracted to light, they fly readily and enter buildings through windows. The females which have short wings are brought indoors with firewood and groceries.

Out-of-doors this species occurs in the hardwood and softwood forests of the east and south-east United States where it hides during daylight in litter and under loose bark. It is only rarely found infesting houses other than as an intruder.

BLATTELLA VAGA

Another species which invades houses under certain conditions is *Blattella vaga* of southern Arizona. It is known as the Field cockroach and is common in irrigated fields where it feeds on plant debris. Its invasions during dry seasons into buildings are temporary although it is known to remain in them on occasions and breed to a limited extent.

Description

It so closely resembles *B. germanica* that mistakes can be made in identification, but *B. vaga* is very slightly smaller and more delicately shaped. The most conspicuous difference, however, is the presence of a black line between the eyes and the mouth of *B. vaga* which is absent in *B. germanica*. In addition, the longitudinal bars on the pronotum of *B. vaga* are darker and more distinct than is usual in *B. germanica*. Additional differences are to be found in examination of the genitalia. The biology of the two species is generally similar, but whereas the number of eggs per ootheca is 37 in *B. germanica* there are only 28 in *B. vaga*. The proportion of hatched nymphs maturing is much less in *B. vaga* and in this latter species the males are much slower completing their development (56 days, with females 45 days). In *B. germanica*, however, both sexes are generally similar (40–41 days).

Chapter 26

Earwigs

DERMAPTERA

Earwigs constitute the order DERMAPTERA and about 1,200 species have been described from most parts of the world. The word 'earwig' is derived from an old Anglo-Saxon word meaning 'ear creature'. Most European countries hold a superstition that the earwig crawls into the ears of sleeping people in order to conceal itself and, indeed, some people believed that it could gnaw its way into the brain.

Throughout the order there is little variation from the usual earwig-like form. They are rather elongate insects, either winged or wingless, almost always nocturnal and range in size from small (3 mm) to large (nearly 80 mm). The forewings are reduced to small leathery pads known as tegmina or elytra, hindwings are large, fan-shaped and intricately folded under the small tegmina when the insect is at rest. The legs are short for running. The tarsi are three-segmented and the long, freely movable abdomen, is terminated by a pair of cerci modified as forceps. This is a popular characteristic of earwigs.

FORFICULA AURICULARIA   EUROPEAN EARWIG

The adult *F. auricularia* must be one of the best known of all European insects. The dark brown body is from 10 to 14 mm in length and the forceps 4 to 9 mm. The head, however, is somewhat darker and the legs paler. The mouthparts are adapted for biting and resemble those of ORTHOPTERA. Compound eyes are present. The forceps are sickle-shaped in the male but straight in the female. Their function in this species is not known with certainty.

The European earwig, *Forficula auricularia*, is almost cosmopolitan. It was first observed in the USA in 1907 and since that time has become widespread throughout a large part of North America. It has also been introduced into Australia where it is said to be virtually cosmopolitan in the cooler regions.

Life-cycle
In Britain mating takes place in the autumn after which the pair hibernates in a cell just beneath the ground surface. In early spring the

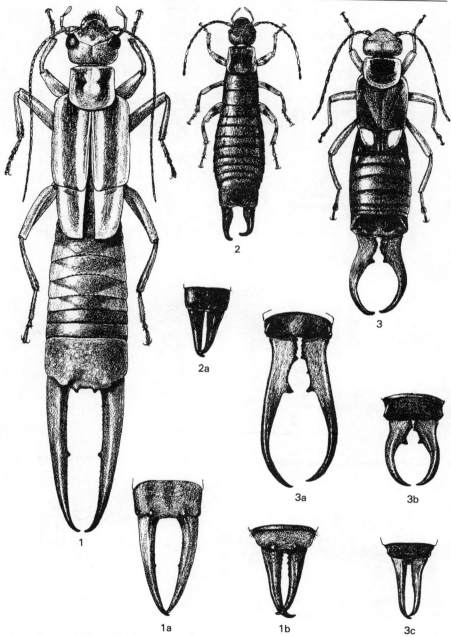

Fig. 26.1 Illustrations of three common earwigs in the United States. 1. Striped earwig, male; 1a. Tip of male abdomen with small forceps; 1b. Tip of female abdomen; 2. Ring-legged earwig, male; 2a. Tip of female abdomen; 3. European earwig, male, with medium forceps; 3a. Tip of male abdomen with long forceps; 3b. Tip of male abdomen with short forceps; 3c. Tip of female abdomen. (Cushman, NPCA)

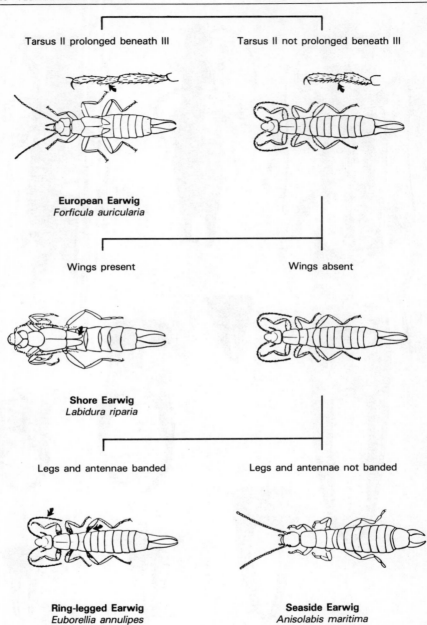

Fig. 26.2 Earwigs. Pictorial key to common domestic species in the United States. (US Department of Health, Education and Welfare)

male leaves the cell and about 30 eggs are laid in it. The female stays with her eggs and not only protects them from predators but tends them in some manner because if she is removed, except in the last few days of incubation they do not develop, either becoming mouldy or desiccated.

The eggs are white and oval about 1.25×1.0 mm and hatch in about 20 days although eggs laid in winter in Washington DC took 73 days. After a few days the young nymphs disperse.

There are four nymphal stages although as many as six have been observed and the nymphs resemble the adults except that the wings are not developed until after the first moult and there are fewer segments in the antennae, the number being 8 in the first instar then increasing progressively in each instar until the 14 of the adult stage are present.

Habits

It very rarely flies and many other temperate species are similarly averse to flight. They are nocturnal and spend the daytime hiding in crevices usually a metre or so from the ground and often deep inside flowers. When disturbed they run quickly and seek another dark retreat. In some years they become exceedingly numerous and enter houses in great numbers especially if the buildings are creeper-clad or surrounded by thick vegetation. They do not breed indoors, however, and can be reckoned only as garden intruders. Their presence, however, is considered objectionable largely on account of their ill-deserved reputation.

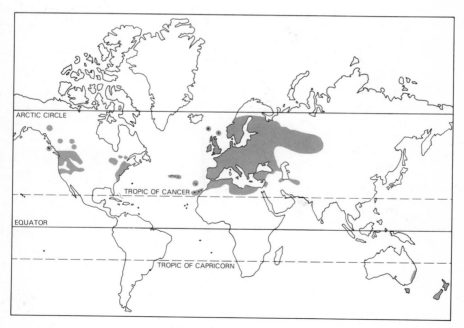

Fig. 26.3 *Forficula auricularia.* World distribution.

Earwigs are carnivorous, catching and eating insects but they are herbivorous also, often eating large holes in the petals of garden flowers.

EUBORELLIA ANNULIPES   RING-LEGGED EARWIG

This species, found in the USA, is about 25 mm in length. It is brown to nearly black in colour with yellowish brown beneath. The yellowish legs are banded with brown around the femora and tibiae. Out-of-doors it inhabits litter and makes shallow nests in rocky places. It is generally common although most abundant in the south and south-west. It occurs also in British Columbia. It feeds on grain as well as preying on the pests of grain; it is one of the most common pests in flour mills and in meat-processing plants and is long-lived.

Other species
A number of earwig species are known to enter buildings in tropical and subtropical areas, being attracted by lights at night, but their presence is accidental and they are of little importance.

Chapter 27

Crickets

ORTHOPTERA

ORTHOPTERA possess typical biting mouthparts. They are generally medium to large in size, indeed some species are among the largest of all insects. The forewings are elongate and narrow and modified into leathery or horny tegmina, while the hindwings are membranous, more delicate and with a large anal area. The pronotum has large, descending lobes. There are many apterous and brachypterous forms. Unjointed cerci are usually present as is an ovipositor. The hindlegs are usually adapted for jumping.

This order of more than 20,000 species is widely distributed throughout the world and includes the short- and long-horned grasshoppers, locusts and crickets, as well as many related groups. Species of ORTHOPTERA have been responsible for a vast amount of damage to Man's crops and have thereby caused famine and devastation on the greatest scale. This was the stimulation for the greatest concentration of effort and research within the field of entomology ever achieved. It is strange, therefore, that members of this order which have played such an important role in Man's economy are of relatively little importance in his buildings. Only in the GRYLLIDAE, the crickets, do we find species that live in buildings.

ACHETA DOMESTICUS   HOUSE CRICKET

This cricket is one of the most widely known insects occurring in buildings in Europe and temperate Asia. The chirruping noise made by the males and their pugnacity have been instrumental in assuring this insect a place in literature. The chirruping is produced by the raised forewings rubbing together. Each bears a file-serrated edge and a ridge or scraper, and the friction of the one on the other of the opposite wings causes the characteristic sound. In addition, there are many vibratory tympana or drum-like structures on the wings which serve to amplify it. Auditory organs are situated on the front legs.

In colour it is a light, yellowish-brown with three dark bands on the head, a dark pattern on the pronotum and a fine reticulation on the forewings. It varies from about 18 to 25 mm in length. The antennae are long and whip-like, and there are large compound eyes. The mouthparts are typically mandibulate, rather similar to those of the cockroach, and it will bite if handled. The hindlegs are much larger than the first two pairs, the femora being exceptionally so, and the tibiae bear heavy spines. The abdomen terminates in a pair of seta-covered cerci which function as receptors for a broad range of air vibration frequencies.

Breeding
Out-of-doors, on refuse dumps and similar situations, the female pushes her ovipositor into the substrate in soft earth with the forelegs and mouthparts (about 35 mm deep) and then lays her eggs. Indoors they are laid singly in dark crevices and behind wainscoting, baseboards, and the like although in the laboratory they appear to prefer damp situations. The eggs are somewhat banana-shaped, white in colour and about 2.4 mm in length and 0.3 mm wide. They are very sensitive to atmospheric conditions. If they become too dry they shrivel and if too damp they become mouldy. The average number of eggs laid was 728 at 28°C and 1,060 at 35°C but another investigator found that from 40 to 170 eggs are laid with an average of about 104 at room temperature. The egg stage lasts about 12 weeks while the nymphal stage lasts from 30 to 33 weeks in which period 9 to 11 moults take place. The first stage nymph is much like the adult in appearance except that the head is proportionately larger and the hindlegs relatively smaller. At the penultimate nymphal stage the rudimentary ovipositor of the female can be easily made out.

Habits
Progression is generally made by a series of short runs, it is only when danger is threatened that a quick escape is made by a series of jumps. The predator can be easily confused by the movement of the cricket's body through an angle of 100° to 170° during the jump so that after landing it jumps off in quite a different direction.

Breeding takes place to an extent throughout the year both in fermenting or rotting refuse dumps and in heated buildings. Bakeries, kitchens and boiler houses and, indeed, anywhere where the temperature does not fall much below 20°C, are the situations where they commonly occur.

In northern Europe and North America crickets often live out-of-doors in refuse tips during the summer but at the onset of colder weather in autumn invade neighbouring properties. The numbers involved are sometimes of extraordinary proportions constituting a veritable plague. In the United States and in Canada there are accounts of thousands of the insects being killed daily in one building and in India a swarm 2 metres long and 100 metres wide is recorded.

Crickets are omnivorous and in buildings eat vegetables, raw or cooked, fruit, bread and dough, meat, raw or cooked and other insects including their own species especially those with a wound. They will also bite holes in textile materials and leather which cannot possibly serve food and will sometimes do considerable damage. They are nocturnal.

Copulation is preceded by a courtship in which the male chirrups and waves his antennae. The chirruping cricket is indeed a sign of the sexual activity of the male and the female moves towards him. The male then vibrates his body in a curious manner in which there are lateral and backwards and forwards movements. The chirruping note then changes and the female mounts on to his back or the male may take the initiative and push his abdomen under hers. He then transfers his sperm in a spherical spermatophore which remains attached to the female for about an hour when it falls off or the female eats it.

GRYLLUS ASSIMILIS   AMERICAN FIELD CRICKET

Distributed throughout North and Central America and the northern part of South America G. *assimilis* is often of importance as an invader of buildings, sometimes in swarm proportions. It is larger and more robust than *Acheta domesticus* and the hindwings project backwards beyond the forewings like pointed tails.

About two weeks after attainment of the adult stages from 150 to 400 eggs are laid over a period of about two months. They are laid singly with no protective secretion and deposited in the soil from 6 to 25 mm deep. According to temperature and rainfall, hatching takes place in May. Immediately after emergence the young cricket can walk, run and jump with great activity. Male nymphs pass through eight instars, the females nine. The length of nymphal life is usually from 78 to 90 days although 65 to 102 days have been recorded.

The species is subdivided into a number of subspecies and occasionally it is referred to as five distinct species. It is sometimes injurious to field crops such as alfalfa, wheat, oats and rye, but also eats dead and dying insects. Indoors, however, a wide variety of materials is liable to be damaged by them. Holes are gnawed in textile materials such as cotton, linen, wool and silk. Furs are damaged and clothing stained with perspiration is especially attractive to them. In addition paper, greasy foods, milk and syrup are recorded as being prone to damage. Other materials suffering injury are given as nylon, wood, plastic fabric, thin rubber goods and leather. The swarming invasions into buildings are said to take place when rainfall follows a period of drought.

GRYLLODES SIGILLATUS   TROPICAL HOUSE CRICKET

This is the cricket most commonly associated with buildings in oriental countries. It feeds on greasy and fatty refuse and is also injurious to paper, cloth and leather bindings of books. *Gryllulus domesticus*, Black-headed cricket of India will eat human food and even clothes when green vegetable matter is scarce.

Chapter 28

Springtails

COLLEMBOLA

COLLEMBOLA are small, primitive, wingless insects rarely exceeding 5–6 mm in length. Their anatomy is remarkable in that the abdomen consists of only six segments, the smallest number found in insects. The antennae mainly consist only of four segments. They are extremely abundant in nature, acting as scavengers in the soil, among dead leaves and other vegetable detritus. They are known as springtails on account of the habit of most species, jumping when disturbed. This is accomplished by means of a special forked spring organ arising from the fourth abdominal segment which allows them to leap 25–50 mm into the air thus enabling them to evade predatory animals. When at rest the springing device is held forwards by another organ on the underside of the third abdominal segment. Over 2,000 species are known.

COLLEMBOLA do not undergo a larval or pupal stage. They usually shun light and most require a high atmospheric humidity. They subsist on a variety of vegetable and animal debris. Although sometimes present in large numbers no damage is caused, and although frequently found in damp areas in buildings such as kitchens, outhouses and cellars where they hide under damp materials or pots, they cause no harm.

COLLEMBOLA are remarkable for their exceedingly large numbers, for as many as 100,000 per square metre are commonly met with in soil of temperate woodland and grassland. Only in very dry situations, such as deserts, are they not abundant. They are the most numerous of all insects. Their lack of ability to withstand desiccating conditions is because the waxy layer of the cuticle, present in almost all other insect groups, is

Fig. 28.1 A Collembolan, Springtail.

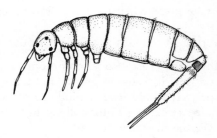

present only on the tips of a pattern of minute tubercles. However, this makes the insect non-wettable, which is most important in the very damp conditions in which it lives. Some COLLEMBOLA resist desiccation by cuticular modifications or developments of setae and scales. Such forms are easily wettable.

Reproduction is generally a simple process. Either a mass of sperms or a spermatophore is deposited near the female and she takes it into her genital opening, or sperms are shed on to any part of the female's body. Reproductive behaviour is sometimes extremely complex, involving strongly dimorphic modifications. The eggs are laid in small groups of 3 or 4 or they may be laid in masses of a 100 or more. The young springtails, which are similar to the adults except in size, undergo from 4 to 10 moults before reaching adult stage. They then continue to moult from time to time during the whole of their adult life, which is unusual in insects. They probably live for about a year, although under laboratory conditions they have lived for the remarkable period of four years.

In North America springtails of the following genera have been recorded as pests in dwellings: *Lepidocyrtus, Pseudosinella, Heteromurus, Tomocerus, Entomobrya* and *Sira*. Two species of *Entomobrya, E. nivalis* and *E. tenuicauda*, have been implicated in cases of pruritic dermatitis in Man, and species of *Orchesella* are known to have infested the heads and pubic areas of a Texas family, probably due to the use of mouldy bedding.

Chapter 29

Bristletails, silverfish and firebrats

THYSANURA

The THYSANURA are among the most primitive of insects. They are wingless, more or less flattened and torpedo-shaped or cylindrical. They are agile, running swiftly for cover if disturbed and the body is often covered with scales and there may be a pigmented pattern. If compound eyes are present, they are much reduced, the ommatidia being isolated. The unspecialised mouthparts are used for biting and chewing. The maxillary palps are five-segmented.

The many-segmented antennae are long and there are leg-like styles present on a number of the abdominal segments, at the most on segments two to nine, but usually on fewer segments. The most usual feature, however, which the non-entomologist would associate with this group, is the presence at the hinder end of a long 'appendix dorsalis' and a pair of cerci – all of which are movable. This has given rise to the common name of 'bristletail'. There are 330 known species, in five families, if we accept the even more primitive ARCHAEOGNATHA with its 250 known species in one extinct and two living families. In this latter order, however, species of economic significance are absent.

Together with the COLLEMBOLA and a few other primitive wingless groups the THYSANURA make up the INSECTA subclass APTERYGOTA (meaning wingless), while all other insect orders form the subclass PTERYGOTA (winged). They are thought to be close to the ancestral form from which all the pterygote insects evolved.

Internal fertilisation takes place and the young, which hatch from eggs, are similar to the adults in shape. After a number of moults, sexual maturity is reached and thereafter the insect continues to moult for the rest of its life. Upwards of 60 moults are recorded for several species. Generally they are long-lived, and in the laboratory have been known to live for as long as seven years, but some species are known not to become sexually mature until they are two years old.

Thysanurans show a remarkable resistance to starvation. One is known to have survived for over 300 days in a laboratory jar without food or water. In nature thysanurans are to be found beneath stones, in the crevices of bark and rocks, and in forest-floor litter. They appear to feed on a variety of decomposing plant and animal matter, together with the associated bacteria and fungi. They also occur in the nests of birds and mammals, and some species are found in ants' nests where they rob the ants of food when it is in the process of regurgitation. In Australia they are also known from the nests of termites.

A number of species of THYSANURA are well-known inhabitants of buildings. In warmer climates they cause considerable damage to a variety of materials. In more temperate regions, even though the harm they cause is minimal, they are unacceptable to the housewife in the kitchen and bathroom, where they usually occur. Several species are cosmopolitan, or nearly so, and in addition some countries have their own endemic fauna.

LEPISMA SACCHARINA   SILVERFISH

Probably the most widely known of all THYSANURA. The silvery-pewter colour, its twisting fish-like shape, as well as its widespread distribution, have given it its common name, known throughout the English-speaking world. It has, so far, not been recorded from South America. Although originating in a tropical or subtropical area, today it occurs widely wherever such conditions exist in the microclimate of a building, especially in dwellings where food is prepared or where warm, humid conditions prevail, such as in bathrooms. Although *L. saccharina* ranges widely around dwellings in darkness, inside it is often trapped in a bath or in a wash basin. Contrariwise, when something is disturbed around the fireplace – where it would be thought especially desiccating conditions would prevail – *L. saccharina* will be seen darting to some alternative shelter. Although this species is found out-of-doors in tropical areas, its truly endemic region in not known.

Lepisma saccharina possesses a silvery sheen imparted to it in the adult stage by the greyish-brown, backwardly-directed scales with which it is covered, and which enable it to slip into small crevices. It reaches a length of 12 mm. Its introduction into new buildings is thought to be due to the transport of the adults, nymphs or eggs in cardboard cartons and papers of various kinds, making their detection difficult. It has a preferential temperature of between 22 and 27 °C, at 75 to 97 RH.

Breeding
The soft eggs, when laid, are white but they quickly turn yellow then brown. From one to three eggs are laid daily or at irregular intervals of several days or weeks, either under various objects or deposited haphazardly. The majority hatch at 22 °C and take 43 days, but take only 19 days at 32 °C. The preferred temperatures for all stages lie between these latter limits. An average of 100 eggs are laid by each female. The first instar lasts from 7 to 10 days, and thereafter each instar takes two or three weeks. There are various estimates of the period from egg to

maturity; from three to four months to two to three years. At the third moult, however, the covering of scales occurs. After each moult following sexual maturity, eggs are laid, and this may occur as many as 50 times, and after $3\frac{1}{2}$ years at 22 °C some are recorded as still laying eggs. Both carbohydrate and protein material are eaten, although the latter is preferred. Cast skins, dead bodies and injured individuals of their own kind are readily devoured. *Lepisma saccharina* is able to digest cellulose, not through the intermediary of symbiont bacteria or other organisms in its gut, but by the secretion of enzymes acting directly.

CTENOLEPISMA QUADRISERIATA FOUR-LINED SILVERFISH

This large silverfish reaches a length of 18 mm, is dark grey in colour, with four lines running the length of the back. It is flatter than *Thermobia domestica*. The nymphs are light brown in colour, often pinkish, until the fourth moult. The eggs are usually laid during warm months, and the resultant nymphs do not reach sexual maturity until the following year. Like several other species they are long-lived; they have lived for five years under laboratory conditions. It appears to have a temperature preference between 27 and 29 °C.

Ctenolepisma quadriseriata is common in the eastern United States as far south as Georgia and Arkansas, and west as far as Missouri and Iowa, but also occurs further west in California. In Canada it is found in the south and centre.

ACROTELSIA COLLARIS

This is one of the largest silverfish, and reaches a length of 16–18 mm without the bristles, and a width of 5 mm. It is well known as inhabiting buildings in many tropical areas. The scales on the dorsal surface are practically black, but if they are lost through abrasion the insect has a pale appearance. It can be readily identified by the triangular tenth abdominal tergum and the presence of a tuft of bristles on the prosternum. Like many other silverfish species, it feeds on the sizes and glues used in bookbinding and other paper manufacture. This species is distributed widely throughout the tropics and has reached northern Australia.

CTENOLEPISMA LONGICAUDATA GIANT OR GRAY SILVERFISH

It was earlier known as *C. urbana*, but some confusion still exists as it is evident that at present *C. longicaudata* consists of two species, one having 2 to 5 sensory papillae on the labial palps and the other from 9 to 12. Maximum size is 19 mm. It is abundant in many warm parts of the world including Australia. In the USA it occurs only indoors and is recorded from North Carolina, Illinois, Missouri, Louisiana and California and it is also found in Hawaii.

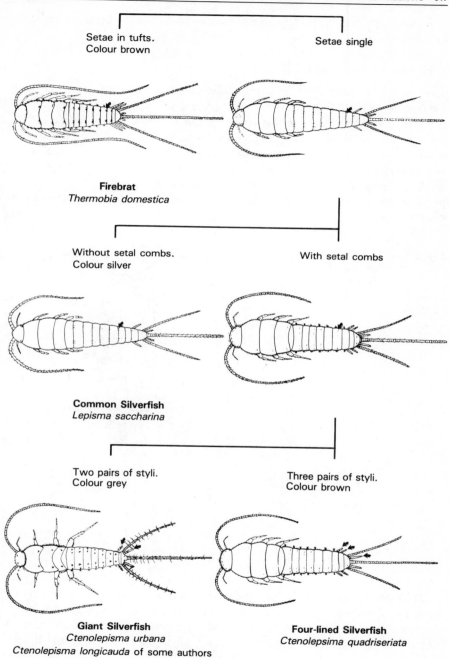

Fig. 29.1 Pictorial key to domestic species of THYSANURA in the United States. (US Department of Health, Education and Welfare)

Among other pest species are *C. lineata* and *C. urbana* both occurring in Australia.

THERMOBIA DOMESTICA   FIREBRAT

This is another cosmopolitan species found in warmer environments than *Lepisma saccharina*. The optimum temperature for development lies between 36 and 41 °C, and this is the reason for its common name throughout the English-speaking world. It is, or was, common around bakeries, cooking ovens, as well as in industrial premises where steam was employed and crumbs and food scraps were commonly left about. With the general tightening up of regulations dealing with the preparation, consumption and disposal of food, it is less abundant than once it was.

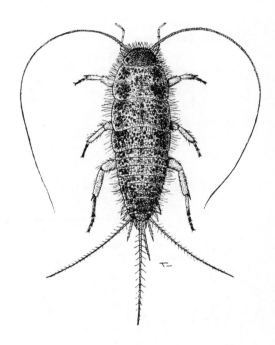

Fig. 29.2 *Thermobia domestica*, Firebrat. (British Museum, Natural History)

Thermobia domestica will feed and thrive on a great variety of materials found in dwellings, but is commonly associated with food particles dropped on to the floor in bakeries and reataurant kitchens. However, it is known to damage many types of textiles, such as cotton, silk, and linen, as well as paper. It is especially injurious to wallpaper, where it is attracted by the starch and dextrin sizes and pastes.

A characteristic of many THYSANURA, but which is shown especially by *Thermobia domestica*, is the ability to withstand the desiccating conditions common in those parts of buildings which they inhabit. Very little water

Fig. 29.3 Damage by *Thermobia domestica*.

is lost through the cuticle, and even so it has no need to drink as it can absorb water vapour from the air down to a relative humidity of 45 per cent. It should be noted that although the name *Thermobia domestica* has been used here because of its universal usage, the insect when first described was named *lepismodes inquilinus* and this latter name may become generally accepted.

Chapter 30

Centipedes

CHILOPODA

The class of ARTHROPODA, known as the CHILOPODA or centipedes, consists of animals with elongate segmented bodies each segment of which bears a single pair of jointed legs. The antennae are filamentous and bead-like and the first pair of legs are modified as jaw-like organs or poison claws. Eyes are present in three of the four orders in which the CHILOPODA are classified. All of the 3,750 or so species are predacious, feeding on insects and other invertebrates, while some of the largest species will attack and consume vertebrate prey, with the exception that species of GEOPHILIDAE sometimes feed on vegetable matter. They are usually nocturnal and in the day hide in the soil, under loose tree bark and other crevice-like situations. When the prey is captured, poison injected through the poison claws immobilises it, although vertebrates caught by the largest centipedes are only partially immobilised.

Centipedes are usually less than 5 cm in length, but some tropical species such as *Scolopendra gigantea*, from Brazil and Trinidad attain a length of 25 cm.

SCUTIGERA

A number of centipede species make their way into buildings, generally fortuitously, when seeking a dark crevice in which to spend the daylight hours. One group, however, is so commonly found indoors in warm climates that they are known as house centipedes.

These are species of *Scutigera* and related genera. They are easily identified, being more like a many-legged spider than the usual image of a short stumpy-legged centipede.

Some, such as *S. coleoptrata*, are cosmopolitan and although apparently originating in the Mediterranean region, they have spread to the United States and elsewhere, but there are only three records for colder Britain. The antennae are long and thin, as are the legs, the hinder ones exceptionally so. The palps of the second maxillae and the poison claws are leg-like, and there are 15 pairs of legs only, which are spotted giving a cryptic effect. Although the legs possess the same number of joints as

Fig. 30.1 *Scolopendra obscura*, Centipede.

Fig. 30.2 *Scutigera forceps*, American House centipede.

other species, in *Scutigera* the distal segment, the tarsus, is secondarily divided into a number of separate parts, but the whole of the tarsus is held flat on the substrate which contrasts with that of other centipedes where only the tip of the tarsus makes contact.

Scutigera is also unique among centipedes in possessing compound eyes similar to those of insects, although it is not thought that they are used in pursuit of prey. The latter consists of the many insect species found in damp situations in buildings where *Scutigera* lurks ready to pounce on them, as it is surprisingly alert, agile and fast in its movements. There are about 130 species of *Scutigera* and of closely allied genera.

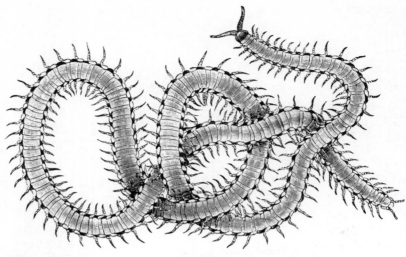

Fig. 30.3 *Notophilus taeniatus*. (After Koch)

Bites

Many centipedes are recorded as biting humans and although the results are usually disagreeable, none is known as being potentially lethal to Man. Many records, unfortunately, do not indicate the centipede species. There is usually intense pain accompanied by localised swelling of the affected part.

In Australia, *Ethmostigmus subripes*, the Common garden centipede, is capable of inflicting painful wounds, with localised swelling of the affected tissue. There are no records of the two other common Australian species, *Schizoribantia aggregatum*, the Ribbon centipede, and *Allothereua maculata*, the House centipede, biting humans.

In the southern states of the United States of America, two fairly large centipedes, *Scolopendra heros* and *S. morsitans*, as well as *S. polymorpha* of the south-western states and Mexico (reaching a length of about 15 cm) can cause severe pain and swelling. In India, Burma and the Andaman Islands, however, certain centipede species bite with more serious consequences, recovery taking sometimes as long as three months. A necrotic process often develops at the site of the bite. Most serious results follow bites by 23 cm long, yellow to dark greenish blue species. The Giant centipede, *Scolopendra subspinipes*, in the Philippines, although giving a bite which causes intense local pain, is not potentially deadly to Man, and the bite is quickly relieved by a local anaesthetic.

In areas where centipedes abound, those sleeping in the open or in tents should ensure that mosquito nets are securely tucked in and should exercise caution in handling clothing and footwear which has been lying on the ground overnight.

Chapter 31

Millipedes

DIPLOPODA

The class DIPLOPODA is so named on account of the fact that nearly all segments of the body bear two pairs of limbs. This is because each segment is composed of two joined tergal plates. There is a nearly circular tergum, one or two small pleural plates, two sternites and two pairs of legs to each segment. There are only two pairs of mouthparts, the first of which, composed of two or three segments, is for biting, while the second pair is fixed to form a broad plate. The reproductive organs have their apertures on the ventral side of the forepart of the body near the head. Eyes are present in some orders but are absent in others.

The distinct head bears a pair of short, unbranched antennae. There are a number of different orders of DIPLOPODA and most are elongated although some are very different in appearance and there is a wide variation in the number of segments of which the body is composed.

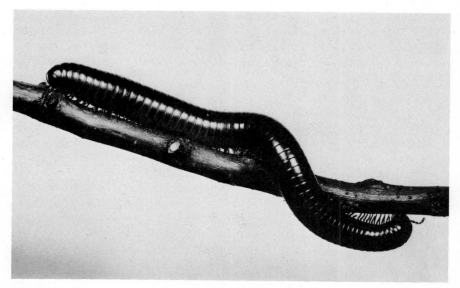

Fig. 31.1 Giant Tropical millipede. (Shell Chemical)

Some are almost tubular in section, while others are flat. When disturbed many of the long species coil themselves into a watch-spring shape. *Glomeris marginata* of Europe, which does not have an elongated body, can curl itself into a ball as in some woodlice. The same habit and shape is shown by the SPHAEROTHERIIDAE of South Africa, Madagascar, southern Asia and Australasia. Some are larger than a golf ball when rolled up.

A number of millipede species are known to enter buildings with some degree of regularity although their generally positive reaction to damp situations would tend against this. In addition, some are able to walk up smooth vertical walls without difficulty so that, as most people view them with some degree of repugnance, they can be considered as undesirable visitors in a house. Most DIPLOPODA possess glands from which a defensive secretion can be discharged either slowly or, in the case of some tropical forms, almost explosively as a fine jet or spray. In the case of *Rhinocricus lethifer* of Haiti, the discharge can be ejected to a distance of nearly a metre. If the fluid touches human skin it first causes a blackening then the skin peels off leaving a wound which heals only very slowly. Many tropical millipedes exhibit warning coloration, generally black and red or yellow, and such colour patterns are often associated with the poisonous nature of the bearer.

Chapter 32

Tapeworms, flukes, roundworms

CESTODA, TREMATODA AND NEMATODA

Cestode, trematode and nematode worms parasitising Man are included in this work as transmission, in a large number of cases, takes place in the kitchen or dining room of a building. Infection often occurs through the consumption of infested food, either through being inadequately washed or insufficiently cooked. Children playing on ground where there has been fouling by the faeces of dogs, rats and Man himself are equally at risk.

Cestoda Tapeworms

Cestodes are internal parasites and are called tapeworms on account of their long, flat shape and whitish colour. The body, or strobila, is divided into a number of similar parts, or proglottides, in each of which is a set of male and female reproductive organs. Each proglottid is connected to the next by muscular tissue, a pair of nerve cords and two pairs of excretory ducts. The head, or scolex, is minute and bears the organs of attachment to the intestinal mucosa of its vertebrate host. There is a narrow, neck-like region immediately behind the scolex where the proglottides are budded off. At first, the latter are small and embryonic, but as they develop they are pushed further and further down the strobila by the production of new ones. The male reproductive system is the first to mature, thus cross-fertilisation between proglottides is ensured. At the hinder end of the strobila the branched uterus is full of eggs. In some species the eggs are shed, passing out with the faeces, and in others the mature proglottides are passed out whole.

The scolex of all tapeworms parasitising terrestial vertebrates, including Man, of course, bears four suckers and one or more rows of hooks encircling the beak-like projection.

Tapeworms do not possess an alimentary canal. They feed entirely by absorbing nutrients directly through their cuticle. These nutrients are in the form of soluble food which has been digested by the host and is about to be absorbed by the host's intestine wall. They are capable of a certain amount of movement, as they move about in the intestine and attach themselves to the gut wall where food is richest. The cuticle is covered with an outer membrane of mucoprotein, which (like the lining of the host's intestine) is resistant to digestive juices.

Cestodes exhibit complicated life-cycles which, in addition to the definitive host, may involve two intermediate hosts.

DIPHYLLOBOTHRIUM LATUM   BROAD TAPEWORM OF MAN

Originally a fish tapeworm, this species has become adapted to parasitising fish-eating carnivores, including Man. It reaches a length of up to 9 m on account of the fact that the proglottides do not become detached and may number as many as 3,000 to 4,000.

The ripe eggs are shed from the uteri and are passed with the faeces. On hatching, a ciliated, free-swimming larva bearing six posterior hooks, is released. This is swallowed by a crustacean, usually a copepod, and it loses its cilia and bores through the gut wall of the copepod into the body cavity. Here it transforms into an elongate larva bearing hooks in the tail region. When the copepod is swallowed by a fish, the larva (proceroid) again bores through the gut wall and develops in the fish's muscular tissue into a hookless larva (pleuroceroid). The process may be repeated several times as small fish are eaten by carnivorous fish such as pike, trout and perch. The pleuroceroid larva may reach a length of several centimetres.

When, finally, the larva is eaten by a warm-blooded animal it develops into the mature tapeworm in the gut. It grows extremely rapidly and may be as long as 1 m and already laying eggs three weeks after being swallowed.

Man becomes infested by swallowing the pleuroceroid larva in raw or undercooked flesh, or the roe of infested fish. *Diphyllobothrium latum* absorbs large amounts of vitamin B from the host's intestine, thus denying it to the host, and as this vitamin plays an important role in the formation of blood, a severe pernicious anaemia is caused in Man.

TAENIA SAGINATA   BEEF TAPEWORM

The adult tapeworm is from 5 to 7 m in length and may consist of as many as 1,000 or more proglottides. The scolex is only 1.5 to 2.0 mm in diameter and there are four suckers but no hooks present. The terminal mature proglottid is approximately 20 mm in length and 6 mm wide, white in colour and contains about 100,000 eggs. The eggs may pass out with the faeces, or the proglottid becomes detached and crawls through the anus, exhibiting a certain amount of agility.

The only recorded intermediate hosts are domestic cattle and their near relatives. The cow or ox grazes on pasture contaminated with human faeces containing the eggs of *T. saginata* and these hatch in the small intestine. The larva is furnished with six hooks and it makes its way via the bloodstream into muscle tissue where it takes about three months to transform into a cysticercus, or bladderworm. This is the stage which infests Man. This is when beef is not adequately cooked, or cooked in such a way that the inside is still raw. When swallowed, the scolex becomes free and attaches itself to the intestinal lining, developing into an adult within two or three months. The length of adult life may be as long as 25 years. It was estimated that in 1947, worldwide, over 38 million persons were infested.

TAENIA SOLIUM   PORK TAPEWORM

Smaller than *T. saginata*, *T. solium* varies in length from 2 to 3 m or a little over and there are always less than 1,000 proglottides. The scolex is approximately 1 mm in diameter and bears four suckers and a circlet of hooks. The gravid proglottides are passed in the faeces of Man singly or in small chains. Pigs swallow the eggs when feeding in fields or yards contaminated with untreated human sewage. In about three months the eggs have developed into the bladderworm (*Cysticercus cellulosae*) stage in the pig's muscular tissue, giving rise to 'measly' pork. If this is eaten the cysticerci develop into adult tapeworms in Man's intestine.

Eggs are sometimes swallowed by Man in contaminated, unwashed

Fig. 32.1 Scolex and anterior proglottides of *Taenia solium*.

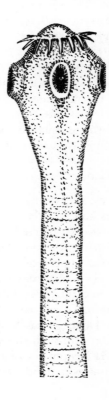

salads, etc., and these give rise to cysticerci. Another method is by anus to mouth contamination as a result of not washing after using the lavatory. The eggs hatch in the upper intestine and, travelling through the lymphatic system and the bloodstream, they may reach various organs of the body. Within two or three months the embryos develop into translucent bladderworms, from 5 to 10 mm in diameter. In order of frequency of invasion are the subcutaneous tissues, muscles, brain, eye, heart, lung and peritoneum. A few hundreds of cysticerci have often gone unnoticed in the subcutaneous tissues, but as many as 20,000 have been found in one person. In the brain, one or two may cause little reaction as long as they remain alive but, when dead, the parasite causes reaction and capsule formation when epilepsy and other disturbances may result. Most cases of cysticerosis concern *T. solium*; seldom has *T. saginata* been involved. When the eye is invaded, detachment of the retina often occurs, the parasite remaining unencapsulated and constantly changing shape.

HYMENOLEPSIS NANA   DWARF TAPEWORM OF MAN

This is the smallest tapeworm commonly infesting Man, being only about 30 mm in length. It is also the commonest cestode parasitising Man and in a heavily infected person several thousand individuals may be present. The scolex bears four suckers and a ring of hooks: the terminal proglottides break up, releasing their eggs which are passed in the faeces. The eggs bear a pair of polar filaments. An intermediate host is not obligatory.

If the egg is swallowed by Man, a six-hooked embryo emerges and enters the intestinal villi, where it transforms to a cysticercus; it then re-enters the intestine to become an adult in about two weeks. Fleas and grain-beetles, however, can function as intermediate hosts.

A form of this species, *H. nana fraterna*, parasitises rats and mice and sometimes occurs in Man.

ECHINOCOCCUS GRANULOSUS   DOG TAPEWORM

This cestode is only from 3 to 9 mm in length. The scolex bears four suckers and a circle of hooks, and there are only three proglottides. A dog becomes infected when it feeds on organs of sheep, pigs, cattle or camel which contain the cysticerci of this species. The tapeworm fastens on to the upper portion of the dog's small intestine. Eggs are passed in the faeces which may contaminate pastures where the intermediate hosts graze, and thus they become infected. Soil around dwellings may also be contaminated and eggs may be transferred to a human host in various ways. The eggs hatch in the duodenum and a six-hooked embryo penetrates through the wall of the intestine and reaches an organ of the host via lymphatics or bloodstream. It then encysts and forms a vesicle known as a hydatid cyst. About 63 per cent are found in the liver.

Hydatid cysts are most abundant in sheep-raising areas where diseased organs are fed to dogs. Specially heavy infections are found in East and South Africa, Australia, New Zealand, the Punjab in India, and Mediterranean countries. The Eskimos of Canada and Alaska also show a high rate of infection. This is due to moose, elk, wolf and dog involvement.

Fig. 32.2 Entire strobilus of *Echinococcus granulosus*.

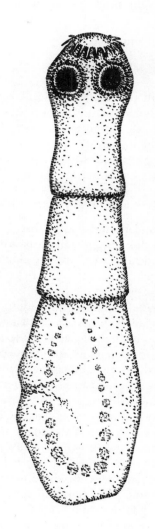

HYMENOLEPSIS DIMINUTA   RAT TAPEWORM

Several hundred cases are known of this tapeworm infecting Man. The intermediate hosts are the larvae and adults of several species of meal moth, grain beetles and fleas. Transmission occurs by Man eating cereals containing these infected insects.

DIPYLIDIUM CANINUM DOG AND CAT TAPEWORM

This cestode is sometimes reported as parasitising Man. The intermediate hosts are fleas and lice, and young children are sometimes the final host. Transmission may take place by a child's face being licked by a dog that has just 'nipped' an infected flea or louse. The latter may then accidentally enter the mouth of an infant.

Trematoda Flukes

Adult trematodes are leaf-like in form and there is a mouth at the front end which is sucker-like. A ventral sucker is usually present with which it attaches itself to the host. They are hermaphrodite with the exception of the genus *Schistosoma*. There is a complicated life-cycle. The adult stage parasitises Man and other mammals, birds and other vertebrates, while the immature stages are parasites of water snails

OPISTHORCHIS SINENSIS   EASTERN LIVER FLUKE

Abundant and widely distributed in China, Indo-China, Korea and Japan, this fluke is generally a parasite of dogs and cats but, wherever fish is eaten raw, smoked or pickled within these regions, human infection can occur.

Life-cycle
The adult flukes are from 10 to 25 mm long and 3 to 5 mm wide. They inhabit the bile ducts of the liver, feeding upon its secretions. Large numbers of eggs are produced which ultimately are passed in the faeces. Further development can only take place if the eggs are transported by some means to fresh water where the fully developed miracidia larvae are released and ingested by certain species of snail and where larval multiplication takes place. Cercariae, which are free-swimming, make their way out of the snail, and finally encyst on fishes of the family CYPRINIDAE, to which many edible species of carp belong. If the raw or preserved fish is only partially cooked then the cysts are liberated in the duodenum, whence the metacercariae migrate to the bile duct and become adult flukes in about a month. The flukes of *C. sinensis* irritate the bile ducts both by their attachment and by their toxic products which cause modifications, principally by proliferation of the epthelium and by thickening the duct walls. The ducts become dilated and filled with the flukes and their eggs. As many as 21,000 flukes have been recovered from the liver of a single person. Heavy infection causes a number of disagreeable symptoms with enlargement of the liver. The flukes are very long-lived, periods of 25 years being reported.

FASCIOLOPSIS BUSKI

The adult fluke inhabits the upper intestine of Man and pig, where it attaches itself to the mucosa. It is of the greatest importance as a parasite of Man in China, the Bengal areas of India and East Pakistan, Thailand, Malaya and bordering regions. Ulcerations and abscesses occur at the attachment points, and if a large number of the flukes are present (a heavily infected person may harbour several thousand of the parasites), there may be bleeding, epigastric pain, severe diarrhoea with nausea and vomiting. Chronic infections are accompanied by oedema, probably caused by the absorption of toxic metabolites excreted by the flukes.

Life-cycle

The hermaphrodite adult flukes are about 3 cm in length and 1.2 cm in width. Thousands of eggs are produced daily by each fluke and are passed out with the faeces. They mature only if they come into contact with fresh water. Here they develop into a ciliated, free-swimming larva known as a miracidium which finds a snail host of certain species, and after piercing the skin, bores into it. Within the snail the parasite multiplies, and after emergence from the host, transformation into tadpole-shaped cercariae takes place. After swimming for a time, the cercariae encyst on the leaves of certain aquatic plants, generally of the genus *Trapa* or *Eliocharis*. These are used as vegetables and unless thoroughly cooked the cysts may be swallowed and develop into adults in the intestine. A common method of infection takes place if the encysted cercariae are present on water nuts when peeled with the teeth and eaten raw.

FASCIOLA HEPATICA   SHEEP LIVER FLUKE

Although only an accidental parasite of Man, the very wide distribution of *Fasciola hepatica* makes it potentially important virtually wherever sheep occur. Herbivorous mammals generally, but sheep more particularly, are the principal hosts. Strangely, up to 1948, one-third of all human infections had reportedly taken place in Cuba.

Life-cycle

The leaf-like adult fluke is about 30 mm in length and 13 mm in width, with a mouth surrounded by an oral sucker, and a ventral sucker. It inhabits the bile ducts, liver and gall bladder. The eggs are large and pass into the intestine. On hatching, the miracidium larva makes its way into a snail of the genus *Limnaea* or of related genera. Here development and multiplication take place until cercariae larvae swim away from the snail and make their way on to wet vegetation such as grass and there encyst. When the cysts are taken into the digestive system of the herbivore, the metacercariae are liberated. They reach the upper part of the small

Fig. 32.3 Adult of *Fasciola hepatica*, Sheep liver fluke.

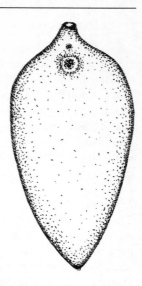

intestine and there bore through the wall, finally reaching the liver. Man usually becomes parasitised by eating infested watercress. Considerable damage to the liver occurs.

SCHISTOSOMA

Some of the most important of human parasites are trematodes, worms of the genus *Schistosoma*. There are three principal species. *Schistosoma mansoni* is distributed throughout Egypt, Africa generally, Madagascar, Yemen, West Indies and the northern part of South America. *Schistosoma japonicum* occurs in central and south China, Japan, the Philippines and

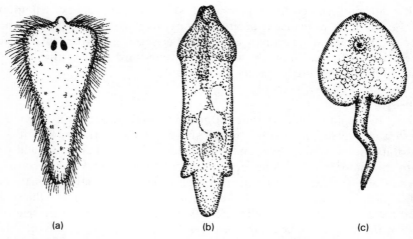

(a)   (b)   (c)

Fig. 32.4 *Fasciola hepatica*, Sheep Liver fluke; (a) Miracidium larva; (b) Redia larva; (c) Cercaria larva.

Celebes. The third species, *S. haematobium*, overlaps the distribution of *S. mansoni* to some extent in Egypt, throughout Africa and additionally in areas of the Near East, Middle East, Madagascar, Portugal and a restricted area near Bombay. Brown (1959) considered that over 100 million persons were infected with schistosomes and that next to malaria it is Man's most serious parasitic infection.

Fig. 32.5 *Schistosoma mansoni*, male and female. (After Loos)

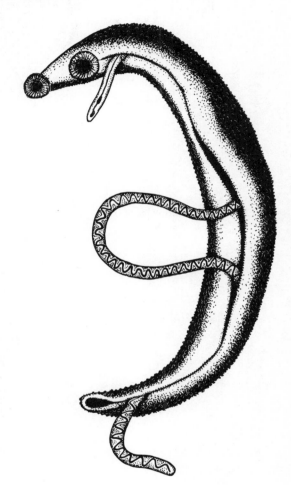

Life-cycle
The adult worms which are up to 2.5 cm in length live in pairs, the female being enfolded within lateral lobes of the males. According to the species up to 3,500 eggs are passed each day into the venules which are ruptured by the egg into the tissue surrounding the intestine or bladder and are finally released. If the egg comes into contact with fresh water, within a short time it hatches into a pear-shaped miracidium larva. This is ciliated and swims about until a specific snail is found. The larva then

Fig. 32.6 (a) Eggs of *Schistosoma heamatobium* (terminal spine); (b) Egg of *S. mansoni* (lateral spine). (Both after Loos)

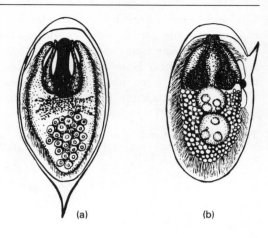

Fig. 32.7 Miracidium larva of *S. haematobium*.

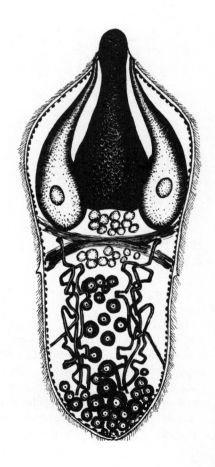

penetrates into the snail's tissue and develops and multiplies asexually through two generations before emerging from the snail as cercaria larvae in large numbers to swim actively aided by their long, slender, forked tail. The snail hosts differ according to species. In the case of the miracidia of *S. haematobium* only snails of the genus *Bulinus* are penetrated while in the case of *S. mansoni* the flat, ramshorn-shaped snails of the genus *Biomphalaria* are selected. Reinfection of man takes place when the person comes into contact with the contaminated water during bathing, working in irrigation ditches, fishing or washing clothes, and the cercariae penetrate the skin. Large numbers cause a tingling sensation. The cercariae then travel via the capillary blood system to the heart and lungs and are then carried into the systemic circulation. The subsequent lodgement of the mature worm-pairs depends on species and takes approximately 20 days from the initial skin penetration. The length of life of the adults is up to 25 years.

Fig. 32.8 (a) Shell of snail *Physopsis globosa*; (b) Shell of snail *Biomphalaria pfeifferi*. (c) A 'right-handed' snail shell.

Schistosomiasis   Bilharzia

The species *Schistosoma haematobium* is the usual cause of vesical schistosomiasis when the adult worms inhabit the venules of the vesical and pelvic plexuses. Most of the eggs are deposited in the bladder and urinary tract. The bladder wall becomes inflamed then ulceration and papilloma occur. The bladder wall becomes thickened and inelastic and there is obstruction of urinary flow with consequent kidney complications. The eggs become enveloped in uric acid, oxalates, phosphates and finally calcium and secondary infections cause severe ulceration. In light infections symptoms may not occur for several years but if the infection is heavy, blood may appear in the urine and later mucus and pus are to be found also.

In areas where *schistosomiasis* is endemic clean water supply is obviously of the greatest importance as is the hygienic disposal of human faecal

Fig. 32.9 Cercaria larva of *S. mansoni*.

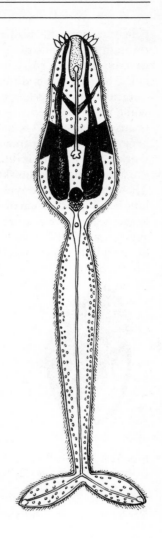

matter and urine. Adequate toilet facilities must be provided within a structure. Snail control over extensive areas has been attempted with such molluscicides as copper sulphate, sodium pentachlorphenate, and dinitro-o-cyclohexylphenol but re-establishment of the snail in previously treated localities has usually occurred.

Several other species of liver-inhabiting trematodes occur in various parts of the world. *Opisthorchis tenuicollis* infests Man throughout the Far East, the USSR, East Germany and Poland. Principal hosts are cats and dogs. Human infestation occurs through the consumption of raw fish. *Opisthorchis viverrini* occurs in Thailand where it is again common in cats and dogs. *Dicroceolium dendritium* is a liver-infesting species of sheep and is cosmopolitan in distribution.

Nemathelminthes Nematodes Roundworms

Nematodes or roundworms are said to be among the most abundant and widely distributed of all invertebrates. Although the greater number of species are free-living, a large number are parasites of plants and mammals. About 50 species have been recorded as parasites of Man; many are rare but two species are present in a substantial proportion of the human race. Some species are minute, only a millimetre or so in length, but others may reach a length of a metre.

Nematodes are characterised by their circular cross-section and by the peculiar nature of their cuticle. The hypodermis is a non-cellular outer covering beneath which is a layer of longitudinal muscle, the cells of which are unique in structure. There is no circular muscle in the body wall, so that there is no contractile system. Locomotion is achieved only by bending movements. Food is taken in through a well-defined mouth, either as fluid or a semi-solid such as intestinal contents. There is usually a muscular pharynx for sucking food into the straight tubular gut, and a rectum opening to the exterior posteriorly. Gaseous interchange for respiration takes place directly through the cuticle; there are no special organs associated with this. The reproductive system mostly takes up a large proportion of the body cavity. Only a few nematodes are hermaphrodite but usually the male gonads mature first. The amoeboid spermatozoa are placed by the male into the female's reproductive opening. The nematode life-cycle is often very complex, involving an alternation of free-living and parasitic generations or taking place alternately in more than one host. Vectors are sometimes utilised for infection of the final host.

ASCARIS LUMBRICOIDES   LARGE ROUNDWORM OF MAN

One of the largest and most widely distributed of Man's parasites, this nematode may reach a length of 30 cm. It is creamy in colour and the general shape recalls that of the earthworm, *Lumbricus*. In some parts of the world nearly three-quarters of the population may be infested with it, the highest incidence being among children. The adults lie freely in the small intestine, and by means of active, looping movements they are able to maintain their position against the rhythmic contractions of the gut forcing food downwards.

Life-cycle
Nearly three-quarters of a million, thick-shelled eggs are produced by one adult female each day. They do not hatch until after being passed out of the intestine of the host with the faeces. A warm, moist environment is necessary for successful development, otherwise the embryonic larva can lie coiled up within the shell for an extended period in moist soil for as long as several years. The egg is swallowed by Man and this probably

takes place when unwashed and inadequately cooked vegetables are consumed or when young children play on the soil around a building where it has been contaminated by human faecal matter. The egg hatches in the small intestine and the larva migrates around the body of its human host before it returns to the intestine. The wall of the intestine is first penetrated, after which the larva is carried through the blood system, first to the liver, then to the heart and then to the lungs. It penetrates the blood capillaries into the lungs where it reaches the bronchi and moves into the trachea. Here it is swallowed and returned to the intestine. This migration lasts only about seven days, and after moulting three times the worm matures in two months.

The migratory phase of the larva is often dangerous to the host when a large number of eggs have been swallowed. This is sometimes fatal, and often an infection is so severe that mental and physical development in children is retarded. A few adults in the intestine may be unnoticed but large numbers can block the intestine or they may migrate to other regions of the host's body where permanent damage may result.

TRICHINELLA SPIRALIS   PORK TRICHINA WORM

The adult trichina is very small and inhabits its host's small intestine. Each female produces about 1,500 living larvae which bore through the intestinal wall, and having found a blood vessel are carried round with the circulating blood. After a time they pierce the wall of the blood vessel and bore into muscular tissue. They then become encysted in a coiled position within a lemon-shaped cyst which is about 0.25 mm in length. No further development takes place until the infected muscular tissue is eaten by another mammal. Man usually acquires the parasites by eating uncooked, infected pork. When digestion takes place the larval trichina worms are liberated into the small intestine and quickly develop to maturity. A common host is the pig which becomes infected through eating uncooked meat scraps, or, from time to time, eating infected rats.

It will be seen that at no time in the life-cycle is there an external phase, as at all times *Trichinella* remains within the host. If the infected muscle tissue is not eaten the cyst eventually becomes calcified and the enclosed worm dies. This, of course, is what occurs when Man is parasitised as his flesh is not eaten – well, seldom! When meat which has been heavily infected is eaten, the number of larvae liberated may run into several millions, often with lethal consequences. The adult worms in the intestine are harmless to Man.

ENTEROBIUS VERMICULARIS   PIN WORM

This is probably the most abundant of Man's metazoan parasites and occurs in all classes of human society and in practically every country of the world. The adult is from 8 to 13 mm in length, very slender and with a pointed tail.

Fig. 32.10 Cyst of *Trichinella spiralis* in muscle. Longitudinal section. Magnified.

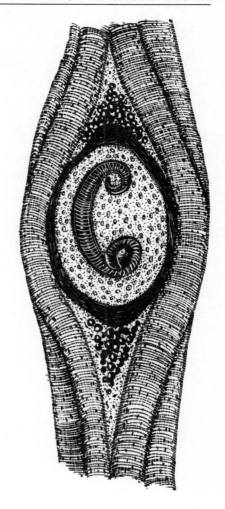

Life-cycle
The adults live in the caecum and adjacent parts of the colon and ileum and are thought to feed on intestinal contents, although the view has been put forward that they may also attach themselves to the wall of the intestine and consume epithelial tissue. The female stores her eggs until the number reaches about 10,000. She then migrates to the anus and emerging to the perianal region then expels her eggs. In the case of young children who feel itching in that region, the eggs may be taken into the mouth from the fingers. In other cases, the eggs become widely disseminated in the environment. They can be swallowed or inhaled in dust. It is recorded that 92 per cent of dust samples taken from households whose members were infected, contained eggs of this parasite.

Eggs are commonly found on lavatory seats, in wash basins and in bedrooms in any community where the worm is abundant. However, the eggs desiccate rapidly and die in from a few hours to a few days in a dry environment. In some cases larvae hatch in the perianal region and re-enter the colon through the anus (retro-infection). Commonly, both adults and children may be infected without showing symptoms, but in many cases there is *pruritis ani*. The usual signs of the parasite's presence are poor appetite, loss in weight, hyperactivity, insomnia, irritability, grinding of teeth when asleep, enuresis, abdominal pain, vomiting and nausea.

Chapter 33

Control methods and substances

The main object of this book is to draw attention to the principal animal species injurious or harmful to Man and his structures so that they may be identified and the nature of their harmful effects gauged. It would be inappropriate, however, not to include an account of the methods available for control. This chapter, therefore, is devoted to this subject.

In deciding on the form which the chapter was to take importance was attached both to the speed at which a reference could be obtained and to brevity. It will be seen, therefore, that this chapter is in alphatical order, not divided into sections. It has been based on the lists given in Hickin (1971) and Cornwell (1973 and 1979) to which the reader is directed for more detailed accounts. This chapter, then, is in the nature of a glossary of pest control terms but it should be used with some reserve on account of the changing attitudes to the balance between conservation of wild animal species on the one hand and control of harmful species on the other. In addition, legislation as regards the use of poisonous substances varies widely from country to country.

It should be noted that a number of words which otherwise would be found in the list are adequately described in the general text of the book. They are, therefore, not included.

Acute poisons

Acute poisons bring about death in a short period of time if the lethal dose is taken. Many rodenticides are acute poisons but some have the disadvantage that the rodent takes only a sub-lethal dose and then leaves the remainder because of its unpalatability or because the onset of the symptoms of poisoning are perceived.

Aerogel (also silica aerogel; 'Drie-die' and 'Drione')

The principle involved in this method of insect control is that of abrasive action and absorption on the insect cuticle. The outer layer of the latter is waxy which prevents water loss but when part of this layer is removed the insect desiccates. It is more successful in hot, dry climates. It has been 'built-in' to buildings, being dusted between partitions and cavity walls for control of termites, cockroaches and fleas. It is mostly used in California where the method was developed and has an important advantage in that it is virtually non-toxic to Man.

Aerosol

In pest control operations, an aerosol consists of an oil solution of an insecticide with a liquefied gas (the propellant) in a pressurised container. It is released as a temporary suspension of minute particles with similar characteristics to a fog or mist. The insecticidal properties may be short-lived or, on the other hand, may have a residual effect. One modification consists of a 'non-stop' canister which empties when released. Its principal use is a space treatment in commodity storage but special precautions must be taken. Another adaptation is for an aerosol to be used as an injector, forcing the insecticide into wood tunnelled by the larvae of wood-boring insects.

Anticoagulants

The development of the anticoagulant type of rodenticide has been of the utmost significance. An anticoagulant is a chemical substance which inhibits the clotting of blood. The most widely-used type is Warfarin which is formulated in cereal baits as well as in liquids and as a contact dust. The action of Warfarin is to cause the rodent to bleed externally and internally until death ensues. Blood-clotting is brought about by the soluble blood constituent fibrinogen being converted to the insoluble fibrin. This takes place in the presence on an enzyme prothrombin, so that the blood does not clot.

Generally anticoagulant baits are left down for a period of from 5 to 10 days. The low concentration of the anticoagulant in such baits reduces 'bait shyness' to a minimum and another reason for the lack of bait shyness is that the poison symptoms are slow to appear and the rodent does not associate them with the anticoagulant-baited food. Although the house mouse is generally regarded as being controlled by this type of rodenticide, access to the bait must take place over a longer period for successful results. Domestic animals have, from time to time, been the accidental victims so that rodenticide baits must be located with this hazard in mind.

Bait block (Rat block)

Rodenticides, in the form of food bait with wax or wax-like substances and with or without an attractant, are available in such blocks. They give a semi-permanent bait station, especially useful in damp situations where otherwise the bait would quickly deteriorate through mould growth.

Bait tray

This is often a special tray of card or plastic which is not easily overturned. It allows rodents to become accustomed to certain feeding stations and also enables success to be ascertained after a treatment programme has taken place.

A bait 'box' is a more permanent form of baiting station. It usually has a hinged lid and is of robust construction for use out-of-doors. It protects the bait from the weather as well as from domestic animals.

Bait tunnel

Where a poison bait is to be isolated from domestic pets or animals it must be placed near the centre of an immovable (and virtually unbreakable) tunnel or pipe, of large enough dimensions to allow access to rodents but small enough to prevent the bait being consumed by dogs, cats, pigs, etc. Ceramic land drains, 0.5 m in length, are generally used but must be held firmly in position against an outer wall be a heavy stone or the like to prevent them being rolled away, when the bait may be displaced to the outside. Indoors, bait tunnels are often of cardboard to give the rodent some 'confidence' in approaching the bait and also to protect it from dust.

Biological control

In some degree all animal populations, whether wild or in buildings, are controlled by other organisms which prey upon them or parasitise them.

The introduction of parasites and predators into populations of harmful species is known as biological control. A number of advantages are offered by this control method. The main one is 'specificity' – that is only the target animal would be affected and, in the absence of chemicals, the potential undesirable effects on the environment would be avoided. The organism *Salmonella enteritidis* (var. Danys) was used from 1928 until 1958 by the British Ratin Company (now Rentokil Ltd) as a bait preparation for rodent control.

Bird proofing

An essential part of any urban programme for bird pest control is to prevent access to roosting and nesting areas in buildings. For this, various forms and dimensions of netting are used. Galvanised chicken wire; expanded aluminium; polythene-coated nylon netting, are all used on occasion. For pigeons the mesh required is 4.5 cm ($1\frac{3}{4}$ in) and 1.2 cm ($\frac{1}{2}$ in) for sparrows. Open windows, caves and niches behind statuary, and girders under bridges are usually denied to birds in this manner.

Chemosterilant

A chemosterilant is a chemical substance capable of causing sexual sterility in animals. Reproduction is either prevented or suppressed. Thus it has no effect on the current populations, but the succeeding generation is reduced in number or may be absent. Chemosterilants may well prove to be the answer to the control of urban populations of birds such as the feral pigeon when there is likely to be a public outcry against their destruction. There are, however, difficulties with the chemosterilants so far known.

Chronic poisons

Chronic poisons require repeated doses over a period of time in order to be effective. A number of rodenticides are formulated at low concentration so that repeated baiting is necessary to achieve control. Most anticoagulant rodenticides are used in this manner and are thus depicted as chronic poisons.

Contact dust

A number of rodenticides are successfully employed as a fine powder at high concentration. They are dusted along the runs and around harbourages and are known as contact dusts – the rodenticidal substance comes directly into contact with the skin of the feet and with the fur of the underside. During grooming the animal transfers the rodenticidal particles directly into the mouth. The principal substances used as contact dusts are Warfarin at 15 per cent concentration and Lindane at 50 per cent concentration. DDT was widely used at concentrations of from 20 to 50 per cent but its use for this purpose is now illegal in many countries. Because the rodenticides are used in high concentrations their toxicity to Man is relatively high so that contact dusts must only be applied to covered areas where there is not the least chance of human contact, nor the possibilty of the dust contaminating foodstuff and other commodities.

Desiccant dusts (see Aerogel)

Dichlorvos (DDVP)

An efficient contact insecticide of the organophosphorus type better known under the trade names of 'Nuvan' (Ciba) and 'Vapona' (Shell). Widely used against commodity pests in storage as well as against most domestic insect pests. In addition to rapid knockdown action it possesses fumigant and flushing out properties. It is in general use for 'slow release strips' but restrictions apply in some countries with regard to the latter where inhalation over long periods could occur such as in hospitals and also where food is exposed. The acute oral LD_{50} (rat) is 40–60 mg/kg.

Dieldrin

This organochlorine compound is one of the most widely-used insecticides of recent years. It possesses high insecticidal activity with a high degree of stability and persistence even in thin films. It is this property, however, which has brought about its banning in a number of countries where it has given rise to pollutant effects. It is highly toxic to fish. The acute oral LD_{50} (rat) is 40–50 mg/kg. It is readily absorbed through the skin and is stored in bodyfat. In general it has been replaced by organophosphorus compounds, although it is still in use for control of subterranean termites.

Dimethyl phthalate (DMP)

A widely used insect repellent for personal protection against biting insects such as mosquitoes and midges. It is applied as a lotion or cream and sometimes formulated with other repellents. It is an irritant to eyes and mucous membranes and care must be taken to prevent contact.

Dioxathion

This organophosphorus insecticide (trade name of 'Delnav' has an acute oral LD_{50} (rat) of 25–40 mg/kg and is in use against dog ticks and fleas and other external parasites of cattle and other domestic animals.

Distress calls

One form of scaring device used against birds where their numbers form a hazard – such as where their roosting places are adjacent to aeroplane runways – is to play previously-recorded distress calls.

Dust gun

Insecticidal dusts and rodenticidal contact dusts are best applied by the use of a dust gun which is hand-operated. The gun consists of a 'blower' containing the dust through which a blast of air is impelled by means of a piston or, in smaller equipment, by bellows or a rubber-bulb. It is possible to reach otherwise inaccessible areas by attaching a piece of tubing to the gun-nozzle.

Dusts

Insecticides and rodenticides in the form of dusts have a number of important applications. A dust in this context consists of a uniform mixture of a low concentration of an insecticide or a high concentration of rodenticide in a low particle-size carrier, such as china clay or talc. Boric acid, however, at about 99 per cent active ingredient is used as an insecticidal dust.

The carrier dust facilitates the uniform distribution of the active component but these dusts can only be applied in dry situations as otherwise they aggregate. An important advantage possessed by dusts is that because they do not adhere readily to even surfaces they are easily picked up by the spines and bristles on an insects legs. They are also economical to use. They may be introduced into locations which cannot be reached by spraying. However, they should never be mixed with water and when in use dust masks should be worn.

Emulsion concentrate

In the context of pest control, an emulsion concentrate is a solution of an insecticide in a non-aqueous solution, together with emulsifying agents, so that it may be diluted with water to the required dilution for use. The concentration of insecticide is usually from about 20 to 50 per cent. When diluted the emulsion is milky-white.

Such emulsions possess a number of important advantages. The water-insoluble insecticides allow water to be the carrier (diluent) which means that they are, therefore, less expensive than oil-based sprays and have also a reduced fire risk. The transport cost of emulsion concentrate is lower than that of oil-based spray. When applied the residue is not so prominent on the treated surfaces as in the case of wettable powders. However, because the emulsion is readily absorbed by porous surfaces, such as brickwork, etc., the residual life is shorter than when wettable powders are used and there is greater water-staining risk to susceptible surfaces.

Light mineral oil is used to dilute certain emulsion concentrates so that they may be used in fogging machines.

Entoleter

This is a centrifuge designed to remove infesting insects from milled products and many modern flour mills have them installed.

Environmental change

In a number of cases a degree of control of a pest species has been brought about by effecting a change in the surrounding environment. The mosquito is a good example. Areas of stagnant water have been drained and the cultivation of plant species with stem-ensheathing leaves which can hold water, thus enabling certain mosquito species to breed, has been prohibited; trees with hollow trunks and branches have been removed for the same reason.

Exclusion screening

It should be a strong element in good housekeeping and good practice to ensure that buildings (domestic and otherwise) do not invite the attention of rodents, bats and other animal pests, by leaving gaps in the outer structure which are sufficiently large for them to enter. Such points of easy access are often to be found under doors and around badly-fitting pipes of various descriptions (but including sewers) where pests may enter or leave the building. Gaps for underfloor ventilation are a well-known access for rodents, while under-eave gaps and broken or badly-fitting roof tiles are the usual place where bats enter to take up residence in roof voids.

Pest animals can only be excluded by filling in these gaps with galvanised wire-netting or metal grids of an appropriate mesh. Pieces of galvanised, metal sheet can be screwed on to the bottom of a door or can cover gaps and holes in walls.

Exclusion can also be applied to the method of preventing mosquitoes and other noxious, flying insects from entering buildings where insects are a hazard. Wire gauze is often used to isolate open windows, large areas (such as verandahs, etc.) from the attentions of biting insects and this method is generally known as 'screening'.

Fluoracetamide

This chemical may be otherwise known as Compound 1081 or 'Fluorakil'. An acute poison, it is used in cereal baits as a rodenticide. It is extremely poisonous to mammals with an acute oral LD_{50} (rat) of 15 mg/kg, but dogs are more susceptible. In Britain its use is restricted by Poison Rules, 1970, to sewers and to drains in certain areas, in docks and in some other establishments where human access is controlled. Fluoracetamide is a crystalline solid, very soluble in water and odourless and tasteless. It is readily absorbed through cuts and abrasions. No human antidotes exist and emphasis must be made on safe handling and storage.

Fogging

This is the method of insecticide application in which various types of machine are employed to break up an insecticide in an oil solution into minute droplets. The fog produced is temporary only and after a short period 'falls out' to give a fine film on upper, horizontal surfaces.

Equipment available is as follows:

1. The TIFA – in which the fog is produced by means of a blast of hot air, using a petrol burner. In this machine the droplet size can be varied and the unit is often mounted on a vehicle.
2. The Swingfog and Dynafog machines produce the fog by using a pulse-jet engine. These machines are portable. Discs rotate at high speed and the fog is produced by the insecticide dropping upon them and breaking up by centrifugal force and then being blown out at high speed.
3. The Microsol generator is also of this type, being portable and electrically operated. The Hi-Fog type of portable machine produces fog by means of a jet of compressed nitrogen gas through a fine nozzle.

Fogging is suitable for use against flying insects – such as several moth species in commodity storerooms but it has little effect on insect larvae infesting stored food within the wrapping. It is also used extensively against mosquitoes and other biting insects out-of-doors.

Fumigation

The application of a toxic substance in the form of a gas, vapour, volatile liquid or volatile solid, in the atmosphere of a closed container in order to control noxious animal species, is known as 'fumigation'. The closed container may be specially constructed – either large for commodities or small (for works of art in the eradication of wood-destroying beetles). The structure itself may be the 'container' – such as a warehouse where stored food products are to be disinfected, or may be the hold of a ship for the extermination of rats and insects. Commodities may be stacked and covered with gas-tight tarpaulins, the ends of which are rolled together and clipped. Termites are eliminated from buildings by covering the whole structure with such tarpaulins rolled and clipped in the same way. This operation is carried out in the USA on a very large scale.

The concentration of the gas (the fumigant) is maintained for a specified period and the concentration × time is known as the CT product. When the fumigation operation is completed the fumigant must be removed by 'airing-off'. Some dense commodities are difficult to fumigate satisfactorily at atmospheric pressure and in such cases a vacuum is first created before the fumigant is introduced.

Fumigation is a dangerous process and should not be attempted without qualified personnel with previous experience or without the appropriate safety equipment.

Good housekeeping

The term 'good housekeeping' can be used in connection with any building – commercial, industrial as well as domestic.

Sanitation (see separate section) must be paramount, so must exclusion screening (see separate section).

It is essential that there are no cracks, crevices or splits in materials, e.g. joins in woodwork, skirting boards (kick boards) or walls. All fixings must be flush. Loose wallpaper should not be tolerated as this is a favourite harbourage for cockroaches, especially in kitchens and for bed bugs in bedrooms. See 'Sanitation' for disposal of waste foodstuff and other organic matter.

Good practice

In the prevention of insect and rodent damage to stored foods and other commodities, good practice during storage is of the highest importance. By this is meant the management of the stocks so that infestations are either maintained at a low level or eliminated altogether. Some of the principal points are as follows:

1. Warehouses and storage rooms must be kept at the highest degree of cleanliness. All spillages must be removed immediately and either rebagged or destroyed.
2. All materials in split or broken containers should be rebagged on receipt.
3. All commodities should be palletised so that they are not in contact with the ground. A minimum distance from the ground of 12 in (30 cm) is advocated by the Public Health Service of the United States Department of Health Education and Welfare. A pallet is a strong, wooden frame of standard size which enables bagged and boxed commodities to be stacked thereon and also makes the movement of stocks by a 'fork-lift truck' economical and minimises damage. Palletised stock should not be stacked to such a height that there is a danger of breakage to the lower packages and it should be kept at a distance from walls so that inspections can be made and rodent harbourage lessened. Correct palletising of stock minimises fire risk.
4. Frequent inspections of stored foodstuffs and other commodities at risk should be made for evidence of infestation. A 6 in (15 cm) band on the ground and outer perimeter of stocks, painted white, facilitates infestation recognition.
5. Good practice requires that all infestations of whatever degree should be dealt with immediately, and appropriate steps taken by trained personnel.

6. Stock should be used in strict rotation. Old stock must not be allowed to accumulate. It is in old stock that small infestations become large ones with consequent contamination of newer stock.
7. Office and factory staff should be discouraged from eating food on the premises. Waste food is liable to be left about which may attract rats and mice. If, however, circumstances make it necessary, then meals should be taken is specified places and nowhere else. The subsequent debris should be meticulously binned.

Grain protectant

Insecticides are often added to grain before any infestation has been observed in order to protect it in storage or during shipment. Malathion is generally used for this purpose at a maximum permissible concentration of 8–10 ppm. The insecticide may be added to the grain as it passes along conveyors before discharge or as a dust added in layers 1 m apart as the grain enters bulk storage. Thorough mixing must, however, take place when the grain leaves storage. Another method is to spray the protectant on to the bagged grain as an emulsion or wettable powder.

High-frequency electric fields (HF)

Insects have been successfully eliminated from stored food on an experimental basis by the method of selective heating by high-frequency electric fields known as HF and it is thought that this method, as well as the use of radiant energy, shows promise as a means of killing insect eggs in bulk-stored grain. (See also Radiant energy.)

Hygiene

Hygiene is the name of the science of maintenance of human health by application of the principles of sanitation. Sanitation has been defined as measures for the promotion of health, especially with regard to drainage, sewage disposal and the creation of conditions favourable for a pest-free environment.

The Public Health Service of the United States Department of Health, Education and Welfare has given some basic rules for sanitation in respect of the control of household insect pests as follows:

1. 'Refuse storage' in a sanitary manner is a fundamental factor in effective household insect control. Wrap garbage in paper and store in durable, rustproof cans which can be kept clean and covered. Provide enough containers so refuse need never be stored in boxes, cartons, bags or on the ground. Keep cans on neat easily cleaned racks, platforms or slabs. Avoid spillage of garbage on the soil. Remove paper, boards, cardboard, loose rock, etc., from basements and yards to control scorpions, spiders, centipedes and other pests and to improve premise appearance. One of the greatest opportunities for household insect control lies in better refuse storage and litter removal.
2. Human waste, loaded with organisms pathogenic to man, is among the greatest hazards to public health. In addition, it provides an excellent breeding medium for flies and other insects. Therefore, all human waste should be disposed of immediately in the most sanitary manner available. Where sewage is water-borne, it is a disposal problem for community sewage plants or community-regulated septic tanks. Where it is not water-borne, it should be disposed of in a sanitary pit privy.
3. Standing water, a source of mosquitoes (vectors of encephalitis, malaria, etc.) and other insects, should be drained, the cavity filled, or maintained in a sanitary condition (clean shore line, no emergent vegetation). Many health departments have regulations governing maintenance of standing water. These should be investigated and followed. Special attention should be given to removing accumulations of water (as in small containers) from premises.
4. Structural harbourage is provided by any item suitable as a hiding place for cockroaches, bed bugs, or other pests. New buildings should be constructed as pest-proof as possible. Older buildings should be modified with screens, etc., so that they present a minimum harbourage for pests. Vacuum cleaning at regular intervals (at least once weekly) plays an important role in insect reduction. It is of special value in household flea control. Vacuum cleaning picks up not only adult fleas, but (of greater significance) eggs, larvae, and pupae in cracks of flooring and fibres of rugs and overstuffed furniture. It thus prevents build-up of large flea population from the few brought into the house by pets. Vacuum cleaning also reduces infestations of ticks, mites, food and fabric insects, and other pests.
5. Weeds are an open invitation for large insect populations to come near human habitations. They provide extensive and varied cover for pests, and prevent adequate control of refuse, standing water, and structural harbourage. Use mowers, clippers, weed burners, etc., when weeds might endanger valuable plant life.
6. Food storage is an important factor in household insect control. Poor storage produces large populations of beetles, cockroaches, moths and other pests. Inspect food periodically. Isolate infested materials immediately from clean stock. Rotate food so the oldest material is

always used first. Keep new stock away from older items.
7. Storage of usable items should be arranged to reduce food and harbourage for pests to a minimum, and to eliminate fire hazards. Stack items on racks 18 in (45 cm) off the ground. Keep objects which might accumulate water under cover. Evaluate stored items critically and dispose of those unlikely to have real future value.

Injection

This term is used in three different contexts in relation to pest control in buildings.

1. Where timber is heavily attacked by Death Watch beetle to the extent that many tunnels run into each other, some of the flight holes may be injected with organic, solvent-type wood preservative. Available for this process are hand injectors which consist of a small pressure chamber containing the preservative (which is pumped up by hand) connected to a spring-loaded nozzle by pressure tubing, the nozzle fitting into the flight hole with a washer. When the nozzle is pressed into the hole the fluid is released under pressure and often penetrates deep into the tunnelled wood.
2. Control of subterranean termites in a badly-attacked building is generally carried out by drilling the walls and floors at specified intervals and then injecting substantial amounts of water-dispersible insecticide known to be toxic to termites (termiticide). The holes are then filled in. Where walls are adjacent to earth, the latter is trenched along the whole length of the walls and specified amounts of insecticide are sprayed into the trench. The equipment required, of course, is considerably larger than described above.
3. Where areas of termite damage in a building are relatively small, powder injection is in general use in Australia and Malaysia. It consists of locating the termite runways in current use and, then, using a small powder blower, injecting a toxic dust. The runways are then carefully closed. The dust usually employed is arsenic trioxide (white arsenic) with graphite or talc. **This substance is extremely toxic to human beings**.

Insecticidal lacquers

In situations where insecticidal dusts and sprays are rapidly removed, the use of an insecticidal lacquer may well prove successful. Consisting of special formulations containing a high concentration of insecticide in a synthetic resin, the lacquer is applied by brush and, when dry, the insecticidal component appears on the surface. Some types of lacquer require the admixture of an 'accelerator' to hasten drying. If Dieldrin is used the surface 'blooms' with fine crystals but Malathion produces a thin, bluish film. If this film is removed by washing or abrasive wear then the insecticide is released again. Such lacquers are effective over considerable periods.

In situ treatment (see Remedial treatment)

Larvicide

An insecticide specially formulated to control the larval stage of a pest insect is known as a larvicide. The best-known example of a target pest insect is the mosquito. The larvae inhabit still water and are controlled by insecticide dissolved in a light oil which has been selected for its ability to spread widely over the water surface. When the mosquito larvae come to the surface to breathe air they must come into contact with the oil film and are exterminated.

Liver baiting

Small pieces of fresh liver are placed on paper or foil and used as attractants so that the size and location of the nests of Pharaoh's ant, *Monomorium pharaonis*, can be found. Minced liver, with an insecticide such as chlordecone is often used to control this species.

Malathion

This is an insecticide of the organophosphorous group distributed under a number of proprietary marks in various parts of the world. Its main use is in the control of stored food pests as a grain protectant and also in household insect and tick pest control. Its residual effect is fair except on alkaline surfaces, and it is corrosive to iron so that lined containers must always be used. It is a good acaricide.

Methyl bromide

Otherwise known as bromomethane, this is a colourless liquid becoming a gas at 4.5 °C. Used as a fumigant on account of its valuable insecticidal properties. It is generally stable, non-corrosive and non-flammable. Natural rubber, however, is degraded by it so that products fabricated of this must be removed before fumigation operation commences. This is especially important when a dwelling is treated against termites, as in the USA. Methyl bromide is highly toxic to Man and as it possesses no odour the gas chloropicrin is usually added so that leaks may be quickly detected. In addition to its use in commodity fumigation in warehouses, it is used in ships and is widely used against termites. Care has to be taken with high oil-content foods such as nuts and soya flour, otherwise tainting may occur. Many countries confine the use of methyl bromide to licensed or trained personnel.

Misting

Textiles are easily stained if sprayed with convential insecticidal sprays to the extent that the textile appears wet. The technique known as 'misting' consists of spraying with minute droplets of the insecticide using a fine nozzle. The droplets are held by the the individual fibres of the cloth, producing an overall film. There are a number of other applications in which misting is recommended practice.

Moisture and temperature control

Many food substances – such as grain – require the moisture content to be maintained within certain limits during storage in order to prevent the growth of moulds which would then favour insect attack. The regulation of moisture content is usually accomplished by ventilation. Where dried foodstuffs are stored, 1 m wide access aisles with 1 m between stacks and walls and 0.6 m top of stack to ceiling, are required in order to provide adequate ventilation.

Doors and ventilators should be open during dry weather and closed during rain or periods of high humidity. It is advantageous, during the cold season, to circulate cold air as low temperatures slow down the development of larvae or insect pests but care must be taken, however, that susceptible foodstuffs are not spoilt by freezing.

The house mouse is remarkably resistant to low temperatures.

Molluscicide

This is a substance used to control snails and slugs. Metaldehyde with a cereal product such as bran has been widely used in gardens. Copper sulphate, sodium pentachlorphenate and dinitro-o-cyclohexylphenol have been used over extensive areas to control the snails harbouring the larvae of schistosomes.

Personal hygiene

Personal hygiene is of the utmost importance.

All parts of the body should be washed daily, using soap to wet thoroughly all hair. Wounds should be protected from flies and other insects by bandaging. Any part of the body becoming soiled should be washed immediately. Only the minimum contact should be made with animals, wild or domestic. Children should have no contact with them whatsoever. Clothing should be changed and washed frequently and no contact made with soiled clothing or linen used by another person. Feminine hygiene requires particular attention, especially with regard to disposal (menstruation).

Pre-baiting

Pre-baiting requires the placing of unpoisoned baits several days before the use of poisoned bait. This accustoms the rodents to eat at certain sites where later the poisoned bait will be laid. Rats show shyness to some acute poisons but pre-baiting tends to make them less so.

Pre-baiting is a recommended preliminary to bird-trapping and bird-narcotising in making the birds less shy of any particular situation.

Pressure impregnation

Impregnation is defined as the diffusion or saturation of an active substance through a material, but in wood preservation it is generally used to describe treatments giving high-loading of preservative in the wood as, for example, in pressure treatments. Pressure impregnation is the usual term used.

It consists of forcing the preservative into the timber by pressure applied in a closed vessel such as an autoclave or pressure cylinder. In a vacuum pressure process, the timber is first subject to a vacuum before the injection of the preservative under pressure. In this process the preservative is applied by evacuation of the air, then filling the cylinder with the preservative and, finally, releasing the vacuum.

Permeable timbers are sometimes treated in this way with preservatives of low viscosity.

One of the principal and most successful wood preservatives in worldwide use applied by pressure impregnation is copper-chrome-arsenate in aqueous solution. These preservatives are available in various proportions.

A feature of these preservatives (known as CCA type), is the high degree of fixation of the components occurring in wood soon after treatment and is mainly due to the chromium content. The biocidal effect is conferred by the copper and arsenic.

Pyrethrum

Pyrethrum is the common name of the plant *Chrysanthemum cinerariaefolium* in the family COMPOSITAE from which the valuable insecticidal pyrethrins are extracted from the flower-heads. It is variable in composition but is usually available as a brown, resinous oil. It is insoluble in water but is formulated in an organic solvent as a spray or aerosol, or may be mixed with an inert powder for use as a dust. It is best known for its very rapid action, giving what is termed 'knockdown' or immobilisation. Kill, however, is not usually adequate unless certain substances referred to as synergists are added. Piperonyl butoxide is the best known of the latter. The principal advantages possessed by pyrethrins (apart from speed of action) are their activity towards a wide spectrum of insect species; lack of toxicity to Man; and their safe use where stored foodstuffs are concerned. On the other hand, they possess a short residual life on treated surfaces and formulations tend to be expensive.

Radiant energy

The eradication of pest species from stored food products by the use of radiant energy techniques has been successful on an experimental scale. It has been used commercially only on a limited scale but is likely to be more widely employed in the future even though a loss of vitamin content may take place. (See also High-Frequency Electric Fields (HF).)

Rat guards

Physical barriers to the climbing activities of rats are often built in to structures. These are usually strips of galvanised iron cemented into place in brick or concrete pillars which would otherwise give access to higher floors. Rat guards are frequently used to prevent the entry of rats into ships by means of their climbing along hawsers when tied, which they are well able to do. The guards are in the form of a truncated cone of metal in which there is a slit through which the rope only may pass.

Red squill

This is an acute poison generally used against the Brown rat. It is extracted as a powder or liquid from the dried bulbs of *Urginea maritima*, Sea squill, in the family LILIACEAE. The bulbs, which are very large (being up to 25 cm in diameter), are obtained from sandy or rocky areas on the North African coast. The active ingredient is scillirocide which is a well-known emetic. Rats are unable to vomit harmful substances once swallowed but cats and dogs are able to do so, thus red squill baits are of a relatively low order of toxicity to them. The LD_{50} (rat) of the pure scillirocide is 2 mg/kg. The Animal (Cruel Poisons) 1962 Act forbids its use in Britain.

Remedial treatment

This is sometimes referred to as *in situ* treatment and is a term applied when exposed wood in a building is treated against wood-boring beetles using insecticides. Heavily-damaged wood may have to be removed and replaced. All wooden surfaces are then sprayed with long-lasting insecticide formulations. Strictly speaking, termite fumigation treatments as practised in the United States and in other parts of the world come under this heading.

Residual spraying

One of the most significant advances in the techniques employed to control noxious insects in recent years has been the development of residual insecticides. These are long-lasting insecticides applied in such a way as to leave a film (the residual film) which remains active for a considerable length of time.

The first long-lasting insecticide was DDT, but subsequently a large number have been produced and used. However, some have proved to be *too* permanent and have been shown to degrade the environment to a serious extent.

The residual insecticide may be applied as a dust or a spray, the latter being formulated as a suspension in water or dissolved in an organic solvent. Water suspensions have an advantage in that they are less likely to taint or contaminate foodstuffs and they are also non-flammable. But, on the other hand, they may discolour and stain wall-coverings and decoration when used in the home.

Residual insecticides have been used on an extensive scale for mosquito control where a toxic film has been sprayed on the inside walls of dwellings. Many countries prohibit (or limit) the use of a number of residual insecticides and advice should be sought from the appropriate government department before residual spraying operations commence.

First of all DDT then later a number of chlorinated hydrocarbon and organophosphate insecticides were found to be long-lasting, i.e. to have a residual effect. Many of the imported species of warehouse insect pests are susceptible to such insecticides if they come into contact with them. However, the chosen insecticide must not be capable of tainting if used where foodstuffs are stored. The risk of fire is greatly reduced if the insecticide is a wettable powder diluted with water, but in such circumstances it must not be toxic to Man.

Residual insecticides are usually applied by spray but dusts are sometimes administered.

Sanitation

Good sanitation, whether in warehouses, factory premises or the home, requires the immediate removal of all waste materials, more especially waste foodstuff and other organic matter. Temporary storage of such material must be in strong bins with lids which can be fixed in position when closed. The bins must be emptied regularly and frequently. If, however, circumstances make this impossible, then waste, organic material (including sweepings) must be carefully burned and any residues from the fire buried deeply.

Floors, walls and shelves must be swept or, preferably, vacuumed and the residues burned. It is essential that all debris (including paper and cardboard and other materials that can be shredded) should be removed so that rats and mice are not encouraged to create their harbourage.

Ceilings, floors and walls, as well as equipment and bins, must be free from cracks, crevices, splits and slits and must be so filled in that immature stages of pest insects cannot utilise them for hiding and for pupation.

All faecal matter and urine from humans should be flushed into sewers immediately or, where a sewerage system does not exist, buried deeply. All animal faecal and urine matter of domestic animals and pets should be disposed of without loss of time and removed from the vicinity of buildings. If this is not done, flies will inevitably be attracted to it, become contaminated and subsequently enter buildings.

Scaring devices

A number of devices have been used, from time to time, to scare birds (usually in flocks) away from areas where they may constitute a hazard such as around airfields (see Distress calls). Recordings of bird distress calls; loud gunshot-like noises; flashing lights and trained hawks, have all been stated to have some measure of success. However, they appear to be limited in value if constantly used as the birds, from experience, learn to ignore them. They should, therefore, be used only sparingly.

Sewer treatment

Throughout the world, sewers are the most important harbourages for rats. 'Elimination of the sewer population can bring about a marked reduction in the number of rats living above ground: of vital importance in modern cities and the first step in any campaign against rats', states Cornwell.

After setting up the organisation procedure with the appropriate authority, the first practical steps will involve sewer treatment. The lifting of manhole covers will require special attention, using equipment according to their design. The sewer is then baited with an acute poison (placed on a benching if present or suspended over the outflow in a muslin bag). Successful rodenticides are sodium monofluoracetate (1080) and fluoracetamide (1081). They are, however, so poisonous to Man that special safety precautions and procedures must be meticulously carried out.* In the United Kingdom the use of these poisons is restricted to the de-ratting of ships and drains in port areas.

The poison 1080 containing nigrosine dye is used as a liquid and 1081 (under the trade name 'Fluorakil' also containing nigrosine dye) is used in cereal baits.

*Legal requirements and safety precautions vary from country to country.

Site hygiene

This is an expression used in termite control to describe the condition of a building site where termites are present. It requires that all wood, odd pieces of timber, concrete shuttering, tree-roots, vegetation. etc., be carefully removed and burned so that no colonies of termites are on the site before concrete is laid for slab of foundations.

Site pre-treatment

Formerly called 'soil poisoning', the application of insecticide, usually as a diluted water emulsion, is made to the area on which a building is to stand in order to inhibit the passage of termites from the soil to overlying wooden structures. This is generally carried out at the stage when foundations have been laid and all utility services have been connected – that is when no further disturbance of the soil will take place.

Slow release strip

This consists of an insecticide (Dichlorvos has been widely used), formulated into a plastic base at 20 per cent concentration. The vapour is only slowly released and thus possesses an extended life. Strips of this nature have been used to control flies indoors as well as to control wool-eating larvae of moths and beetles in clothes cupboards and wardrobes. Made into a collar, they are served to control dog and cat parasites. In some countries the use of slow-release strips is subject to regulation where food is exposed and where humans may be exposed to prolonged vaporisation.

Smoke generator

A smoke consisting of finely-divided particles which, when produced by a pyrotechnic mixture on ignition, forms a cloud which then distributes itself widely. Insecticides (Lindane has been in wide use) can be formulated in this way. The particles thus produced are solid – *not* gases – and eventually come to rest, particularly on upper, horizontal surfaces. Smoke generators are especially useful in applying an insecticide film in inaccessible places, e.g. the roof of a church.

Sodium monofluoracetate

This is also known as compound 1080. It is used as an acute rodenticide in liquid form but WHO standards require the commercial product to be dyed with 0.5 per cent nigrosine dye. This extremely poisonous substance which has an acute oral LD_{50} (rat) of 1.5 mg/kg, is rapidly absorbed through the intestinal tract as well as being absorbed through cuts and abrasions. In Britain restrictions apply and it may only be used in elimination of rats from ships and from dockside drains, as for fluoracetamide.

Soil poisoning (see Site pre-treatment)

Space spraying

A number of techniques come under this general description. Fogging machines and mist machines fill the space as well as covering exposed surfaces – in particular upper surfaces. However, space spraying should not be confused with the far more efficient fumigation method.

The insecticide is in the form of minute droplets able to kill flying insects as well as those walking on surfaces of pallets and packaging. If the insects have already penetrated the packaging then space spraying is not applicable. For such spraying only insecticides of low human toxicity are suitable and their employment may be regarded as preventative maintenance rather than elimination techniques. A number of suitable machines are marketed for the purpose of space spraying.

Sticky board

Otherwise known as a glue board, this is essentially a piece of hardboard or cardboard (about 30–60 cm square) coated on one side with a tacky, non-drying varnish or a thixotropic gel about 3 mm thick. This is formulated with a strong enough 'tack' to hold a rodent coming into contact with it and is usually sited on rodent 'runs', being sometimes baited around the edge. It is generally used to clear up survivors or stragglers after a large-scale bait treatment.

Strychnine

Strychnine is an acute poison of extremely rapid action now generally considered to be too dangerous for widespread use. A number of countries have imposed legal restrictions on its use. It is an alkaloid extracted from the plant *Strychnos nux-vomica* of the family LOGANIACEAE and sometimes available as the water-soluble sulphate. The acute oral LD_{50} (rat) is 5 mg/kg but it is not considered successful as a rodenticide on account of bait shyness as far as rats are concerned. It has proved more successful against mice. Today, however, its use is considered to be cruel due to the convulsive spasms it causes. The Animal (Cruel Poisons) 1962 Act permits its use in Britain only against moles.

Sulphuryl fluoride

A colourless, odourless gas used as a fumigant against drywood termites under trade name 'Vickane'. It is not to be used on foodstuffs.

Synthetic pyrethroids

Similar to natural pyrethrins in chemical structure but of varying degrees of knockdown and kill efficiency.

Tactile repellent

Various devices and chemical substances are widely used to prevent feral pigeons from roosting in particular places on buildings. The birds must then find other roosting areas. Thick gel and various plastic compositions have proved to be successful when carefully applied to the roosting ledge. A caulking gun is generally used to lay long 'strings' of the repellent and when the birds' feet touch the tacky substance they apparently feel insecure and fly off.

Wires and metal strips are far more difficult to install and corrosion often takes place fairly rapidly; their use is usually confined to window sills and other small lengths of roosting ledge. Electrically charged wires have been found to possess a relatively short life also.

Test baiting

The use of unpoisoned baits as a preliminary to a treatment programme gives an important picture as to the size of a rodent population and its principal concentrations. These can be estimated with a fair degree of accuracy from the amount of bait consumed.

In a similar way the degree of success of a treatment programme can be gauged by a test baiting afterwards. This method is also known as census baiting.

Tracking dust

Tracking dust (or powder) is a non-toxic dust used to detect the presence of rodents and to give information on their numbers and the location of their harbourages. Flour or talc is commonly used and the dust is laid overnight and inspected the following morning when the footprints and tail marks will be visible. Tracking dust is also used after a bait treatment to detect the presence of survivors.

Trapping

Perhaps the oldest method of rodent control is trapping. There are many types of trap involving the capture of the animal either dead or alive. Spring-loaded 'break-back' traps are widely employed for rats as well as for mice. Meat and fish are used as baits for *Rattus norvegicus*, while fresh fruit is more acceptable to *Rattus rattus*. Traps should be set on well-used runs and tied to beams or pipes; they may be set without bait in such situations.

In the case of *Mus musculus*, milk chocolate or half a sultana is recommented as bait. Although in common use, cheese is not so attractive as a bait as is supposed. Traps should be inspected every day. Traps for birds such as feral pigeons are made of galvanised steel mesh with funnel-shaped openings into which the birds are lured with mixed grain. They are best located on flat roofs near to the usual roosting places and out of sight of the public so as not to offend.

In some countries there are regulations concerning the disposal of trapped birds.

In Australia and New Zealand the Brush-tailed possum is caught in steel mesh traps having a spring-loaded gate. In Australia the animal must be set free in the nearest national park but in New Zealand law requires them to be killed as pests.

Ultrasonics

Claims have been made that high frequency sound has the effect of repelling rodents to the extent that they are forced to leave the building which they are infesting. The air vibrations produced by commercial equipment are above the human audible range (greater than 20,000 Hz). There are several practical limitations, however, one being the rapid absorption of the vibrations by commodity stacks and thinly-partitioned walls. Tests have shown that the effect of the 'sound' is temporary and limited in extent.

Ultra-violet light

Light immediately beyond the blue end of the visible light spectrum attracts many insects including a number of pest species. This phenomenon is employed in a number of types of fly-control equipment when, after being drawn to the ultra-violet light, the flies are killed when they come into contact with electrically-charged and earthed grilles. Such equipment should be used indoors only; if suspended outdoors it will kill many harmless or beneficial species of insect.

Warfarin

This is the best-known 'anticoagulant' rodenticide of the hydroxycoumarin type. It has been approved and used worldwide. Developed by the Wisconsin Alumni Research Foundation, on which the name is based, it was originally used at concentrations as low as 0.005 per cent in baits. Due to resistance by rats in some localities, however, the concentration has been increased to 0.2 per cent, i.e. 40 times as concentrated as that originally found to be successful. Repeated baiting over about five days is required. It is considered to be a 'safe' rodenticide although cats and dogs and especially pigs are sometimes accidentally poisoned by it through consuming a large quantity at one time. The chronic oral LD_{50} (rat) is 1 mg/kg per day (for five days) and the acute oral LD_{50} (rat) is about 60 mg/kg. Warfarin is available in concentrated form, liquid concentrates for dilution with water, contact dusts at 1 per cent and as a gel. Anticoagulants are chemicals which inhibit blood clotting so that the poisoned rodent bleeds to death through internal and external haemorrhages.

Wettable powder

Sometimes referred to as 'water dispersible' powder, this is a powder of small particle size containing (in pest control applications) a substantial percentage of insecticide, usually between 30 and 50 per cent. This is formulated to be readily wetted by water and to form a suspension. Such suspensions require to be agitated from time to time as otherwise settling takes place. The wettable powder is generally regarded as the cheapest form of insecticidal spray and the preferred form for spraying porous surfaces when the water is absorbed and the insecticidal particles all remain on the surface.

Zinc phosphide

An acute, quick-acting rodenticide which in some countries is confined to professional use or is licensed. The acute oral LD_{50} (rat) is 45 mg/kg. Usually used as a 2 per cent bait with oils or fats. Pre-baiting is necessary for rat control but not for mice. It is stable in dry conditions but decomposes when damp when the odour of garlic can be detected indicating release of phosphine.

References

Aitken, A. D. (1975) *Insect Travellers*, Vol. 1, COLEOPTERA. *MAFF Tech. Bull.*, **31**, HMSO.

Barnett, S. A. (1963) *A Study in Behaviour*. Methuen, London.
Bezant, E. T. (1956) Further records of the Australian Carpet Beetle, *Anthrenocerus australis* (Hope). (COL.: DERMESTIDAE) in Britain, *Ent. mon. Mag.*, **92**, 401.
Bezant, E. T. (1957) The Australian Carpet Beetle, *Anthrenocerus australis* (Hope (COL.: DERMESTIDAE) in Britain, *Ent. mon. Mag.*, **93**, 207.
Blake, G. M. (1958) Diapause and the regulation of development in *Anthrenus verbasci* (L.) (COL.: DERMESTIDAE), *Bull. ent. Res.*, **49**(4), 751–75.
Blake, G. M. (1958) Length of life, fecundity and oviposition cycle in *Anthrenus verbasci* (L.) (COL.: DERMESTIDAE) as affected by adult diet, (Bull. ent. Res., **52**(3), 459–72.
Blake, G. M. (1972) Photoperiodic effects regulating the internal clock in *Anthrenus verbasci* (L.) (COL.: DERMESTIDAE), *Proc. Roy. Ent. Soc. Lond.* (Journal of Meetings), **37**, 17, 25.
Brendell, M. J. D. (1975) COLEOPTERA, TENEBRIONIDAE (Handbooks for the Identification of British Insects, Vol. V, part 10). Roy. Ent. Soc. Lond.
Broadhead, E. and **Hobby, B. M.** (1945) The Booklouse, *Discovery*, May 1945, 142–7.
Broadhead, E. (1954) The infestation of warehouses and ships' holds by psocids in Britain, *Ent. mon. Mag.*, **90**, 103–5.
Brown, H. W. (1959) In Cecil, R. L. and Loeb, R. F., *Textbook of Medicine*. Saunders & Co, Philadelphia.
Busvine, J. R. (1951 and 1966) *Insects and Hygiene*. Methuen, London.

Cecil, R. L. and **Loeb, R. F.** (Eds) (1959) *Textbook of Medicine* (10th edn). Saunders, Philadelphia.
Chitty, A. J. (1904) *Ptinus tectus* and *Lathridius bergrothi* in Holborn, *Ent. mon. Mag.* **40**, 109.
Cloudsley-Thompson, J. L. (1958) *Spiders, Scorpions, Centipedes and Mites*. Pergamon Press, London.
Cloudsley-Thompson, J. L. and **Sankey, J.** (1961) *Land Invertebrates*. Methuen, London.
Coombs, C. W. and **Woodroffe, G. E.** (1955) A revision of the British species of *Cryptophagus* (Herbst) (COL.: CRYPTOPHAGIDAE), *Trans. Roy. ent. Soc.* **106**, 5, 237–82.
Cornwell, P. B. (1968) *The Cockroach*, Vol. 1. (1976) Vol. 2. Hutchinson, London.

Cornwell, P. B. (1973 and 1979) *Pest Control in Buildings:* A guide to the meaning of terms (Rentokil Library). Hutchinson, London.
Crowcroft, P. (1966) *Mice all Over.* Foules, London.

Edwards, R. (1980) *Social Wasps, Their Biology and Control.* Rentokil, East Grinstead.
Evans, H. E. and **Eberhard, M. J. W.** (1973) *The Wasps.* David and Charles, Newton Abbot.

Freeman, J. A. (1962) The influence of climate on insect populations of flour mills, *Proc. XI Int. Congr. Ent.* (Vienna 1960), **II**, 301–8.
Freeman, J. A. (1965) On the infestation of rice and rice products imported into Britain, *Proc. XII Int. Congr. Ent.* (London 1964), 632–4.
Freeman, J. A. (1967) On the infestation of almonds with special references to those imported into Great Britain from Mediterranean countries, *EPPO Pubns Ser A*, **46A**, 121–4.
Freeman, J. A. (1974) A review of changes in the pattern of infestation in international trade. European and Mediterranean Plant Protection Organisation, *EPPO Bull.*, **4**(3), 251–73.

Giles, P. H. (1965) Control of insects infesting stored sorghum in N. Nigeria, *Journal of Stored Products Res.*, **1**(2), 145–8.
Gillett, J. D. (1971) *Mosquitos.* Weidenfeld and Nicolson, London.
Gillett, J. D. (1972) *Common African Mosquitos and Their Medical Importance.* William Heinemann Medical Books, London.
Ghouri, A. S. K. (1965) Physio-ecology of *Gryllodes sigillatus* (Walker) (ORTHOPTERA: GRYLLIDAE), *West Pakistan Journal of Agricultural Research*, **3**, 181–99.

Harris, W. V. (1961 and 1971) *Termites, Their Recognition and Control.* Longman, London.
Hickin, N. E. (1963 and 1975) *The Insect Factor in Wood Decay.* Hutchinson (1963) and Associated Business Programmes (1975), London.
Hickin, N. E. (1971) *Termites: A World Problem.* Hutchinson, London.
Hinton, H. E. and **Corbet, A. S.** (1972) Common insect pests of stored food products. *BM(NH) Econ. Ser.* 15, (5th edn).
Howe, R. W. (1959) Studies on beetles of the family PTINIDAE. XVI: Conclusions and additional remarks, *Bull. ent. Res.*, **50**(2), 287–326.
Howe, R. W. and **Burges, H. D.** (1955) Studies on beetles of the family PTINIDAE. II: Some notes on *Ptinus villiger* Reit., *Ent. mon. Mag.*, **91**, pp. 73–5.

Keegan, H. L. and **Macfarlane, W. V.** (Eds) (1963) *Venomous and Poisonous Animals and Noxious Plants of the Pacific Region.* Pergamon, Oxford.

Levi, W. M. (1941 and later editions) *The Pigeon.* Levi Publishing, Sumter, SC, USA.

Mallis, A. (1945 1st edn; 1969 5th edn) *Handbook of Pest Control.* MacNair-Dorland, New York.
Mattingley, P. F. (1969) *The Biology of Mosquito-Borne Disease*, (Science of Biology Series 1). Allen &Unwin.
Monro, H. A. U. (1961) *Manual of Fumigation for Insect Control.* FAO United Nations, Rome.
Moore, B. P. (1957) The identity of *Ptinus latro* auct., *Proc. Roy. ent. Soc. Lond.* (B), **26**, 199–202.
Mphuru, A. N. (1974) *Araecerus fasciculatus* De Geer (COL.: ANTHRIBIDAE): A

review. *Tropical Stored Products Information*, **26**, 7–15. Tropical Stored Products Centre, Slough.
Munro, J. W. (1966) *Pests of Stored Products* (Rentokil Library). Hutchinson, London.
Murton, R. K. and **Wright, E. N.** (Eds) (1968) *The Problems of Birds as Pests* (Institute of Biology Symposium 17). Academic Press, London.

National Pest Control Association (1971) *Biology and Habits of Commensal Rodents*, 8–71. Elizabeth, New Jersey, USA.
New, T. R. (1971) An introduction to the natural history of the British PSOCOPTERA, *Entomologist*, **104**, 59–76.

Oldroyd, H. (1964) *The Natural History of Flies*. Weidenfeld & Nicolson, London.

Pearson, T. G. (Ed.) (1936) *Birds of America*. Doubleday & Garden City Books, New York.
Pollitzer, R. (1954) *Plague*. World Health Organisation Monograph 22, Geneva.

Robinson, R. (1965) *Genetics of the Norway Rat*. Pergamon, Oxford.

Sayed, T. El. (1935) On the biology of *Araecerus fasciculatus* with special reference to the effects of variation in the nature and water content of the food, *Ann. appl. Biol.*, **22**(3), 557–77.
Scott, Sir Ronald (Ed.) (1966) *Price's Textbook of the Practice of Medicine* (10th edn). Oxford University Press, London.
Shorten, M. (1962) *Squirrels, Their Biology and Control* (Bulletin 184 Min. Ag. Fish.). HMSO, London.
Shrewsbury, J. F. D. (1970) *A History of Bubonic Plague in the British Isles*. Cambridge University Press.
Smith, K. G. (1956) The occurrence and distribution of *Aphomia gularis* (Zell.) (LEP.: GALLERIIDAE) a pest of stored products, *Bull. ent. Res.*, **47**(4), 655–67.
Smith, K. G. (1960) Insect infestation associated with French shelled walnuts with particular reference to the occurrence of *Aphomia gularis* (Zell.) (LEP.: GALLERIIDAE), *Bull. ent. Res.*, **50**(4), 711–16.
Smith, K. G. (1961) Recent occurrences and a review of the status of *Aphomia gularis* (Zell.) (LEP.: GALLERIIDAE) as a storage pest in Britain, *Ent. mon. Mag.*, **96**, 167–8.
Smith, K. G. (1968) Some aspects of the biology of *Paralipsa (Aphomia) gularis* (Zell.) (LEPIDOPTERA) in relation to its distribution, *Proc. XII Int. Congr. Ent.* (London 1964), 626.
Smith, K. G. V. (Ed.) (1973) *Insects and Other Arthropods of Medical Importance*. British Museum (Natural History).
Southern, H. N. (Ed.) (1964) *Handbook of British Mammals*. Blackwell, Oxford.

Thearle, R. J. P. (1968) Urban Bird Problems, in *The Problems of Birds as Pests*, R. K. Murton and E. N. Wright (Eds). Inst. of Biology and Academic Press, London.
Tooke, F. G. C. (1949) Beetles injurious to timber in South Africa, *Department of Agriculture, Science Bull.*, **293**, Union of South Africa, Pretoria.

Wilkinson, J. G. (1979) *Industrial Timber Preservation*. Associated Business Press, London.
Williams G. C. (1964) The life history of the Indian meal-moth, *Plodia interpunctella* (Hubner) (LEP.: PHYCITIDAE) in a warehouse in Britain and on different foods, *Ann. appl. Biol.*, **53**, 459–75.

Witherby, H. F., Jourdain, F. C. R., Ticehurst, N. F. and **Tucker, B. W.** (1940) *The Handbook of British Birds*, Vols 1–5. Witherby, London.

Woodroffe, G. E. (1951a) A life history study of the Brown House Moth, *Hofmannophila pseudospretella* (Staint.) (LEP.: OECOPHORIDAE), *Bull. ent. Res.*, **41**(3), 529–53.

Woodroffe, G. E. (1951b) A life history study of *Endrosis lactella* (Schiff.) (LEP.: OECOPHORIDAE), *Bull. ent. Res.*, **41**(4), 749–60.

Woodroffe, G. E. and **Southgate, B. J.** (1954) An investigation of the distribution and field habits of the Varied carpet beetle, *Anthrenus verbasci* (L.) (COL.: DERMESTIDAE) in Britain with comparative notes on *A.fuscus* Ol. and *A.museorum* (L.), *Bull. ent. Res.*, **45**(3), 575–83.

Zinsser, H. (1935) *Rats, Lice and History*. George Routledge, London.

Index

Page numbers in **bold** refer to illustrations. References to countries have been combined under continents e.g. Canada and United States are combined under North America. Only Britain and the Soviet Union are separate.

Acanthomyops interjectus (ant), 171
Acanthoscelides obtectus (weevil), **107**
Acari *see* mites; ticks
Acarus siro (flour mite), 54, **55**, 56–7, **58**, 70
Acheta domesticus (cricket), 309–11
Acrotelsia collaris (silverfish), 316
adders, 45
Aëdes (mosquitoes), **190**, 191–4, 196
  A. aegypti, 191, 193–4, 196
  A. africanus, **190**, 193
  A. albopictus, 191, 194
  A. scutellaris, 191, 194
  A. simpsoni, 191, 193
Aerogel, 342
aerosol, 342
Africa
  ants, 164, 167, 169, 172
  beetles, 97, 99, 102, 108–9, 111–12, 114, 119, 125, 141
  birds, 34, 36
  bugs, 244, 249
  carnivores, 1–2, 5
  cockroaches, 286, 290–1, 293–4, 298–300
  earwigs, 307
  fleas, 179
  flies, 186, 189–94, 201, 203, 206, 214, 219, 221, 223
  millipedes, 324
  moths, 228–9, 234
  rodents, 7, 21
  scorpions, 85–6, 88, 90
  snakes, 45
  tapeworms and flukes, 329, 331–3
  termites, 273, 276, 282
  ticks, 72, 78
African eye worm, 205–6
Aglais urticae (butterfly), 227
agricultural crops, damage to, 37, 68
  see also stored goods
ahasverus advena (beetle), 141
alate termites, 274–5
Allochernes italicus (false scorpion) 87
Allodermanyssus sanguineus (mite), 21
Amblycera (feather lice), 260
Amblypygi (scorpions), 86
Amblyomma americanus (tick), 20
American cockroach, 286, **289**, 291–3, 295
American field cricket, 311
American house centipede, **321**
Amitermes (termites), 277
anaemia, 75
Anatoecus dentatus (louse), 261
Andrenidae (bees), **154**
Androctonus australi (scorpion), 90
Angoumis grain moth, **228**, 229, **233**
Anisolabis maritima (earwig), **306**
Anilus cristatus, **240**
Anisopodidae (Window gnats), 204
  Anisopus fenestralis (Window gnat), 204
Anobiidae (beetles) 93–4, **95–7**, 98–9, **100**, 101–3
  Anobium pertinax, 101
  A. punctatum, 94, **95**, 96–100
Anopheles (mosquitoes), 188–9, 192–3, 195–6

A. freeborni, 196
A. gambiae, **189**
A. quadrimaculatus, 196
Anoplura *see* sucking lice
ant-like wasp, 99
Anthrenocerus australis (beetle) 129
Anthrenus (museum and carpet beetles), 123, 127, 128, 129
  A. verbasci, **123**, 128, **129**
Anthribidae (beetles), 102–3
anticoagulants, 342–3, 365
ants, **154** 164, **165**, 166, **167–70**, 171–2, **173**, 174, 249, 315
  bites and stings, 169–70
Apidae (Honey bees), **154**
Aplonis panayensis (starling), 38
Apocrita (wasps) 153–74
Apoidea (Social and Solitary bees) **154–8**,159–62
  Apis dorsata, 161
  A. indica, 161
  A. mellifera, 160–1
Arachnida *see* false scorpions; false spiders; mites; scorpions; spiders; tail-less whip scorpions; ticks
Araecerus fasciculatus (weevil), 103
Aranea *see* spiders
Argas (ticks), 72–4
Argentine ant, 166, **167**, 249
Aridius nodifer (beetle), 130
Arilus cristatus (bug), 240
Armadillidium (woodlouse), 91
Ascaris lumbricoides (roundworm), 337–8
Asia
  beetles, 103–5, 108, 119, 123, 138–9, 141
  birds, 34, 36, 38
  bugs, 249
  carnivores, 1, 5
  centipedes, 322
  cockroaches, 293–4, 299–300
  control methods, 353
  fleas, 177
  flies, 186, 191, 194–5, 203, 206, 209, 214, 219, 221
  insectivores, 27
  millipedes, 324
  moths, 232
  rodents, 9, 15–16, 22
  snakes, 45–7
  spiders, 80
  tapeworms and flukes, 329–33, 336
  termites, 276
  ticks, 73, 77–8
  wasps/ants/bees, 161, 164, 172

Assassin bugs, 239, **240**
asthma, 57, 201
Astigmata (mites), 54, **55**, 56–7, **58–9**
Atrax robustus (spider) 81–2
Attagenus (Fur and carpet beetles) 122–4
  A. pellio, **123**, 124
Auchmeromyia luteola (maggot), **222**
Australasia
  ants, 167, 169
  beetles, 96–7, 102, 104–5, 108, 112–14, 120, 129, 135–7
  birds, 36, 38, 41
  centipedes, 322
  cockroaches, 287, 289, 293–4, 297, 300
  control methods, 353, 364
  earwigs, 307
  flies, 191, 194–5, 201, 203, 206, 210
  marsupials, 30
  millipedes, 324
  mites, 66
  rodents, 9, 15
  silverfish, 315–16, 318
  snakes, 45–6
  spiders, 80–1
  tapeworms, 329
  termites, 276, 282–3
  ticks, 78
Australian carpet beetle, 129
Australian cockroach, 286, **289**, 293–4
Australian spider beetle, 136–7
Austrosimulium pestilens (fly), 203
Aves *see* birds

Bacon beetles, 126–7
badger, 177
bait, 343
baiting, test, 363–4
Bald-faced hornet, 153
Bamboo borers, **104**
Bancroftian filariasis (disease), 192
Bandicota (bandicoots) 6, 13–15, 27
  B. bengalensis, 15–16
Bark beetle, 101–2
Barn swallow bug, **241**, 250
Bartonella bacilliformis, 187
Bartonellosis (disease), 187
bats, x, 5, 23–6
  bugs, **241**, 249
bed bugs, **241**, **244**, 245–50
Beef tapeworm, 326–8
bees, **154**, **159**, 161–2, **163**, 164
  stings, 161–2, 166
beetles, 93–145, 158, 226

INDEX 373

see also Anobiidae; Anthribidae;
  Bostrychidae; Brenthidae;
  Bruchidae; Buprestidae;
  Cerambycidae; Cleridae;
  Cryptophagidae; Cucujidae;
  Curculionidae; Dermestidae;
  Lathridiidae; Lyctidae;
  Lymexylidae; Nitidulidae;
  Oedemeridae; Ptinidae; Silvanidae;
  Tenebrionidae; Trogositidae
Big-headed ants, 169
Bilharzia (disease), 335–6
biological control, 343–4
Biomphalaria (snails), **335**
birds, 33–8, **39**, 40–3
  bugs, **241**, 250
  control of, 344, 346, 360, 363–4
  diseases and, 42–3
  fleas, 176, 178–9, 182
  lice, 251, 260–3
  mites, 53
  ticks, 72
Biscuit beetle, 94–5, **96**
bites *see* diseases; stings and bites
Biting house fly, 218–19
biting midges, 200–1
biting lice *see* Mallophaga
Blaberus (cockroaches), **296**, 301–2
Black bug, **240**
Black flies, 202
Black rat *see Rattus rattus*
Black salt-marsh mosquito, 196
Black widow spider, **80**, 81
Blackwater fever, 200
bladderworm, 327
Blaps mucronata (beetle), 141
Blattodea see cockroaches
  Blattella germanica, 286, **287–8**,
    289–90, 303
  B. vana, 288, 303
  Blatta orientalis, 286, **289**, 290
blowflies, 219–21, **222–4**
Blue-barred rock pigeon, 34
Bluebottles, 219–20
Bodega black gnat, 201
body louse, 253, **254**, 255–6
Bombidae (Bumble bees), **154**, 159
Bombycoidae (irritating hairs) 236
booklice, 264, **265**, 266–7
boomslangs, 45
Boophilus (ticks), 74
borers, **104**, 105–6, 110, 113, 120,
  135, 136, **226**
Bostrychidae (beetles), 103, **104**, 105,
  106

Bostrychopsis parallela, **106**
Bothrops (pit vipers)
Boutounese fever, 77–8
Brachycera (flies), 204–6
Braconid wasp, 99
Brenthidae (beetles), 106–7
Brill–Zinser disease, 259
bristletails, 314–19
Britain
  ants, 172
  bats, 26
  bedbugs, 248–9
  beetles, 98–9, 104, 109, 113,
    116–17, 120, 123–4, 129, 135–9
  birds, 35, 37, 40–1
  booklice, 264–6
  cockroaches, 290–1, 293, 295, 298
  control methods, 344, 348, 358,
    360, 362–3
  fleas, 177–8
  flies, 200–1, 219
  insectivores, 27
  moths, 228, 232, 234, 236
  rodents, 6–7
  spiders, 79
  wasps, 152, 163
Broad tapeworm of Man, 326
Bromomethane, 354–5
Brown cockroach, 286, **289**, 294–5
Brown dog tick, 78
Brown house moth, 230–1
Brown rat *see Rattus norvegicus*
Brown recluse spider, **82**, 83
Brown spider beetle, 138
Brown-banded cockroach, 286, **288**,
  297
Bruchidae (beetles), **107–8**
  Bruchus ervi, **107**
  B. pisorum, **106**
Bryobia praetosia (mite), 68, **69**, 70
Buffalo gnat, 202
bugs, 238–9, **240–4**, 245, **246**, 247–50
  diseases and bites, 239, 242–4, 247
Bulinus (snails) 335
Bumble bees, **154**, 159
Bungarus (kraits), 45
Buprestis aurulenta (beetle), 108–9
Bushmaster (snake), 47
Bush-tailed possum, 364
Buthus occitanus (scorpion), 90
butterflies, 225–7

Cabbage white (butterfly), 226
Cadelle beetle, **145**
Calliphoridae (Blowflies, Blue-

bottles, Greenbottles), 219–21, **222–4**
Camponotus (carpenter ants), **173**, 174
  C. herculaneus, **173**
Canadian porcupine, 16
Canicola fever, 18
carnivores, 1–2, **3**, 4–5
  rabies and, 4–5
Carpenter ants, 172, **173**, 174
Carpenter bees, 162, **163**, 164
Carpet beetles, 122–4, 127–9
Carpoglyhus lactis (mite), 57
Carpophilus (beetles),134–5
  C. dimidiatus, 134, **135**
  C. hemipterus, 134
Carrion's disease, 185–7
Cartodere constrictus (beetle), 130
Caryedon gonagra (beetle), **108**
Case-bearing clothes moth, 231–2
cats, x, 3–4, 15
  fleas, 4, **178**, 180
  flukes and tapeworms, 315, 330, 336
  lice, 262–3
  mites, 63–4
cattle parasites, 75, **224**
Cediopsylla simplex (flea), **180**
Centipedes, 320, **321–2**
Cerambycidae (beetles), 109–11, **112**, 113–14
Ceratophyllus (fleas),179, 182
Ceratopogonidae (midges and sandflies) 200–1
Cerobasis annulata (booklouse) 265
Cestoda *see* Tapeworms
Chaga's disease, 239, 242–4
Chaoboridae (Phantom midges), 201–2
Cheese mite, 56
Cheese skipper, 209–10
Cheirisium museorum, 87
Chelonethi (false scorpions) 86–7, 249
  Chelifer cancroides, 87, 249
chemosterilant, 344
Chequered beetle, 114, **115**
Chewing lice *see* Mallophaga
Cheyletidae (mites), 70
  Cheyletus eruditis, 56, 70, 230
Chilopoda (centipedes) 320, **321–2**
Chimney swift, 249
Chiracanthium (spiders), 81
Chironomidae (midges), 201
Chiroptera *see* bats
Chlamydozoon trachomatis (virus), 214
Chloropidae (Eye flies), 208–9
cholera, 211
Chrysomia (blowflies) 220–1

Chrysops dimidiata (fly), 205–6
Chrysops discalis (fly), 20
Churchyard beetle, 141
Cigarette beetle, 96, **97**
Cimicidae (bed bugs), **241**, **244**, 245–9
  Cimex adjunctus (bat bug) **241**, 249
  C. columbarius (pigeon bug), 249
  C. hemipterus **241**, 249–50
  C. lectularius, **241**, **244**, 245–9
  C. pilosellus (bat bug), **241**, 249
  C. pipistrellus, 249
Cinereous cockroach, **296**, 299–300
Clear Lake gnat, 202
cleg, 205
Cleridae (beetles), 114, **115**
Clostridium (micro-organism), 220
Clover mite, 68, **69**, 70
Cluster fly, 215, **216–17**
Clytus arietis (beetle), 113
Cnemidocoptes (mites), 64
cobras, 45–6
Cochliomyia (Blowfly), 221
cockroaches, 285–6, **287–9**, 290–5, **296**, 297–300, **301**, 302–5
Cocoa moth *see* Warehouse moth
Codiosoma spadix (beetle), 121
Coelopidae (Seaweed flies), 209
Coffee bean weevil, 103
Coleoptera *see* beetles
Collembola (Springtails), **312**, 313
colonies
  ants, 164–5
  bees, 160–1
  wasps, 153–8
Colorado tick fever, 76–7
Columbidae (Doves and Pigeons), 34–5
  Columba liva, 33–5, 179
Common bean weevil, **107**
Common bird mite, 53
Common chicken louse, 261
Common clothes moth, 231
Common furniture beetle, 94, **95**, 96–100
Common garden centipedes, 322
Common house fly, 53, 211–14
Common malaria mosquito, 196
Common silverfish, 315–16, **317**
Common wasp, **156–7**
compounds, poisonous 348, 362
Cone-nose, 249
Congo floor maggot, **222**
Coninomus nodifer (beetle), 264
contact dust, 345
control methods and substances, 341–66

INDEX 375

Coptotermes (termites), 277, 282–3
Corcyra cephalonica (Rice moth), **229–30**, 232–3
Cordylobia anthropophaga (fly), 221–2
Corn moths, 231–2
Corsair, **240**
Cossonus (beetles), 121
Crab louse, 253, **254**, 256
Crane flies, 185
Crazy ant, 172
Crickets, 309–11
crops *see* agricultural
Crotalidae (pit vipers), 46
Crotalus (rattlesnakes), 47
Crustacea (Isopoda) *see* woodlice
Cryptolestes ferrugineus (grain beetle) 117, **118**
Cryptophagidae (fungus and plaster beetles), 115, **116**, 117, 264
Cryptostigmata (mites), 53–4
Cryptotermes brevis (termite), **273**, 282
Ctenocephalides (fleas), 4, 177, **178**, **180**
Ctenolepisma (silverfish) 316, **317**, 318
Cucujidae (grain beetles), 117, 118
Culcidae (mosquitoes), 187–8, **189–90**, 191, 192, 195–6
  Culex pipiens fatigans, 189, **190**, 192, 195–6
  C. quinquefasciatus, 195–6
Culicoides (midges), 200–1
Culiseta melanura (mosquito), 196
Curculionidae *see* Weevils
Cyclorraphaga (flies), 206–10, **211**, 212–14, **215–18**, 220–1, **222–4**
Cysticerus cellulosae (Bladderworm), 327

Daddy Long-leg spiders, **50**
Damon (tail-less whip scorpions), 86
Dampwood termites, 274–7
DDT, 358–9
Death watch beetle, 99–100, 267, 353
Debris bug, 230, 250
Deer fly, 20
Deladenus siricidicola (nematode), 152
Demodex (dog mange mite), **64**, 65
Dengue (disease) 187–8, 194–5
Dermacentor andersoni (fever tick), 20, **75**, 76
Dermacentor variabilis (tick), 20, 76
Dermanyssus gallinae (mite), 53
Dermaptera (earwigs), 304, **305–6**, 307–8
Dermatitis, 56–7
Dermatophagoides (mites), 57, **58–9**

Dermatosis, pruritic, 58
Dermestidae (beetles), 121–2, **123**,124, **125–9**, 130
  Dermestes (Hide, Larder, Bacon beetles) **126**, 127
Desmondontidae (bats), 23, 25, **26**
  Desmondus rotundus (Vampire bat), 25, 26
dessicant dusts, 345
Diamanus montanus (flea), 181
diarrhoea, infantile, 211
Dichlorvos (DDVP), 345, 361
Dicroceolium dendritium (fluke), 336
Didelphis marsupialis (opossum), 5, 32
Dieldrin, 345
Dimethyl phthalate (DMP), 346
Dinoderus (Bamboo borer), **104**
Dioxathion, 346
Diphyllobothrium latum (tapeworm), 326
Diploda (Millipedes), **323**, 324
Diptera *see* flies
Dipylidium caninum (tapeworm), 177, 254, 262, 330
diseases transmitted or caused by
  beetles, 99
  birds, 42–3
  booklice, 266–7
  bugs, 239, 242–4
  fleas, 178, 182, *see also* rodents *below*
  flies, 183, 185–206 passim, 209, 211, 213–16, 221, 223–4
  flukes, 335–6
  lice, 256–9, 262–3
  mites, 22, 51, 53, 56–7, 62–7
  rickettsia, 21, 53, 66, 257–9
  rodents, 7, 16–31, 178
  tapeworms and roundworms, 325–33, 337–40
  ticks, 20, 72–8
  see also stings and bites
dogs
  fleas, 177, **180**, 186
  lice, 253–4, **255**, 261, **262**
  mites, 63–5
  tapeworms and flukes, 177, 254, 262, 315, 320, **329**, 330, 336
  ticks, 76
Dolichovespula (wasp), 155
Dried currant moth, 234–5
Dried fruit mite, 57
Drone flies, 207
Drosophilidae (fruit and vinegar flies), 207–8
Drug store beetle, 94–5, **96**

Dryobates major (woodpecker), 42
Drywood termites, 272, **273**, 274–6
Duck louse, 261
Dung flies, 210
dusts, 346–7
dysentery, 211
Dwarf tapeworm of Man, 262–3, 328

earwigs, 304, **305–6**, 307–8
Eastern liver fluke, 320
Echidnophaga gallinacea (flea), 178–9, **181**
Echinococcus granulosus (tapeworm), 328, **329**
Ectobius livens (cockroach), **296**
Ectopsocus vachoni (Booklouse), 266
Egyptian mongoose, 1–2
electricity, high frequency, 351
Elephantiasis, 183, 189, 192
emulsion concentrate, 347
Encephalitis, 53, 77
Endrosis sarcitrella (moth), **229**
Enicmus minutus (beetles), 264
Enterobius vermicularis (worm), 338–40
entoleter (centrifuge), 347
environmental change, 347
Ephestia (moths)
   E. cautella, 234, **235**
   E. elutella, **230**, **233**, 234, **235**, 236
   E. kuehniella, **233**, 234, **235**
epidemic louse-borne typhus, 256–8
Eremotis (beetles), 121
Ergates spiculatus (Pine-borer), 113
Erinaceidae (hedgehogs), 27, 177
Eristalis tenax (fly), 207
Erithizon (porcupines) **16**, 17
Ernobius mollis (beetle), 94, 101–2
Ethmostigmus subripes (centipede) 332
Euborellia annulipes (earwig), **305–6**, 308
Euophryum (weevils), 120
Europe
   beetles, 96, 99, 101, 103, 108–9, 111–12, 114, 121, 133–5, 138–9, 141
   birds, 33–4, 36–9, 41–2
   bugs, 249
   centipedes, 320
   cockroaches, 287, 290, 298–9
   crickets, 310
   earwigs, 304–8
   fleas, 178, 182
   flies, 195, 206, 210, 214, 219–20, 223
   millipedes, 324

mites, 54, 67, 69, 77–8
moths, 228, 232
rodents, 9–10
scorpions, 85–6, 90
snakes, 46
spiders, 80
tapeworms and flukes, 329, 333, 336
termites, 280, 284
ticks, 77–8
wasps/bees/ants, 147, 152–9, 163, 170, 172
woodlice, 91
European earwig, 304, **305–6**, 307–8
European hornet, **155**
European social wasps, 153–4, **155–8**, 159
European starling, 33, 37–8, **39**
Eurycotis floridae (cockroach), **296**, 302–3
Eutrombicula (mites), 66
exclusion screening, 348
Eye flies, 208–9
Eyeless tampan, 72, **73**

Face fly, 214–15
False scorpions, **50**, 87, 249
False spiders, 85–6
False stable fly, 219
Fannia canicularis (house fly), 216–17, **218**
Fannia scalaris (latrine fly), 218
Fasciola hepatica (liver fluke), 331, **332**
Feather lice *see* Mallophaga
Felidae *see* cats
   Felis subrostratus (cat parasite), 263
Feral pigeon, 33–5, 363–4
Field cockroach, 288, 303
Filaria bancrofti (fly), 191
filiarial worm *see* nematodes
Filariasis (disease), 187–8, 191–3
Filter flies, 185
Fire ants, 170
Firebrats, 317–19
fleas, 18–19, 21, 175, **176–8**, 179, **180–2, 226**
   diseases and, 178, 182, *see also* rodents
   species attacking Man, **177–8**, 179, **180–2**
Flesh fly, 222
flies, 158, 183–224, 230
   illustrations, **189–90, 204, 215–18, 222–4**
   see also diseases; mosquitoes
Florida cockroach, **296**, 302–3

Flour beetles, **143–4**
Flour mite, 54, **55**, 56–7, **58**, 70
Flukes, 330–6
  diseases and, 335–6
Fluoracetamide (poison), 348
fogging, 348–9, 362
follicle mite, **64**
food *see* agricultural; stored products
Forficula auricularia (earwig), 304
  305–6, 307–8
Formicidae *see* ants
Four-lined silverfish, 316, **317**
fruit flies, 207–8
fumigation, 349, 355, 358
fungus beetles, 115–17
Funnel-web spider, 81–2
Fur and Carpet beetles, 122, **123**, 124
Furniture beetles, 94, **95**, 96–100
Furniture mite, 56–7

Galleriinae (moths), 232–3
Gamasina (ticks), 52–3
gangrenous ulcer, 82–3
Garden ants, 171–2
Gasterophilus intestinalis (bot), **224**
geckos, 47–8
Gelechidae (Grain moths), 227, **228**, 229
German cockroach, 286, **287–8**, 289–90
Giant cockroach, **296**, 301–2
Giant silverfish, 316, **317**, 318
Giant tropical millipede, 323
Gibbium psylloides (spider beetle), 139
Glomeris marginata (millipede), 324
Glossina (tsetse flies), **211**
Glossy tree starling, 38
glue board, 362
Gluvia dorsalis (false spider), 85
Glycyphagus domesticus (mite), 56–7
Gnataocerus (grain beetle), 144
gnats, 201–2, **204**
Golden spider beetle, **138**, 139
good housekeeping, 350
good practice, 350–1
grain beetles, 117, **118–19**, 124, 140–1, 144–5
grain moths, 227, **228–30**, 231–2, **233**, **235**
grain protectant, 351
grain weevil, 118–19
Gray silverfish, 316–18
Great Indian scorpion, **89**
Great spotted woodpecker, 42
Greenbottles, 220

Ground Squirrel flea, 181
Gryllodes sigillatus (cricket), 311
Gryllus assimilis (cricket), 311
gulls, 39–41

Haematobia irritans (fly), 215
Haematosiphon inodorus (bug), **241**, 250
Haemogamasus (mites), 53
Haemogogus (mosquito), 193
Haemorrhagic diseases, 18, 77
hairs, irritating, 236
Hairy spider beetle, 138
Halarachnidae (mites), 53
Hard ticks, 74–5
Harlequin roach, 300, **301**
Harvest mite, 65
Harvest sickness, 18
Harvester termites, 275–6
Head louse, 252–4, 256
Hedgehogs, 27, 177
Hemiptera *see* bugs
Henfleas *see* Poultry fleas
Henoticus californicus (beetle), 117
Hermetia illucens (fly), 206
Herpestes (mongooses), 1–2, 5
Heteroptera *see* bugs
Hide or Skin/Larder/Bacon beetles, 126–7
hides/furs/textiles, damaged by
  beetles, 122–4, 127–9
  crickets, 311
  moths, 230–2
  silverfish, 318–19
High-frequency electric fields (HF), 351
Hippelates (eye flies), 209
Histolysis (disease), 65
Hodotermitidae (termites), 275–6
Hofmannophila pseudospretella, 230–1
Holoconops kerteszi (gnat), 201
Honey ant, 171
Honey bee, **154**, 160–1
Horn fly, 215
hornets *see* Wasps
Horse bot, **224**
Horse flies, 204–5
House ant, 171
House cricket, 309–11
House dust mites, 57, 58, **59**
House flies, 53, 210–14, 216–17, **218**, 219
House longhorn beetle, 110
House moths, **229–30**, 231
House mouse, 6, 13, **14**, 15, 27, 355, 364

flea, 179, **180**
mite, 21
House sparrow, 33, 35–7
Hover flies, 207
Human flea, 19, 175, **176–7, 181**
hygiene, 351–5, 356, 361
Hylecoetus dermestoides (beetle), **133**
Hylotrupes bajulus (beetle), 110–11, **112**
Hymenolepsis (tapeworms) 262–3, 328–9
Hymenoptera *see* ants; bees; wasps; woodwasps
Hypoderma bovis (warble), **224**

Ibalia leucospoides (woodwasp), 150–1
in situ treatment, 354, 358
Indian bee, 161
Indian grey mongoose, 5
Indian meal moth, **233**, 235–6
Indian mongoose, 1
injection of preservatives and poisons, 353
insect repellant, 346
insecticides, 342, 345–9, 351, 353–5, 357–9, 361–2, 365–6
insectivores, 27–8
Iridomyrex humilis (ant), 166, **167**, 249
Ischnocera (feather lice), 260
Isoptera *see* termites
Itch mite, 60, 61–2
Ixodidae (hard ticks), 74–5

Jewel beetles, 108–9
jiggers, 179, **182**

Kala-azar (disease), 185–6
Kalotermitidae (drywood termites) 272, **273**, 274–6, 284
Khapra beetle, **124**, 125–6
Kissing bugs, 239, **240, 244**
Korynetes caerulus (beetle), 99, 114–15
kraits, 45

Labidura riparia (earwig), **306**
Lacertilia (lizards), 47–8
Lachesilla pedicularia (booklouse), 265–6
Lachesis mutus (snake), 47
Laelaps (mites), 53
Larder beetles, 126–7
Large roundworm of Man, 337–8
Large yellow ant, 171
Laridae (gulls), 39–41
larval stages of common pests, **226**
Lasioderma serricorne (beetle), 96, **97**
Lasius (ants), 171–2
Latheticus oryzae (beetle), 144

Lathridiidae (plaster beetles), **116**, 129–30
Latrine fly, 218
Latrodectus curacaviensis (spider), 81
Lactrodectus mactans (spider), **80**, 81
Lead cable borer, 105–6
Leather jackets, 185
Leishmania donovani (protozoan), 186
Lepidoglyphus destructor (mite), 57
Lepidoptera *see* butterflies; moths
Lepinotus (booklice), 265
Lepisma saccharina (silverfish), 315–16, **317**
Leptotrombicula (mite), 22
Leptoconops torrens (gnat), 201
Leptopsylla segnis (flea), 179, **180**
Leptospiroses (disease), 17–18
Leptotrombidium scutellaris (mite), 66–7
Lesser bandicoot, 15–16
Lesser dung flies, 210
Lesser grain borer, 105
Lesser house fly, 216–17, **218**
Leucophaea maderae (cockroach), **296**, 298–9
lice
  diseases and, 256–9, 262–3
  see also booklice; Mallophaga; sucking lice
Lieurus quinquestriatus (scorpion), 90
Linognathus setosus (louse), 253–4, **255**
Liposcelis (booklouse), **265**, 266–7
Liptocera caenosa (fly), 210
Little black ant, **168**
Little fire ant, **169**, 170
liver baiting, 354
liver flukes, 330–1, **332**
lizards, 44–5, 47–8
Loa loa (worm), 205
Loaiasis (disease), 205–6
Longhorn beetles, 109–11, **112**, 113–14
Lophocateres pusillus (beetle), 145
Loxosceles reclusa (spider), **82**, 83
Lucilia (Greenbottles), 217, 220–1
Lycosa (spiders), 81
Lyctidae (powder-post beetles), 130, **131–2**
  Lyctus brunneus, **131**, 132
Lyctocoris campestris (bug), 230, 250
Lymexylidae (beetles), 133–4
  Lymexylon navale, 134

Macrocheles muscaedomesticae (gamasid), 53
Madeira cockroach, **296**, 298–9

malaria, x, 183, 187, 196–200
Malathion, 354
Mallophaga (Feather/Biting/Chewing/Bird lice), 251, 260–1, **262**, 263
Mange mites, 63–4
Mango fly, 205–6
Mansonia (mosquitoes) 192–3
marsupials, 29–30, **31**, 32
Masicodamon allanteus (scorpion), 86
Masked Hunter, **240**
Mastotermiditdae (termites), 274–6
  Mastotermes darwiniensis, 276, 282
mealworms, **226**
Mediterranean flour moth, **233**, 234, **235**
Megachiroptera (bats), 23–4
Megadermatidae (bats), 24
Melanolestes picipes (bug), **240**
Meles meles (badger), 177
Menopan pallidum (louse), 261
Mephitis mephitis (skunk), 2, **5**
Merchant grain beetle, 140, **141**
Mesium affine (Spider beetle), 139
Mesostigmata (mites), 52–3
metal, damage to, 7, 105, 113
Metaldehyde, 355–6
Methyl bromide, 354–5
Metoecus paradoxus (beetle), 158
Mezium americanum (Spider beetle), 139
Microchiroptera (bats), 23–5
Microscerotermes (termites), 277
Micotermes (termites), 271
midges, 200–2
Mill moth, **233**, 234, **235**
millipedes, **323**, 324
misting, 355
mites, 21, 49, **50**, 51–7, **58–60**, 61–3, **64**, 65–7, **68–9**, 99, 230
  disease and, 22, 51, 53, 56–7, 62–7
moisture control, 355
moles, 27, 363
molluscicide, 336, 355–6
mongooses, 1–2, 5
monkeys, 190
Monomorium minimum (ant), **168**
monomorium pharaonis (ant), 167, **168**, 249, 354
mosquitoes, 187–8, **189–90**, 191, 347
  see also malaria
moth flies, 185–6
moths, 225, **226**, 227, **228–30**, 231–2, **233**, 234, **235**
  see also clothes moths; grain moths; house moths

mouse *see* House mouse
Mummucia variegata (sun spider), 85
Murine typhus, 21–2, 53
Mus musculus see House mouse
Muscidae (House flies), 53, 210–15
  Musca autumnalis, 214–15
  M. domestica, 53, 211–14
  M. stabulans, 219
  M. scorbens/humilis, 212
Muscoid fly, **226**
Museum beetles, 127–9
Mutillidae (ants), **154**
Myiasis (disease), 185, 204, 213–14, 219, 222

Nacerdes melanura (borer), **135**, 136
Nasutitermes (termites), 277, 283
Nauphoeta cinerea (cockroach), **296**, 299–300
Necrobia rufipes (beetle), 115–16
necrotic spider bite, 82–3
Nemapogon granella (moth), 232
Nematocera (flies), 185–8, **189–90**, 191–203, **204**
  diseases and, 186–7, 191–5, 197–204
nematodes (roundworms), 22, **152**, 189–93, 202–3, 205, 337–8, **339**, 340
Neostylopga rhombifolia (cockroach), 300, **301**
Neotoma (Wood rats), 16
Neotrombicula autumnalis (mite), 65
Niptus hololeucus (Spider beetle), **138**, 139
Nitidulidae (beetles), **134–5**, 136
Noctilionidae (bats), 24
noise problems, birds, 34–5, 38, 40, 42
Non-biting stable fly, 219
North America
  beetles, 97, 99, 101–3, 105, 108–9, 112, 115, 117, 123, 125, 127, 135–9
  birds, 34–6, 38, 42
  booklice, 266
  bugs, 239, 249–50
  carnivores, 2–3, 5
  centipedes, 320–2
  Chaga's disease, 242, 244
  cockroaches, 287–9, 291–303
  control methods, 342, 350–1, 355, 365
  crickets, 310–11
  earwigs, 304–6
  fleas, 178, 180–2

flies, 186, 195–6, 201–2, 210, 214, 216, 219, 221
marsupials, 32
mites, 66–7, 69
moths, 228, 231–3, 235
rodents, 7–10, 15–16, 19–21
sanitation rules, 351–3
scorpions, 85, 90
silverfish, 316–17
snakes, 46
sowbugs, 91
spiders, 80–4
springtails, 313
tapeworms, 329
termites, 271–4, 281–2, 284
ticks, 73, 75–8
wasps/ants/bees, 147, 152–3, 158, 163, 166, 168–73
Northern rat flea, 19, **181**, 182
Norway rat *see Rattus norvegicus*
Nosopsyllus fasciatus (rat flea), 19, **181**, 182
Nostril flies, 223
Notoedres (mites), 63
Notophilus taeniatus (centipede), **322**
Notostigmata (mites), 52
Nutmeg weevil, 103
Nymphopsocus destructor (booklouse), 265

Oak longhorn beetle, 113
Odorous house ant, 171
Oeciacus (swallow bugs), **241**, 250
Oecophoridae (house moths), **229–30**, 231
Oedemeridae (beetles), 134, **135**, 136
Oestridae (warble/nostril flies), **223**
Oestrus ovis (nostril fly), 223–4
'Old house borer' (beetle), 110
Onchocerca volvulus (worm), 202–3
Onchocerciasis (disease), 202–3
Oniscus asellus (woodlouse), 91, **92**
Opilo domesticus (beetle), 99
Opisthoglypha (snakes) 45
Opisthorchis (flukes), 320, 336
opossums, 5, 30–2
Orchopeas howardi (flea), **181**, 182
Oriental cockroach, 286, **289**, 290
Oriental rat flea, x, 18–19, 21, 178, **181**
oriental sore, 185
Ornithodorus (ticks), 72, 73, 78
  O. moubata (tampan), 72, **73**
Ornithonyssus bacoti (mite), 21, 53
Ornithosis (disease), 42–3

Oroya fever, 185–7
Orthoptera (crickets), 309–11
Oryzaephilus (grain beetles), **140–1**
  O. mercator, 140, **141**.
  O. surinamensis, **140–1**
Otodectes (mites), 64
Owl midges, 185
Ox warbles, **224**

Palmetto bug (cockroach), **296**, 302–3
Palorus (beetles), 144
Pandinus imperator (scorpion), 88
Panstrongylus megistus (bug), 239
Papataci fever, 185
Paper wasps, 159
Paracharon caecus (scorpion), 86
Paralipsa gularis (moth), 232
parasites, 150–2, 197–8
  biological control with, 343–4
  see also diseases; fleas; flies; flukes; lice; mites; nematodes; tapeworms; ticks
Paratrichina longicornis (ant), 172
Paravespula vulgaris (wasp), 153
Parcoblatta (cockroaches), **288–9**, 303
Paridae (tits), 42
Passer domesticus (sparrow), 333, 35–7
Passer montanus (sparrow) 37
Pasturella pestis (bacterium), 18–19
Pasturella tularensis (bacterium), 77
Pavement ant, 170
Pediculidae *see* Sucking lice
  Pediculus humanus capitis (head louse) 252–4, 256
  P. humanus corporis (body louse), 253, **254**, 256
Pedipalpida (whip-scorpions), **50**
Pentarthrum huttoni (weevil), 120–1
Periplaneta (cockroaches)
  P. americana, 286, **289**, 291–3, 295
  P. australasiae, 286, **289**, 293–4
  P. brunnea, 286, **289**, 294–5
  P. fulginosa, 286, **289**, 297
Peucetia viridans (spider), 81
Phalangida (Daddy Long-leg spiders), **50**
Phantom midges, 201–2
Pharoah's ant, 167, **168**, 249, 354
Phauloppia lucorum (mite), 54
Pheidole (ants), **169**
Phidippus audax (spider), 81
Phlebotomus (sand flies), 185–6
phoresy, 53
Phoridae (flies) 206–7
Phormia (Greenbottles), 220

Phorocantha (beetles), 114
Phrynichus ceylonicus (scorpion), 86
Phthiraptera (lice), 251
  Phthirus pubis (Crab louse), 253, **254**, 256
Phycitidae (moths), **233**, 234, **235**, 236
Phyllostomatidae (bats), 24
Phymatodes testaceus (beetle), 113
Physopsis globosa (snail), **335**
Picinae (woodpeckers), 34, 41–2
Pieris (butterflies), 226
pigeons, 33–5
  bug, 249
  control of, 363–4
  flea, 179
  tick, 72
Pin worm, 338–40
Piophila casei (skipper), 209–10
Pit vipers, 46–7
plague, x, 7, 18–19, 178
Plasmodium (protozoa), 197–9
  P. falciparum, 199–200
  P. vivax, 197–9
Plaster beetles, 115, **116**, 117, 129–30, 264
plastic, damage to, 7, 17, 282–3
Platypodinae (beetles), 121
Ploceidae (weaver birds), 35–7
Plodia interpunctella (moth), **233**, 235–6
poisons
  bee, 166
  snake, 45–6
  see also control methods; diseases; stings and bites
Polistes (paper wasps), 159
Pollenia rudis (fly), 215, **216–17**
Polynesian rat *see Rattus exulans*
Polypax spinulosus (louse), 21
Porcellio scaber (woodlouse), 91
porcupines, **15**, 16–17
Pork tapeworm, **327**, 328
Pork trichina worm, 22, 338, **339**
possums, 5, 30–2
poultry
  bug, **241**, 250
  fleas, 176, 178–9, 182
  mites, 53
  tick, 72
Powder-post beetles, 130, **131–2**
pre-baiting, 356
pre-treatment, site, 361
predators, 53, 79, 85–7
  on bed bugs, 249
  bugs as, 239
  on beetles, 99

biological control with, 344
  on flies, 115
Prenolepsis imparis (ant), 171
preservatives, 353, 356–7
pressure impregnation, 356–7
Priobium carpini (beetle), 101
Prostigmata (mites), **60**, 61–2
Proteroglypha (snakes), 45
Protozoa, 277–8
  see also Plasmodium
Pseudeurostus hilleri (beetle), 139
Pseudocanthotermes (termites), 271
Pseudocleobis morsicans (sun spider), 85
Pseudoscorpionidae see False scorpions
Psittacosis (disease), 42–3
Psocoptera (booklice), 264, **265**, 266–7
Psoroptes communis (mite), 63
Psychodidae (flies), 185–6
Ptilinus pectinicornis (beetle), 94, **100**, 101
Ptinidae (Spider beetles), 136–7, **138**
  Ptinus clavipes/hirtellus, 138
  P. fur, 137–8
  P. raptor, 139
  P. tectus, 136–7
  P. villiger, 138
Pulex irritans (flea), 19, 175, 176–7, **181**
Pymotidae (mites), 67, **68**
Pyemotes ventricosus (mite), 67, 68, 99
Pynoscelus surinamensis (cockroach), **296**, 298
Pyralidae (moths), 232–3
Pyralis farinalis (moth), **235**
pyrethroids, synthetic, 363
Pyrethrum, 357

Rabbit fleas, 178, **180**
Rabies, x, 4–5
radiant energy, 357
Rasahus biguttatus (Corsair), **240**
rats, 6–7, **8**, 9, **10**, 11, **12**, 13, 16, 178, 364
  control and poisoning, 341–8, 356–8, 360, 362–6
  fleas, 18–19, 21, 176, **181**, 182
  mites, 21, 53, 63
  Rattus exulans, 6–7, **12**, 13
  R. norvegicus, 6, **10**, 11–13, 15–16, 178, 358, 364
  R. rattus, 6–7, **8**, 9–10, 13, 15, 178, 364
  tapeworm, 263, 329
  see also diseases, rodents
rattlesnakes, 47

Red fox, 5, 177
Red poultry mite, 53
Red squill, 358
Red-back spiders, **80**, 81
Reduviidae (bugs), 238–9, **240**
 Reduvius personatus, **240**, 249
Relapsing fever, 72–4, 78, 256
remedial treatment, 358
reptiles, 44–8
residual spraying, 358–9
Reticulitermes (termites), **272**
 R. flavipes, **269**, 281, 284
 R. lucifugus, **280**, 284
Rhinotermitidae (subterranean termites) 274–5, 277
Rhipicephalus sanguineus (tick), 78
Rhizopertha dominica (grain borer), 105
Rhodnius (bugs), 239
Rhyncolus culinaris (beetle), 121
Rhyssa persuasoria (woodwasp), 150, **151**
Rice moths, **229–30**, 232–3
Rice weevil, 118, 119
Rickettsia
 disease, 21, 53, 66, 257–9
 R. akari, 21
 R. conori, 77–8
 R. mooseri, 21
 R. prowazeki, 256, 259
 R. rickettsii, 75
 R. tsutsugamushi, 22
Ring-legged earwig, **305–6**, 308
River blindness, 202–3
Rock bee, 161
Rock dove, 34, 179
Rocky Mountain Spotted fever, 75–6, 78
 tick, 20, **75**, 76
rodents, 6–7, **8**, 9, **10**, 11, **12**, 13, **14–15**, 16–22
 see also Bandicoota; diseases; House mouse; porcupines; rats
Roof Rat *see Rattus rattus*
Roundworms *see* nematodes
Russell's viper, 45, 47
Rust-red flour beetle, **143–4**
Rust-red grain beetle, 117, **118**

Salmonella (bacterium), 20, 220, 344
Salmonellosis (disease), 20
Sand fleas, 179, **182**
Sand flies, 185–6, 200–1
sanitation, 359–60
Sarcophaga (fly), 222
Sarcoptes scabiei (mite), **60**, 61–3

Sarcoptidae (Mange mites), 63–4
Saw-toothed grain beetle, **140–1**
Scabies, 62–3
scaring devices, 360
Scenopinidae (Window flies), 115, 206, 230
Schistomiasis (disease), 335–6
Schistosoma (trematodes), 330, **333–4**, 335
 S. haematobium, 333, **334**, 335
 S. japonicum, 332
 S. mansoni, 332, **333–4**, 335–6
Scobicia declivus (beetle), 105–6
Scolopendra (centipedes), 320, 321, 322
 S. obscura, **321**
Scolytinae (beetles), 121
scorpions, **50**, 88–90
 see also False scorpions; Tail-less etc
scrub typhus, 22, 66–7
Scutigera (centipedes), 320, **321**
Seaside earwig, **306**
Seaweed flies, 209
Sepsidae (flies), 209
Serpentes (snakes), 45–7
Seven-day fever, 18
sewer treatment, 360
sheep
 flukes, 331, **332**, 336
 nostril fly, **223–4**
Ship rat *see Rattus rattus*
Shore earwig, **306**
Short-circuit beetle, 105–6
shrews, 27–8
Silica aerogel, 342
Silvanidae (Grain beetles), **140–1**
Silverfish, 314–15, **317–19**
Silvicola fenestralis (gnat), **204**
Simuliidae (Black flies/Buffalo gnats), 202
Simulium (flies), 203
Sinoxylon anale (beetle), 104
Siphonaptera *see* fleas
Siphunculata *see* Sucking lice
Siricides (Woodwasps), 146–7, **148–52**, 153
 Sirex cyanea, 148–9, 152
 S. juvencus, 149, 152
 S. noctilio, **148**, **150**, 151–2
Sistruna (rattlesnakes), 47
site treatment and hygiene, 361
Sitophilus (grain weevils), 118, **119**
Sitotroga cerealella (Grain moth), **228**, 229, **233**
skunks, 2–3, 5
Sleeping sickness, 183

slow release strip, 361
slugs, 355–6
smoke generator, 361
Smoky-brown cockroach, 286, 289, 297
snails, **335**, 336, 355–6
snakes, 44–7
sodium monofluoracetate, 362
'soil poisoning', 361
Soldier flies, 206
Soldier termites, 275–6
Solenopsis (Fire ants), **170**
Solenopsis molesta (Thief ant), **170**, 171
Solifugae *see* False spiders
Solupugidae (Sun spiders), **50**
Soricidae (shrews), 27–8
South and Central America
  ants and bees, 166, 172
  bats, 24, 26
  beetles, 103, 105, 108, 119, 127, 135
  bugs, 239, 250
  carnivores, 5
  centipedes, 320, 322
  cockroaches, 291, 294, 297–302
  crickets, 311
  fleas, 179
  flies, 186–7, 191–3, 203, 209–10, 214, 221, 223
  flukes, 331–2
  marsupials, 32
  millipedes, 324
  moths, 228, 236
  rodents, 6, 9
  scorpions, 85, 90
  snakes, 46–7
  spiders, 80, 82, 84
  termites, 276
  ticks, 76, 78
Soviet Union, 21, 77, 101, 112, 223, 336
Sowbugs, 91–2
space spraying, 362
sparrows, 33, 35–7
Spathius exarator (wasp), 99
Sphaeroceridae (flies), 210
Sphaerotheriidae (millipedes), 324
Sphecidae (wasps), **154**
Spider beetles, 136–8
spiders, 79, **80**, 81, **82**, 83–4, 249
  see also False spiders
Spilogale putorius (skunk), 2–3, 5
Spilopsyllus cuniculli (flea), 178
Spined pine-borer, 113
Spirillary rat-bite fever, 20–1
Spirillium minus (spirochaete), 20

Spirochaete, 20, 72–4, 78
Spotted Mediterranean cockroach, **296**
Spotted skunk, 2–3, 5
Springtails, **312**, 313
Squamata (lizards and snakes), 44–5
Squirrel flea, **181**, 182
Stable fly, 218–19
starlings, 33, 37–8, **39**
'Steamfly', 286–90
Steel-blue beetle, 99, 114–15
Stegobium paniceum (beetle), 94–5, **96**
Stegomyia fasciatus (fly), 191
Stephanapachys rugosus (beetle), 105
sterility, chemically induced, 344
Sticktight flea, 178–9, **181**
sticky board, 362
stings and bites
  ants, 169–70
  bees, 161–2, 166
  bugs, 247
  centipedes, 322
  flies, 201, 215
  rodents, 7
  scorpions, 89–90
  snakes, 45
  spiders, 80–4
  ticks, 72
  wasps, 158–9
  see also diseases
Stomoxys calcitrans (fly), 218–19
stone, damage to, 33, 168, 170
Stored nut moth, 232
stored products, damaged by
  ants, <u>167, 171–2</u>
  beetles, <u>93, 96</u>, 98, 103, 105, 115–19, 122–30, 134–45
  birds, 33, 35, 42
  crickets, 310
  mites, <u>54–7</u>
  moths, <u>226–30, 232–6</u>
  rodents, <u>11, 15</u>
  see also grain
Stored products, protection of, 350–3, 355, 357, 359–60
Stratomyidae (flies), 206
Straw itch mite, 67–8, 99
Streptobacillus moniliformis, 21
  rat-bite fever, 21
Striped earwig, **305**
Striped skunk, **5**
Stromatium barbatum (beetle), 110
strychnine, 362–3
Sturnidae (starlings), 37–9
  Sturnus vulgaris, 33, 37–8, **39**
subterranean termites, 274–5, 277

sucking lice, 251–3, **254–5**, 256–7
  diseases, 256–9
Sulphuryl fluoride, 363
sun spiders, **50**
Suncus murinus (shrew), 27–8
Supella supellectilium/longipalpa (cockroach), 286, **288**, 297
Surinam cockroach, **296** 298
Swallow bug, **241**, 250
Swamp fever, 18
Sweat bees, **154**
Swineherd's disease, 18
symbiosis, 148–9, 277–8
Symphata *see* Woodwasps
Syrphidae (Hover/Drone flies), 158, 207

Tabanidae (horse flies), 204–5
tactile repellant, 363
Taenia saginata (tapeworm), 326–8
Taenia solium (tapeworm), **327**, 328
tail-less whip scorpions, 86
Talpidae (moles), 27
tapeworms, 254, 262–3, 325–6, **327**, 328, **329**, 330
  diseases, 325–33, 335, 337–40
Tapinoma (ant), 171
Tasmanian bush-tailed possum, 30, **31**
temperature control, 355
Tenebrio molitor (mealworm), **142–3**
Tenebrionidae (beetles), 141–4
  Tenebroides mauritanicus (beetle), 145
termites, 268–84, **269**, 270–1, **272–3**, 274–9, **280**, 281–4
  beneficial role, 280–1
  control of, 353, 355, 361, 363
  economic loss and cost, 279–84
  feeding, 277–9
  groups, 272–7
  organisation, 268–72
  world distribution, **273**, 284
Termitidae (termites), 274–5, 277
Termopsidae (dampwood termites), 274–7
test baiting, 363–4
Tetramorium caespitum (ant), 170
Tetranychidae (mites), 68–70
Tetrastigmata (mites), 52
Thanasimus formicarius (beetle), 114, **115**
Thanatus flavidus (spider), 249
Thaumatomyia notata (fly), 217
Theocolax formiciformis (wasp), 99
Thermobia domestica (Firebrat), **317–19**

Thief ant, **170**, 171
Three-day fever, 185
Thrips (Thysanoptera), 237
Thysanura (Bristletails/Silverfish/Firebrats), 314–16, **317–19**
ticks, **50**, 71–2, **73**, 74, **75**, 76–8
  diseases, 20, 72–8
Tinea (moths), 231–2
  Tineola bisselliella (moth), 231
Tipnus unicolor (beetle), 139
Tipulidae (flies), 185
tits, 42
Tityus (scorpion), 90
Tobacco moth *see* Warehouse moth
Toxochernes panzeri (False scorpions), 87
Trachoma, 214–15
tracking dust, 364
trapping, 364
trees *see* wood
Trematoda *see* Flukes
Tremex (Woodwasps), 147
Trench fever, 256, 258–9
Triatoma (bugs), 239, **240**, 244
Tribolium castaneum (beetle), **143–4**
Tribolium confusum, **143**
Trichinella spiralis (worm), 22, 338, **339**
Trichinosis (disease), 22
Trichodectes canis (louse), 261, 262
Trichosurus vulpecula (possum), 30, **31**
Trigona (bees), 161
Trigonogenius globulus (beetle), 139
Trilobium (flour beetle), 143–4
Trimeresures (pit vipers), 47
Trogium pulsatorium (booklouse), 264, 266–7
Trogoderma (beetles)
  T. granarium, **124**
  T. inclusum, **126**
Trogostidae (beetles), 145
Trombiculidae (mites), 65–6
Trophallaxis, 158
Tropical bed bug, **241**, 249–50
Tropical house cricket, 311
Tropical rat flea *see* Oriental rat flea
Tropical rat mite, 21, 53
Tropical warehouse moth, 234, **235**
tropophallaxis, 277–8
Trypanosoma cruzi (bug), 239, 243
Trypanosomiasis *see* Chaga's disease
Tsetse fly, **211**
Tularaemia (disease), 20, 77
Tumbu fly, 221–2

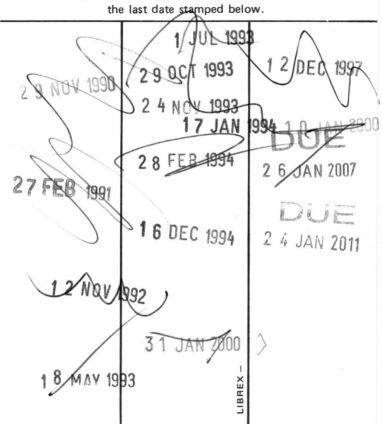

PEST ANIMALS IN BUILDINGS